L. Ashok Kumar
K. M. Janani

Nanosensors for Healthcare and Environmental Monitoring

Fabrication, Characterization, Simulation, and Applications

nova
science publishers
www.novapublishers.com

https://doi.org/10.52305/XVNH0074

NOTICE TO THE READER

Library of Congress Cataloging-in-Publication Data

ISBN: 979-8-89113-316-7

Published by Nova Science Publishers, Inc. † New York

Contents

List of Figures

List of Tables

Chapter 1

Introduction to Nanotechnology and Nanomaterials

Learning Outcomes

- The reader will have a clear overview of nanotechnology and nanomaterials, how and where they are used throughout history, and their evolution till the modern technological era.
- The reader will be able to identify the naturally occurring nanoparticles and engineered nanoparticles as well as their significance and hazardous effects on the environment.
- The dimensional classification of nanomaterials as 0D, 1D, 2D, and 3D NPs can be understood in-depth and the reader will be able to identify the different range of nanomaterials for the desired real-time application.
- Various nanoparticle classifications based on their material, mechanical, electrical, optical, chemical, and magnetic properties are understood in detail so that the reader can make the proper material choice before fabricating a device.
- The top-down and bottom-up methods of the nanofabrication process and their types are explained in detail. Hence, at the end of this chapter, the reader will have thorough knowledge about the nanomaterials, their occurrence, their types, properties, and their fabrication.

Abstract

Today's world of electronics revolves around the new age of scaling down and miniaturization of devices. This demands the use of materials that are in a very smaller scale leading to the discovery of various nanomaterials and the evolution of nanotechnology, the process of incorporating those materials into novel devices of daily usage. In this chapter, an introduction to nanotechnology and its evolution dated back from the prehistoric period are discussed. The scale of nanomaterials, their occurrence in nature, incidental and engineered nanomaterials along with their environmental impact are intensely discussed. To help readers understand how to recognise different nanomaterials for desired applications, the categorization of nanomaterials based on their electron confinement dimensionality and kind of source material is more clearly explained in a well-organized sequence. Every nanomaterial will possess unique physical, chemical, optical, magnetic and electronic properties. Engineering of the nanomaterials to control such properties plays a crucial role in the outcome of a novel device for which the understanding of the fundamental properties is significant and is clearly explained in this chapter. After having thorough knowledge about the fundamental classification and properties of nanomaterials another essential aspect is nanofabrication, by which the nanomaterials are synthesized. There are two fundamental methods for nanofabrication: top-down and bottom-up. This chapter clearly discusses these manufacturing methods as well as the numerous synthesis processes used.

Keywords: nanotechnology, engineered nanomaterials, quantum dots, nanocomposites, semiconductors, carbon-based materials, low-dimensional nanomaterials, nanofabrication

1. Introduction to Nanotechnology

Nanotechnology focuses on the integration of materials and electronics at that size, and the consequences of manipulating matter at that scale lead to devices with new and unusual features that are impossible to achieve in bulk or at bigger scales. Proteins and enzymes, which are examples of nanomaterials engineered by living organisms, are similar in size to nanoscale devices. The study and engineering of materials at the nanoscale, i.e., at the atomic and molecular level, is known as nanotechnology. Engineered materials are used to create miniature devices that operate effectively. Carbon nanotubes are by far the most extensively employed nanomaterials. Nanomaterials generally have a size range of 0.1 nm to 100 nm (1 nm = 10^{-9} m). We may use the size of well-known things we come across in everyday life to get a sense of nanometers. A strand of human hair, for example, is roughly 80,000 nanometers broad, and nanometers are 1000 times smaller than micrometers. The minimal size necessary for a free electron to function as a quantum particle is believed to be around 1 billion nanometers (Table 1).

The word "nano" is derived from a Greek word that means "dwarf." At this scale, the fundamental processes of the living world really take place, and substances at this size exhibit extraordinary chemical properties. The rise in surface area relative to volume as particles go smaller, together with the understanding of unusual quantum phenomena at the atomic scale, are the causes of these profoundly divergent properties. All matter experiences property destruction and creation at these sizes (on a different scale). This final fact has paved the way for novel materials, techniques, and equipment. There are an unlimited number of novel "unlikely-to-be" combinations that may be created as a consequence of these because of the varied forms and structures compared to their bulk equivalents.

Table 1. Range comparison table of nanometer scale with real-time examples

S. No.	Comparative Item	Size in nanometers (nm)
1.	1-inch	25,400,000
2.	Red blood cells	~7,000 in diameter
3.	White blood cells	~10,000 in diameter
4.	A hydrogen atom	1
5.	Nanoparticles	1 to 100
6.	Fullerenes	1
7.	Quantum Dots	8
8.	Dendrimers	~10
9.	DNA strands	2.2
10.	Proteins	5 to 50
11.	Viruses	75 to 10
12.	Bacteria	1,000 to 10,000
13.	Atoms	0.1 to 0.2

Humanity has unknowingly used certain peculiar properties of nanoscale materials for aeons. Tiny gold particles, for instance, have the ability to appear red or green; this property has been used to colour stained glass windows for more than 1000 years. Carbon hardens when it is tightly packed into a nanoscale structure called a nanotube. Graphite, carbon's native form, is rather pliable. Metallic copper is transparent at the nanoscale. Nanotechnology is currently present in a broad variety of goods, from pigments in sunscreen and lipstick to nanometre-thick coatings on "self-cleaning" windows.

One aspect of nanotechnology is the creation of functional mechanisms using nanoscale components (MNT), such as bacteria-sized supercomputers with the capacity of today's MIPS, supercomputers the size of sugar cubes with the power of a billion laptops, or a regular-sized desktop model with the power of trillions of today's PCs. The second involves scaling down existing technologies to the nanoscale; examples may be seen on our Current Uses page. Many of the most exciting potential applications of nanotechnology are governed by the principles of quantum physics. At this size, quantum physics principles take control, creating new opportunities in the domains of computers, catalysis, magnetic storage, optics, and electronics. People that make it a point to stay up to date on technical developments all agree that nanotechnology, which is still in its infancy, has the potential to revolutionise every aspect of life.

2. Basics of Nanomaterials

The foundation of nanoscience and nanotechnology is nanomaterials. Chemical substances or materials created and used on a very small scale are known as nanomaterials. According to ISO, a "nanomaterial" is a substance with "any external dimension in the nanoscale (size range from about 1 to 100 nm) or having a structural properties or surface morphology in the nanoscale" (2015). The unique optical, physical, chemical, magnetic, electrical, and other characteristics that arise at this size make nanomaterials interesting. Electronics, medicine, and other disciplines stand to benefit greatly from these emerging features.

Some nanoparticles occur naturally, but engineered nanomaterials (EN), which are intended for and are now being used in many commercial goods and cycles, are highly persuasive. They may be found in a variety of everyday items, including sunscreen, personal care items, sporting goods, clothing that resists stains, tyres, and electronics. They are also used in medicine for diagnosis, imaging, and drug delivery purposes.

In order to make advantage of their tiny size and novel qualities not present in their normal mass equivalents, designed nanomaterials are being created at the atomic or nanoscale level. The two primary theories as to why materials somewhere at nanometer scale could have various properties are an expanded relative surface area and unique quantum impacts. Compared to normal materials, nanomaterials have a substantially larger surface area to volume ratio, which may enhance chemical reactivity and affect cohesion. The qualities and features of a material may be even more influenced by quantum phenomena at the nanoscale, leading to distinctive optical, electronic, and magnetic behaviours.

Nanomaterials are currently used commercially, and some of them have been available for years or even for a very long period. Only a few examples of the commercial goods available now include stain- and kink-free textiles, sunscreen, hardware, paints, and stains. A broad variety of consumer products, including windows, sports goods, bicycles, and cars, employ nanocoatings and nanocomposites. Tennis balls made of butyl-elastic/nano-earth composites endure longer, while new UV-impeding coatings on glass bottles shield liquids from sunlight. Nanoscale silica is used as a filler in a variety of items, including cosmetic products and dental sealants. Nanoscale titanium dioxide, on the other hand, is used in cosmetics, sunscreen lotions, and self-cleaning windows.

3. Evolution of Nanomaterials

Nanoparticles are not truly delivered by modern amalgamation research facilities, although they have undoubtedly existed in the nature for a long period, and so their employment may be traced back to antiquity. The Big Bang event, which resulted in the genesis of the universe, is regarded to be the origin of nanoparticles and nanostructures (Figure 1). The development of the Earth has so far been made possible by the presence of naturally occurring nanoparticles. From the origin of the first living cell to the development of the human being, nanoparticles have been crucial to the evolution of life on Earth (Barhoum et al. 2022). Nanoparticles like fullerenes, carbon nanotubes, and other combustion by-products were found in the smoke from human-made prehistoric fires. These carbon nanostructures and waxes were employed by our forefathers to create cave paintings (Figure 2). The earliest examples of human artistic expression have been found to date on hand stencils from Sulawesi, Indonesia. These cave paintings were created about 40,000 BCE using fat, charcoal, and plant pigments, according to carbon dating evidence (Youssef et al. 2017; Heiligtag and Niederberger 2013).

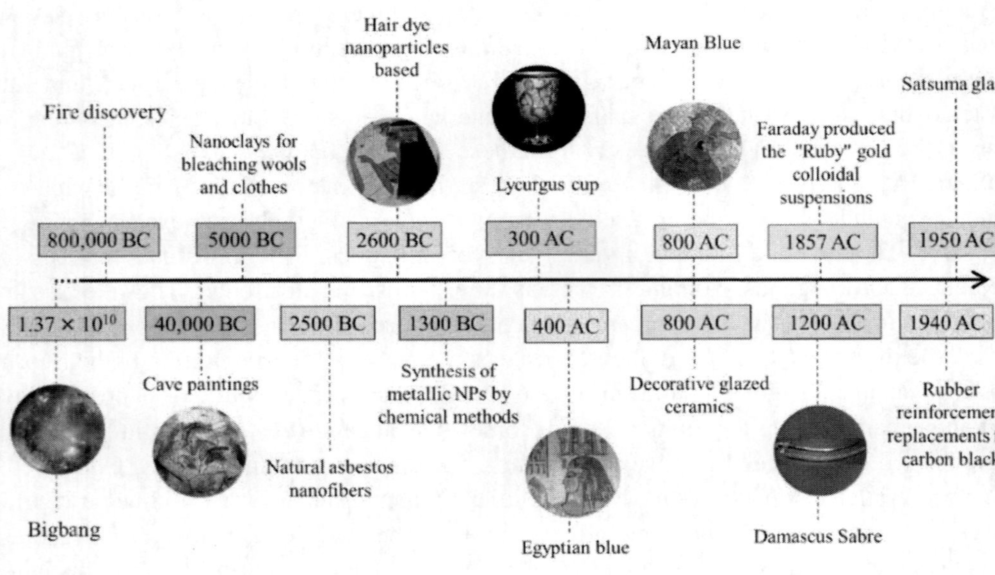

Figure 1. Evolution of nanotechnology.

Figure 2. Paintings found in Lascaux caves located in southern France © UNESCO ("Image 129614 - UNESCO World Heritage Centre" n.d.).

Figure 3. The Lycurgus Cup ("Lycurgus Cup: A Piece of Ancient Roman Nanotechnology | Amusing Planet" n.d.; "The History Blog" "Blog Archive" Lycurgus Cup Inspires Cool New Sensor Technology" n.d.)

Nanomaterials have been managed to produce and even used since ancient civilizations, despite the reality that the term "nanomaterials" was only coined at the end of the 19th century. For writing on papyrus, high-opacity, stable black pigments were produced in ancient Egypt using oil lamp soot. At the time, Egyptians were not aware that soot contained carbon nanomaterials. One of the most well-studied illustrations of natural nanomaterials used in antiquity is clay minerals, which are composed of nano-platelets with a thickness of about 1 nm (5000 BC). Humans first worked with porcelain composite material that were reinforced with asbestos nanofibers more than 4,500 years ago (50–200 nm in diameter) (Mubeen et al. 2021).

Glass can be coloured in an extraordinary way by metal nanoparticles. One of the strongest indicators of ruby glass is found in the British Museum's Lycurgus Cup (Figure 3). Glass has long been painted a stunning crimson colour using gold. It was made by the Romans in the 4th century and looks green in natural light before becoming red when lighted inside (Gartia et al. 2013).

The first scientific study used colloidal gold particles created by Michael Faraday in 1857. Additionally, for more than 70 years, nanostructured catalysts have been researched. By the early 1940s, precipitated and fumed nanomaterials were being manufactured and marketed as extremely fine carbon black substitutes for rubber reinforcement material in Germany and the United States. Metallic nanopowders for magnetic storage cassette tapes were created in the 1960s and 1970s. The first nanocrystals made using the now-common inert-gas evaporation method were published by Granqvist and Buhrman in 1976.

A lecture by physicist Richard Feynman exploring the idea of creating things at the atomic and molecular scale gave rise to the concept of nanotechnology in 1959. However, it wasn't until 1981 when the first scanning tunnelling microscope was created by IBM scientists that experimental nanotechnology truly came into its own. This enables us to see individual atoms by gliding a tiny probe across the silicon crystal's surface.

In 1990, IBM researchers used an STM to manipulate individual xenon atoms on a nickel surface. Since then, more methods have been created to take pictures at the subatomic level. These include magnetic resonance imaging (MRI) and the atomic force microscope (AFM). A 60-carbon molecule with the shape of a soccer ball was created in 1985 by chemists and named Buckminsterfullerene. Carbon nanotubes, extremely durable rolls of carbon atoms, were developed in 1991. These are 100 times stronger than steel while being six times lighter.

About 20–30 years ago, the advancement of synthesis methods was the clear focus of nanoparticle research. Whereas chemically synthesized practise is still important today, a rapidly developing area of study is the use of nanoparticles as key components, similar to LEGO bricks, and their integration into larger structures (Talapin 2008). It is clear that such a toolkit offers up exciting possibilities for the creation of novel materials when you take into account how many nanoparticles of various compositions and qualities are accessible, as well as how many techniques to mix and arrange them. For the regulated assembly of nanoparticles, it is essential to comprehend and manage interatomic interactions such as van der Waals, electromagnetic, molecular, and entropic forces (Bishop et al. 2009).

Strong plastics, microchips, harmful gas detectors, and many other novel materials have all been made with nanotubes. In the distant future, a space elevator might be constructed using nanotubes. In a rapidly expanding range of inorganic and biological structural and functional materials, nanophase engineering is being used to alter the physical, catalytic, electric, magnetic, optical, and electronic properties. It is common to separate microscopic clusters, fuse them together to form a bulk-like material, or embed them in modular liquid or mechanically strong solid materials to produce nanophase or cluster-assembled materials.

4. Occurrence of Nanomaterials

Nanomaterials have always coexisted in this Earth in various forms naturally since its life began over several billion years ago. After the evolution of human beings, the activities carried out by the particular species unintentionally led to the form of many kinds of nanomaterials. It is only in this modern scientific era these nanomaterials are specifically identified, examined, manufactured, and engineered to a greater extent leading to a new field of nanotechnology (Figure 4). Based on the origin and occurrence of each nanomaterial, they can be classified as natural, incidental, and engineered nanoparticles (Hochella et al. 2019).

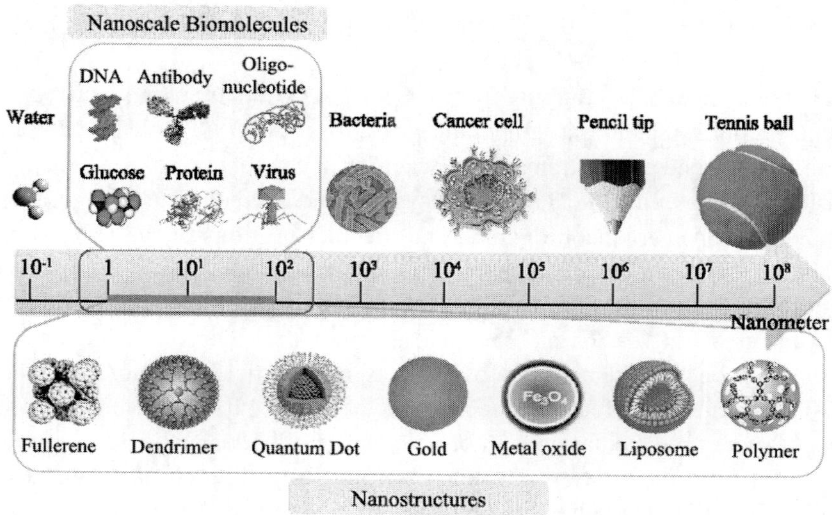

Figure 4. The comparison scale of various nanomaterials.

Because particles cannot be observed in isolation, environmental behavior is influenced by a variety of interacting factors. Nanoparticles interact with a wide range of abiotic and biotic components via a variety of interconnected pathways, resulting in changes in environmental compartments independent of their origin (Baalousha, Lead, and Ju-Nam 2011; Wagner et al. 2014).

4.1. Natural Nanomaterials (NNMs)

Naturally occurring nanomaterials will be formed from various environmental processes such as sedimentation, chemical changes, living cell mutations, animal/other creature's evolution, biological and ecological changes, aggregation, ocean ecosystem, mineralization, oxidation, reduction, etc (Figure 5). These NNMs play a vital role in balancing the earth's ecosystem and are present in all spheres (i.e., in the atmosphere, hydrosphere, lithosphere, and biosphere) (Sharma et al. 2015).

The NNMs can be in the form of either nanoparticles or colloids. The colloids that results from the chemical and biological changes during the decay of earthly matters are a kind of nanomaterial which ranges from a 1 nm to 1 μm. these are generally called as environmental colloids. Chemically separated fractions of total natural organic carbon with good trace metal binding properties are known as humic substances (HS) (Filella 2008). HS can exist as a scattered substance in colloids with diameters of 5 nm. The HS can also create the composite by aggregating with mineral colloids such as Iron and Manganese oxides and aluminosilicates (Stolpe and Hassellöv 2007). Such nanoparticle colloids have a noticeably high surface-to-volume ratio as a result of this aggregation.

Astronomical phenomena and direct examination of "stardust" have revealed that stars are composed of dust and gas, most notably hydrogen. This dust contains oxide, silicate, carbide, nitride, carbonaceous, and organic nanoparticles (Hochella, Spencer, and Jones 2015). Nanodiamonds are a fascinating example of NNPs with a size range of 1-10 nm and an average diameter of 3-4 nm. They have been found in micrometeorites like Murchison and Orgueil's pieces (Dai et al. 2002).

NNPs in the form of ashes ranging from 100 nm to 200 nm are closely attributed to erosion and volcanic eruption reaching a temperature of above 1400° C. Volcanic ash particles can reach the high troposphere and stratosphere, where they can remain suspended and "travel" for years, causing harm to human health, the ecosystem, and even climate change. Due to particle deposition in the upper, tracheobronchial, and alveolar portions of the respiratory tract as a result of this suspension in the atmosphere, inhaling nanoparticles made from volcanic ash can cause serious respiratory disorders. ("What Nanomaterials Exist in Nature?" n.d.; Griffin et al. 2017; Ermolin et al. 2018). Every chemical element that exists as such or in one of its combinations can be found in the atmosphere as nanoparticles during eruptions or shortly afterward. This is due to the fact that the majority of metal and non-metal ores throughout the Earth's mantle were produced as a result of volcanic activity during geological time.

Dust storms from huge deserts, which are a consistent source of nanoparticles, tend to transport them into the atmosphere via air currents (Péwé and American Association for the Advancement of Science. 1981). Recent studies show that both mineral mining dust and manmade urban pollution on continents move over great distances, with roughly half of the aerosols emissions in the troposphere being minerals coming from deserts (Shi et al. 2005). Large amounts of Si, Al, Ca, and Fe are found in the fine particulates throughout a dust storm bordering South Korea and China according to analytical tests ("Northern Dust Brings Dirty Skies to Shanghai -- China.Org.Cn" n.d.; Chun et al. 2008; Strambeanu, Demetrovici, and Dragos 2015). While graphite and silica particles have been discovered in the diffuse interstellar medium, interstellar dust clouds have been found to include amorphous calcium silicate (ACS), CO, SiC, polynuclear aromatic hydrocarbons (PAH), or other simpler organic structures (Strambeanu, Demetrovici, and Dragos 2015).

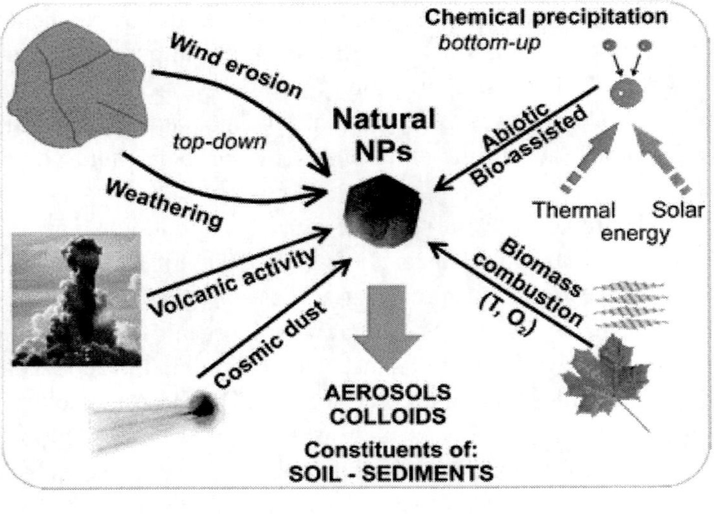

Figure 5. Natural reasons behind the production of nanomaterials (NNPs).

Intracellular structures like magnetosomes and external assembly like lipoproteins and viruses are examples of biological nanoparticles (Besford et al., n.d.). They engage in a variety of tasks, from intercellular signalling to mineral stockpiling. Exosomes are among the countless vesicles and macromolecules that cells create, ranging in size from 3 to 200 nm (Xiong et al. 2021). In the watery inner environment of primates and insects, lipid-protein complexes called lipoproteins carry lipids that are not soluble in water. Lipids and apolipoproteins combine to generate core-shell, spherical or discoidal nanoparticles with sizes between 7 and 80 nm (Stanley 2014; Shih et al. 2007).

Inorganic nanoparticles are also produced by biological organisms, especially iron-containing nanoparticles like ferrite nanoparticles and magnetite (Geppert and Himly 2021). Ferritin is produced by eukaryotes, archaea, and bacteria. Its main functions are to produce and preserve iron oxides as well as to act as a protein nanocage to sequester potentially hazardous iron ions (Bai et al. 2015). While most bacteria have ferritin, magnetotactic bacteria have evolved a second nanoparticle that is iron-rich (Stanley 2014). They include a specialised organelle called a magnetosome, which is composed of a lipid bilayer and harbours magnetic iron-bearing minerals like magnetite or greigite. Bacteria are able to swim more efficiently when aligned with the earth's magnetic field, thanks to magnetosomes that have a diameter of 50 to 80 nm and can form one or more networks (Komeili 2012).

4.2. Incidental Nanomaterials (INMs)

Incidental nanomaterials are unintentional by-products of human activity such as manufacturing and other mechanical processes. These nanoparticles are very asymmetric in size and shape, with no identifiable size or shape. These INMs are discharged into the environment and interact with other particles in the air, causing toxicity and increased health hazards. Vast number of daily activities lead to the increased release of INMs.

Vehicles emit nanoaerosols with diameters of roughly 20 nm as a result of the combustion of particles in fuels (Czerwinski et al. 2017). These have a high concentration of aggregates of primary pollutant particles, which are easily absorbed by the body and can induce respiratory and olfactory health problems. Vehicle nanoparticles are nucleated metal oxides generated as a consequence of metal additions mixed with the lubricating oil utilised (Buchholz, Dibble, and Rich 2003; Sonntag et al. 2012; Hadler et al. 2015).

Among the many types of nanoparticles, automobile tyres release silica, brake pads release metallic dust, engine lubricating fluids release calcium carbonate, and black soot releases partially burned hydrocarbons. The wastewater discharges of the semiconductor sector and other industries, such as photography and healthcare, contribute to the environmental release of nanoparticles.

4.3. Engineered Nanomaterials (ENMs)

Researchers are already discussing the creation of generations of nanoparticles. Coatings, individual particles, and other passive nanostructures are examples of first-generation nanostructures. Some consumer goods already include manufactured nanoparticles of this sort. A number of nanostructures are now being developed; these are 2nd-generation structures which either carry out an active function or react adaptively. By 2010, it is anticipated that third-generation synthetic nanomaterials would be 3D systems that self-assemble or might be used to target medicine delivery to certain body areas. Fourth-generation nanomaterials will, by definition, have molecular structures.

When compared to bulk materials or micrometer-sized particles, fundamental electrical, mechanical, chemical, biological, optical, and other features may be substantially different at the nanoscale. Surface area could be a factor in one explanation. It is significant because the majority of chemical reactions involving solids occur at their surfaces, where there are insufficient chemical contacts. Solids have a surface area per cubic centimetre of 6 cm^2, or about the same as one side of a half-stick of gum. However, an ultrafine powder with particles as small as 1 nm has a surface area per cubic centimetre of 6,000 m^2, or exactly one-third the size of a football field.

Collections of NSPs with high surface areas may be particularly reactive since around one-third of an NSP's chemical connections occur there. Since it has been shown that silver nanoparticles are effective bactericides, for instance, several companies have developed reusable silver fibre water filtration devices. The features of many materials, including magnetic moment, permeability, electrical conductivity, luminescence, opacity, and other properties, start to change considerably at around 100 nm, when quantum-size effects start to overwhelm thermal effects. Quantum-size impacts on electrical properties increase inversely with decreasing particle size.

At certain diameters, however, additional distinguishing features of certain materials become noticeable. For example, gold nanoparticles exhibit significantly improved catalytic capabilities at 3 nm. Characterizing material effects at various scales is a field of basic research that is currently engaged.

There are several structural forms of carbon and a few other elements, including oxygen, tin, and sulphur, together known as allotropes, each of which has distinctive properties. For instance, crystalline forms of pure carbon are found in various sizes of Buckminster fullerenes, which are very soft and hard, respectively.

Despite having the same chemical composition, engineered nanomaterials may take on many morphologies such as planes, rings, fibres, tubes, and spheres. Each of these forms may also have unique physical characteristics. This is due to the fact that despite having the same atoms, molecular bonds may have different patterns.

NSPs have unique properties due to their morphologies, however the Toxic Substances Control Act (TCSA) prohibits the classification of manufactured nanoparticles as new substances unless they also have a unique composition. For instance, despite the fact that both types of TiO_2 have distinct properties, TiO_2 nanoparticles are controlled in the same way as bulk TiO_2.

A few investigations have shown that the toxicities of materials with the same chemical composition but different sizes and forms often vary, and that this connection is not linear as one would anticipate. For example, a research indicated that 10-nm particles were more harmful than 1.6-m particles, despite the fact that nanoparticles between 50 and 130 nm across quartz-crystalline silica (a poisonous material) were less dangerous. The dosage and method of entrance into the body, however, also have an impact on toxicity.

5. Classification of Nanomaterials Based on Dimensional Confinement of Electron Movement

Nanomaterials in general are very tiny scale materials that are classified based on their occurrence in each dimension. Considering the dimensional confinement of electron movement they are categorized into four types – zero-dimensional (0D), one-dimensional (1D), two-dimensional (2D), and three-dimensional (3D) nanomaterials (Figure 6). The borders between nanostructures, which serve as the fundamental building blocks of microstructures with tailored chemical and physical characteristics, may be coherent or incoherent. Grain boundary engineering is the process of modifying the nature of the borders that already exist between the nanostructured building blocks and changing the characteristics of the nanostructure, particularly in heterogeneous systems.

Nanomaterials may take on spherical, tubular, or irregular shapes and can be isolated, fused, aggregated, or agglomerated. Fullerenes, single- or multi-walled nanotubes, dendrimers, and manufactured quantum dots are a few examples of typical nanomaterials. Nanomaterials, which differ physically and chemically from ordinary materials like CNTs, fullerene, Au and Ag nanoparticles, and silica, are used in nanotechnology.

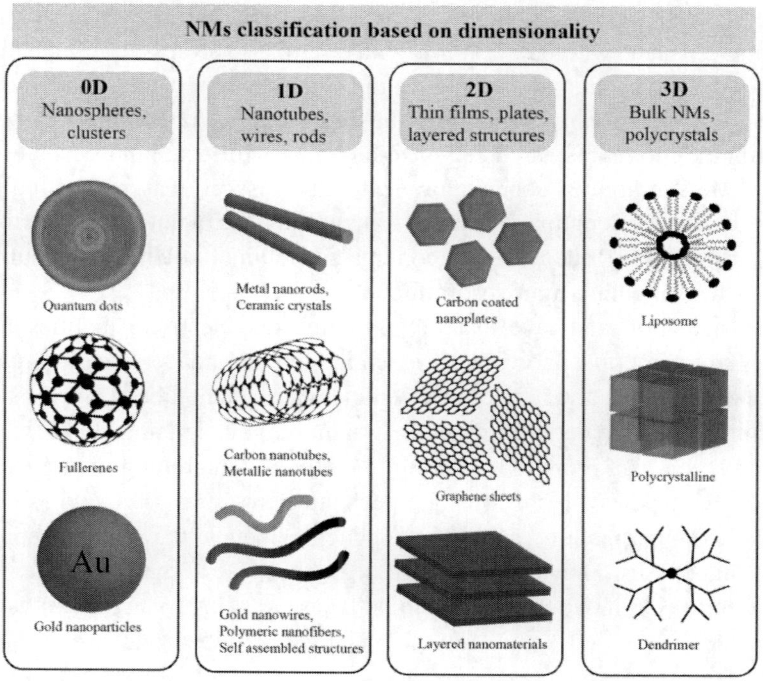

Figure 6. Classification of Nanomaterials based on their dimensions (Poh et al. 2018).

5.1. 0D Nanomaterials

The majority of zero-dimensional nanomaterials are synthetic nanostructures, which are created by restricting the flow of the structure's electrons to within 100 nanometers in all three dimensions. The electrons existing in the nanostructure will be deeply entrapped in a dimensionless region, meaning that the nanostructure will have all three dimensions within the nanoscale range. Quantum dots, fullerene, and gold nanoparticles are examples of zero-dimensional materials. The most common forms of these 0D nanomaterials are nanospheres and nanoclusters.

Nanoclusters are gold alloys that have a metal iota core and organic clusters that are covalently attached to the surface gold molecules. These nanoclusters are so small that they naturally glow in the near infrared (NIR). Groups are formed by combining atoms and iotas. Little refers to very little issues with nanometer-sized widths made up of a few to many units. Nanoclusters have a constrained size distribution with at least one dimension between 1 and 10 nm. Nanoparticles are larger clusters of at least 1000, whilst nanoclusters are smaller clusters of up to 100 iotas in size.

5.2. 1D Nanomaterials

These materials only have one dimension over 100 nm. Excellent electron sources with minimal electric field emission include metal, metal oxides, and high aspect ratio carbon-based 1D nanomaterials (such as SWNTs, MWNTs, nanorods, nanowires, and nanofibres). Polymer nanofibre mats, veils, and webs are 1D NMs with enormous surface-to-volume ratios, improved porosity, and small holes that are utilised for decontamination, catalysis, filtration, and super-absorbent and scaffold nanomaterials for tissue engineering and wound dressings.

By altering the rod radius and using the quantum size effect, nanorods, a one-dimensional structure that offers a directed conduit for electrical transport, may be utilised to regulate the bandgap. By altering the length and breadth of nanorods, the quantum confinement effect may increase the efficiency of quantum dot conjugated polymer solar cells, resulting in smaller devices with better incoming light absorption.

Nanotubes are sturdy and have a variety of thermal and electrical qualities that chip makers demand. They are made in a lab. One day, semiconductors made of carbon nanotubes could take the place of silicon in a variety of electronic devices. Nanotube characteristics include the quantity of concentric cylinders, the cylinder's radius, and its length. Chirality, which displays longitudinal twisting, is a property of certain nanotubes. Combining multiple nanotubes may result in nanomachines, which are compact mechanical systems.

Carbon, silicon, germanium, and other conductive metals like gold and copper may all be used to create nanowires. They are good conductors because to their tiny size, which makes it possible for electrons to move through them with ease, leading to significant advancements in computer science.

5.3. 2D Nanomaterials

These materials come in two sizes greater than 100 nm. They resemble plates and feature thin layers that are at least one atomic layer thick. Some examples of 2D nanomaterials include graphene, graphene oxide, reduced graphene oxide, silicate clays, layered double hydroxides, transition metal dichalcogenides, transition metal oxides, black phosphorus, graphitic carbon nitride, hexagonal boron nitride, antimonite, boron nanosheets, and tin telluride nanosheets.

Their distinctive physical, chemical, optical, and biological properties account for their uniform forms, high surface-to-volume ratio (in contrast to the bulk material), and surface charge.

Graphene, a two-dimensional sheet of carbon atoms organised in a chicken wire pattern, is a unique material with a range of fascinating features, including its mechanical strength and remarkable thermal, electrical, and optical capabilities. Graphene is a material that has been the focus of much study and development, but its present high cost continues to be a hurdle.

Graphene oxide is a kind of graphene with features not seen in graphene and comprises oxygen functional groups. By reducing graphene oxide, these oxidised functional groups are taken out, resulting in graphene. Reduced graphene oxide, or rGO, is the name given to this kind of graphene. Another method for producing rGO is to reduce graphite oxide, which is made up of multiple layers of graphene oxide, to graphene oxide and ultimately to rGO (Figure 7) (Priyadarsini et al. 2018).

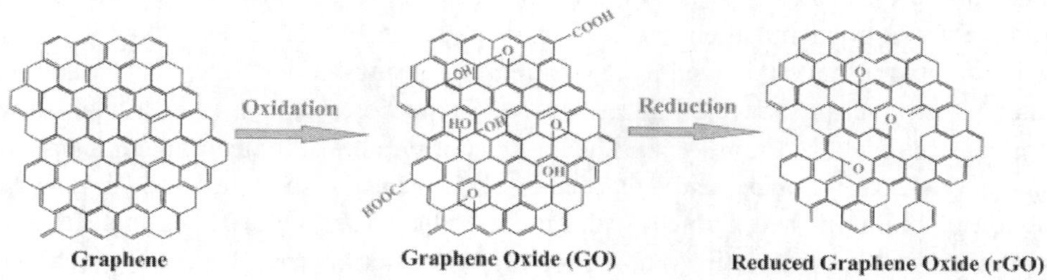

Figure 7. Two-Dimensional structure of Graphene, Graphene Oxide, and Reduced Graphene Oxide.

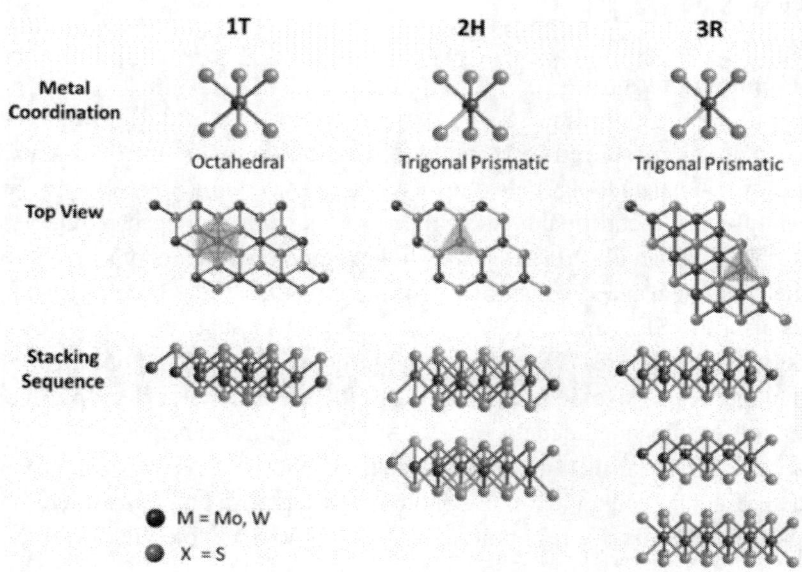

Figure 8. Atomic structure and layered arrangement of TMDCs (Freedy and McDonnell 2020).

2D TMDCs, like graphene and its variations, are perfect for loading a range of molecules, including chemical compounds and genes, via van der Waals interactions or covalent bonding because of their ultrathin structure and high surface-to-mass ratio (Figure 8). Compared to graphene, 2D TMDCs have a very different electrical structure. Graphene and 2D TMDCs have different electrical structures, making graphene a direct band gap semiconductor and 2D TMDCs an indirect band gap semiconductor. Furthermore, a number of chemical modifications are the only ways that graphene may acquire a broad range of characteristics. 36 different kinds of materials fall under the broad category of two-dimensional TMDCs.

5.4. 3D Nanomaterials

These materials have three dimensions that are more than 100 nm. Box-shaped graphene nanostructures, bundles of nanowires, and nanotubes are all included in this class, although they are not the only ones. The design, manufacture, and evaluation of 3D nanostructures as electrodes for electrochemical energy conversion, storage like batteries and super capacitors, and water treatment have received a lot of attention during the last ten years. Materials with intricate 3D structures may exhibit distinctive properties, such as the crucial shift from mechanical strength to biostability in a square twist origami. Microelectromechanical systems, healthcare devices, robotics, and solar cells all depend on complex 3D structures. The production of buildings and gadgets with an unparalleled degree of functional integration is made possible by the synergistic combination of nanomaterials and 3D printing technology.

6. Types of Nanomaterials

6.1. Metal Nanoparticles

Metal nanoparticles are produced by reducing metal-ion precursors in solution using chemical reducing agents. Metal nanoparticles are synthesised chemically, electrochemically, or photochemically using metal precursors. Since they have a high surface energy, they might possibly adsorb tiny molecules for research, biomolecule sensing and visualization, as well as biological and ecological analytical applications. For instance, the substance is coated with gold nanoparticles prior to a SEM analysis. This is often done to enhance electrical flow and generate the best possible SEM pictures ("What Are the Different Types of Nanoparticles?" n.d.).

Metal nanoparticles' size-dependent optical characteristic is surface plasmon resonance (SPR). Silver and AuNPs are the main areas of SPR's attention. The plasmon will interact with excitation source when a metallic nanoparticle is photoactivated, increasing the atom's electromagnetic field. When an incoming photon interacts with an electric field, it scatters and swallows light. Surface plasma resonation research is fairly reliable in conventional analyzers and industry.

The most reliable metal nanoparticles are AuNPs. Investigations into size-dependent electrical and optical characteristics are extensive. As was previously shown, the size-subordinate hiding difference in AuNPs accounts for the surface plasma band (SPB). The SPB is anticipated to be absent in AuNPs with an estimated size of less than 2 nm. Popular uses for AuNPs include surface coatings, solar cells, inks, sensors, optics, catalysis, clinical diagnostics, and solar cells. They are also used as experts in electron-dense naming in cytochemistry and histochemistry.

AgNPs: When compared to other metal nanoparticles, AgNPs have greater antibacterial activity. It is anticipated that bacteria, illnesses, and other eukaryotic microorganisms will be very resistant to AgNPs. The cytotoxicity of AgNPs to healthy microbes and the human body has not yet been proved, despite their usage in several commercial goods. Ag nanoparticles' capacity to fight bacteria is being researched. Some believe that the disintegration of the bacterial cell wall is caused by an interaction between Ag nanoparticles and the membrane. Reactive oxygen species are allegedly produced when AgNPs block a bacterial respiratory enzyme, according to another idea. AgNPs have been demonstrated to be absorbed by localised SPR at wavelengths between 380 and 450 nm. SPRs were cited as being crucial for antibacterial properties.

CuNPs: Cu is a plentiful element that is easily and cheaply produced. CuNPs' oxidation-induced instability restricts their use in a number of industries. Scientists have created more intricate CuNP structures, such as core/shell CuNPs, to address this problem. CuNPs may be linked to a range of substrates, such as iron oxides, silicon dioxide, carbon-based nanomaterials, and polymeric materials, to produce extremely active, sensitive, and stable nanocatalysts. Cu is a transition metal in three dimensions with a variety of mechanical and physical characteristics. Cu has four oxidation states, each of which opens a one- or two-electron pathway: Cu^{0+}, Cu^{1+}, Cu^{2+}, and Cu^{3+}. CuNPs have a specific set of features that make them useful in nanocatalysis.

6.2. Semiconductor Nanoparticles

Semiconductor nanoparticles have characteristics with both non-metals and metals. They may be located in the periodic table's groups II–VI, III–V, or IV–VI. These particles may display a variety of behaviours when adjusted because of their wide bandgaps. They can be used in electrolysis of water, photo-optics, semiconductors, and photocatalytic degradation.

Examples of semiconductor nanoparticles include GaN, GaP, InP, and InAs from groups III to V, as well as ZnO, ZnS, CdS, CdSe, and CdTe from groups II to VI, silicon, and germanium from group IV.

6.3. Carbon-Based Nanomaterials

Carbon is an essential element in nature because of its capacity to connect in a variety of ways. The most crucial biological building blocks, including DNA and proteins, may be identified by looking at their structures as being mostly made of carbon bonded to nitrogen, hydrogen, and oxygen. Since it could exist in *sp hybridization*, the carbon atoms with an electrical arrangement of $1s^2\ 2s^2\ 2p^2$ can take on a variety of crystallographic and amorphous shapes. For an exceptionally long time, graphite and diamond were considered to be the only two allotropic forms of carbon.

In 1985, by coincidence, Kroto and Smalley noticed an unlikely outcome in the mass spectra of dissolving carbon that was dissolving. The strongest fullerene is considered to be

C60. C60 is composed of sp-hybridized 60 identical carbon iotas. The detailed C60 has an estimated width of 0.7 nm.

CNTs are hexagonal, one atom thick layers of graphene that have been wrapped up. The benzene rings found in carbon atoms are comparable to these hexagonal rings. The chirality of the carbon atom determines this substance's metallic and semiconducting properties. CNTs were discovered in 1991 by Sumio Lijima, a NEC Corporation employee who produced them using the arc-discharge technique. There were CNTs with a diameter of 4 to 30 nm and a length of 1 m.

Graphene is made up of hexagonally organised sheets of sp-bonded carbon atoms that are one atom thick. This structure serves as the basis for all other carbon nanomaterials. For example, it might be coated with carbon nanotubes or laminated with graphite (CNTs). It may acquire pentagons to transform into a fullerene. Graphene's electrical zero-gap semiconductor behaviour is closer to that of metals. Due to its unique characteristic, various electrical phenomena, such as pseudospin chirality, conductivity without charge carriers, and ballistic transport, have been discovered (the movement of charge carriers across a medium).

6.4. Quantum Dots

Quantum dots (QDs) were created as a result of the 1980 finding that the band gap is dependent on particle size. When smaller nanocrystals ranging between 2 to 10 nm were stimulated into these states, scientists began to investigate them. When the outcomes of these experiments were compared to their bulk equivalents, it was discovered that these small nanocrystals had different electrical properties than bulk semiconductors. A high surface-to-volume ratio and quantum confinement both have an impact on this characteristic fluctuation. Quantum confinement happens when a molecule is smaller than the distance between electron-hole pair i.e., the excitonic Bohr radii. As an illustration, the band gap of bulk PbSe is 0.28 eV, and the Bohr exciton radius is 46 nm.

In contrast to the mass of PbSe, the band hole moves to 0.82 eV, yielding a solid confined bathochromic shift of.500 m eV, as if the thickness of PbSe were shrunk to 4.8 nm. The greatest distinguishing feature of these materials is their fluorescence, which may exhibit various frequencies depending on the size of the molecule. If the molecule is smaller than the Bohr span, the energy level will become quantized (Pauli's avoidance rule). Instead of using building materials, this energy level technique may measure down to the atomic level.

6.5. Ceramic Nanoparticles

Ceramic nanoparticles are inorganic solids formed of oxides, carbides, carbonates, and phosphates. These nanoparticles are distinguished by their higher temperature resistance and excellent corrosion resistance. They might be used for photocatalysis, dye degradation, imaging, and drug delivery.

By adjusting their dimensions, surface area, permeability, surface to volume ratio, and other features, ceramic nanoparticles function effectively as drug carrier agents. Medications for conditions including cancer, glaucoma, and infectious diseases have all been effectively delivered by these nanoparticles.

6.6. Lipid-Based Nanoparticles

Lipid nanoparticles are generally spherical in form and range in diameter from 10 to 100 nm. It is made up of a soluble lipophilic molecular matrix and a stable lipid core. Emulsifiers and surfactants help to stabilise the outer core of these nanoparticles. These nanostructures are being utilized in the field of biotechnology as cancer therapy agents, drug delivery systems, and RNA discharge agents.

In light of this, the field of nanotechnology is not yet completely established and is, like the statistic suggests, set for exponential growth. It talks about the state of biotechnology in the 1980s compared to the 1960s state of information technology. Therefore, it is straightforward to predict that this topic will have the same exponential growth as the two prior technological domains.

6.7. Polymeric Nanoparticles

Organic-based polymeric nanoparticles have a small size. These have nanocapsular or nanosphere-like structures, depending on the technique of production. While a nanocapsular particle has a core-shell shape, a nanosphere has a matrix-like structure. In the latter, as opposed to the former, where they are uniformly distributed throughout the polymer, the active chemicals are confined and enclosed by a polymer shell.

Polymeric nanoparticles have various benefits, including controlled drug release, therapeutic drug macromolecular protection, the ability to integrate therapy with imagery, accurate targeting, and many more. They are used in diagnostics and medicine delivery. Polymeric nanoparticle-based medication delivery systems are very biocompatible and biodegradable.

6.8. Nanocomposites

This category includes multi stage materials with at most one nanoscale element (between 1 and 100 nm) or nanometric phase boundary. The basis of matrix-based hybrid nanocomposites may be made of a polymer, ceramic, hydrogel, metal, or other substance. In nanocomposites, polymers serve as a medium for both organic and inorganic nanostructures of diverse shapes. Depending on whether organic or inorganic nanoparticles are used as reinforcement, the composite has different characteristics. For instance, professional ceramics are ceramic matrix composites, which are fiber-reinforced structures designed for specialised applications in aerospace sector. Hydrogels exhibit easy interactions with both polymeric chains and nanostructures and possess bioactive properties that may be tuned to certain tissue engineering applications. Metal nanocomposites are composed of several metals. Examples include intermetallic complexes as well as alloys including nanoscale metal particles and core-shell nanoparticles. Carbonaceous materials used in the aircraft sector may be present in the metal matrix.

7. Properties of Nanomaterials

Nanomaterials have structural traits that are halfway between atoms' and bulk materials'. Even while most microstructured materials share features with the corresponding bulk materials, the material properties with nanoscale dimensions are quite unique from those of atoms and bulk materials. This is mostly caused by the materials' nanoscale size, which makes them: (i) a significant portion of the surface atoms are present; (ii) the surface energy is high; (iii) there is spatial confinement; and (iv) there are fewer defects than in the analogous bulk materials.

7.1. Mechanical/Physical Properties

Numerous industries, including surface engineering, and nano-manufacturing or nanofabrication, are made possible by the special mechanical characteristics of nanoparticles. As a particle's size approaches the nanoscale or the atomic density on the surface of an amorphous particle changes, both of which have characteristic length scales that are comparable to or smaller than the de Broglie wavelength of the charge carrier or the wavelength of light, the periodic boundary conditions of the crystalline particle are also destroyed.

7.2. Chemical Properties

Nanoscale differences in the chemical characteristics of nanomaterials also exist. Because there are more surface atoms in nanoparticles than in bulk materials, nanomaterials have higher reactivities than bulk materials. One important factor influencing changes in material behaviour at the nanoscale is surface dominance. Aspects like electrical transport of charge carriers are no longer governed by solid-state electronic bulk processes since the surface atoms present will make up around half of all the atoms in nanoparticles. Since nanomaterials contain a larger percentage of surface atoms than lengthier structures do, their atoms have a higher average energy. When a catalyst is made smaller to the nanoscale, for instance, it exhibits enhanced chemical activity per exposed surface atom. The structure and the nature of chemical bonding between the atoms present at the surface may influence interactions between particles at such tiny scales, and surfaces and interfaces can draw defects and impurities. Molecular monolayers may be utilised to alter or control surface characteristics and also facilitate interactions between nanoparticles ("Chemical Properties of Nanomaterials| Winner Science" n.d.).

7.3. Electrical Properties

The principles of electrical conductivity in various nanomaterials like nanorods, CNT, photoconductivity and electrical conductivity of nanorods and composites are examined in "Electrical Properties of Nanoparticles." Mechanically weakening a nanowire to estimate the electrical flow for a constant voltage applied is an innovative method for illustrating the changes in conductance.

A single electron wave mode carries the electrical flow in carbon nanotubes that are electrically leading. Given that carbon nanotubes come into touch with the outer layer of mercury at different periods due to their variable lengths and orientations, two sets of data are produced: (i) the influence of CNTs dimension (especially length) on opposition; and (ii) the protections of various nanotubes. Since the nanotubes' lengths vary, as the fibre group grows, more carbon nanotubes will come into touch with the mercury drop's outer layer and help carry electrical current.

7.4. Optical Properties

Enthralling features of semiconductor and metallic nanoparticles and nanocomposites include their linear absorption, photoluminescence emission, and nonlinear optical characteristics. Due to the quantum confinement phenomenon, nanomaterials with tiny particle sizes have increased optical emission and nonlinear optical characteristics.

By adjusting its form, size, and surface functioning, a variety of optical effects may be produced for a number of purposes. Depending on the makeup, size, and orientation, this manipulation may be done in a number of ways. Numerous applications of nanoparticles depend on their optical characteristics. With modifications in form, size, or type, they may limit their electrical characteristics to induce quantum effects, which affects the colour emitted by the nanoparticle. While a 100 nm spherical gold nanoparticle emits orange colour and a 25 nm gold nanoparticle appears green. Similar to how spherical silver nanomaterial with a 100 nm diameter looks yellow and spherical silver nanomaterial with a 100 nm diameter appears orange (Table 2).

Table 2. Comparison of change in optical properties (colour) of nanoparticles corresponding to the change in its size

Nanoparticle	Size and Shape	Color
Gold	10–20 nm	Red
	2–5 nm	Yellow
	> 20 nm	Purple
Silver	40 nm	Blue
	100 nm	Yellow
	Prism shape	Red

The optical properties of nanomaterials are impacted by elements, for example, include size, shape, surface characteristics, and different factors, for example, doping and contact with the general climate or other nanostructures. Essentially, structure might fundamentally affect the optical qualities of metal nanostructures. The optical qualities of CdSe semiconductor nanoparticles are modified by a basic change in size. Whenever metal nanoparticles are amplified, their optical qualities change pretty much nothing, as displayed in the different gold nanosphere tests. Whenever an anisotropy is brought into a nanoparticle, the optical qualities of the nanoparticles modify considerably (nanorods development for example).

8. Fabrication of Nanomaterials

Engineering at the nanoscale is a difficult task, and scientists must devise whole new ways to construct from the "bottom-up" rather than the typical "top-down" method (Figure 9).

Some nanomaterials, like nanowires and other fundamental devices, have been shown to self-assemble under the right circumstances, while other investigations at larger sizes are seeking to exemplify the core concepts of self-assembly. Like trees, microelectronics may be encouraged to develop from the ground.

New ways are being discovered to incorporate protein molecules, DNA, viruses, bacteria, and various microorganisms in the construction of nanomaterials, as well as drawing inspiration from the natural world.

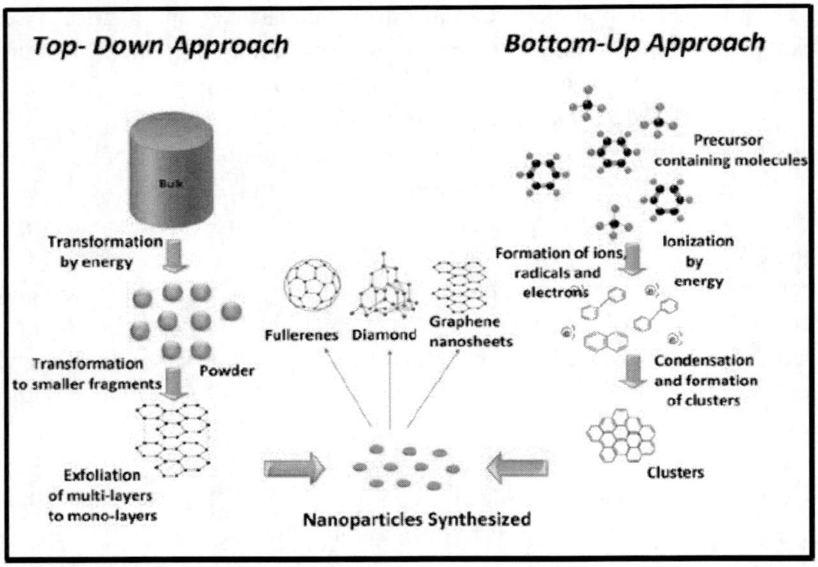

Figure 9. Schematic of Top-down and Bottom-up approach involved in the nanofabrication process.

8.1. Top-Down Approach

Using top-down techniques, nanomaterials may be created by constructing minuscule structures from larger material fragments. Etching may be used to construct circuits on the surface of silicon microchips, for instance.

8.1.1. Mechanical Grinding

Bulk material may be mechanically milled down to a nanoscale in size. This approach may be used to combine many stages. This approach is better suited for large-scale manufacturing. The energy that is transferred from the balls to the sample throughout the process is the fundamental tenet of this procedure. Steel and tungsten are often utilised ball materials because dense materials are needed for milling (Figure 10). The kinetic energy of the ball and the properties of the sample powder determine the temperature that is produced throughout the procedure.

The amount of strain experienced during the balls' impact also affects the final particle's form and structure ("High Energy Ball Milling Process for Nanomaterial Synthesis" n.d.).

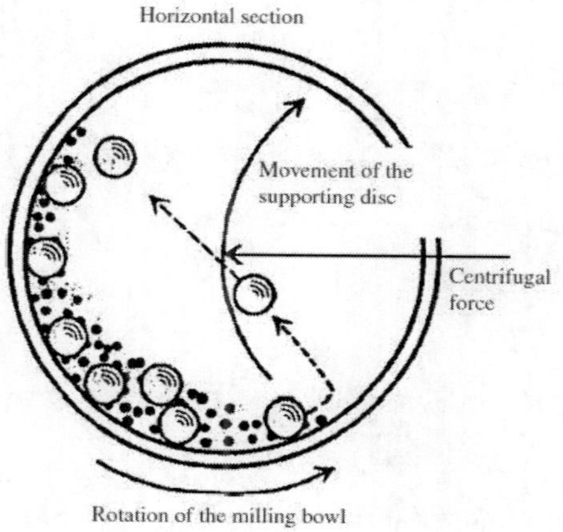

Figure 10. Schematic of mechanical grinding of nanoparticles.

8.1.2. Mechanochemical

If the circumstances aren't correct, milling might result in powder deformation, fracture, and welding. The powder begins to bend plastically as a result of the production of shear bands, which eventually break down into subgrains. Continued milling results in a subgrain size reduction to the nanoscale (Figure 11). The processing speed at which mechanochemical processing yields to the nanoscale is its distinctive characteristic. The reaction kinetics are accelerated by the nanoscale powder, causing chemical reactions that essentially need high temperature (Moores 2018).

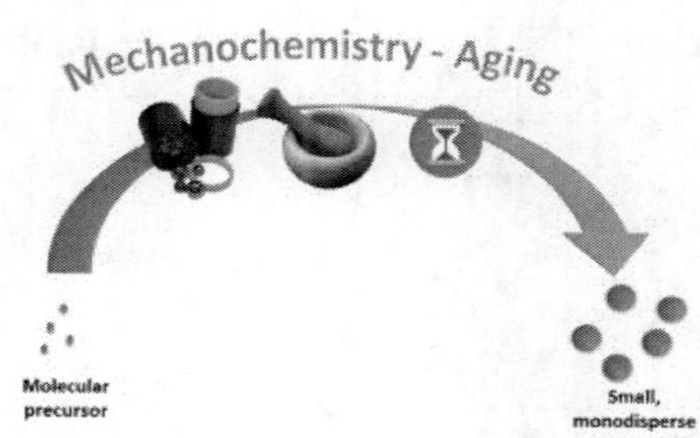

Figure 11. Schematic of mechanochemical aging.

8.1.3. Electroexplosion

The aerosols of various metals like gold, silver, aluminium, copper, etc, are produced using the electroexplosion technique developed by Karioris and Fish in 1962. The produced particles ranged in size from 30 to 50 nm. This essentially represents the electrical explosion of metal wires that produced highly active nanoparticles. (Figure 12). A metal wire is subjected to a high-density electric pulse (104–106 A/mm), which causes a rise in heat upto 30,000°C and subsequent explosion and generation of particles that are transformed into nm sized nanoparticles by passing the appropriate gas further. This technique makes it feasible to create powders with special qualities that are difficult to create using other techniques.

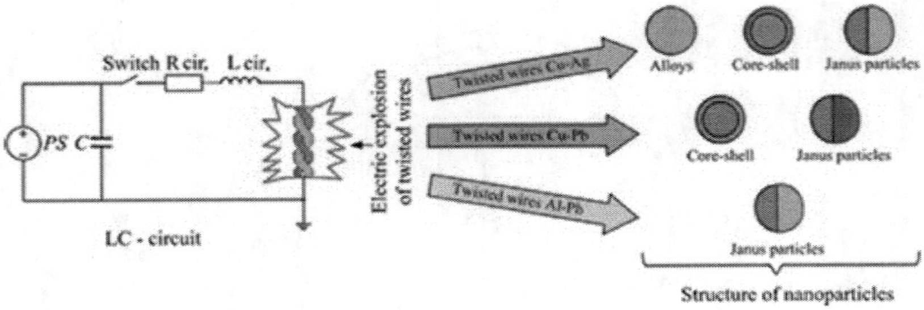

Figure 12. Electroexplosion of various nanoparticles (Lerner et al. 2016).

8.1.4. Sputtering

Grove made his first observations of the material being deposited within the discharge tube and the energetic ions spitting on the tube's cathode surface in 1852. These days, this method involves bombarding a sample surface with powerful gaseous ions, which causes surface atoms or tiny clusters to be ejected. There are three distinct methods of sputtering: DC-diode, RF-diode, and mag-neutron sputtering. The main use of argon plasma is sputtering. There is no melting of the samples as there is with other vapour phase procedures (Figure 13).

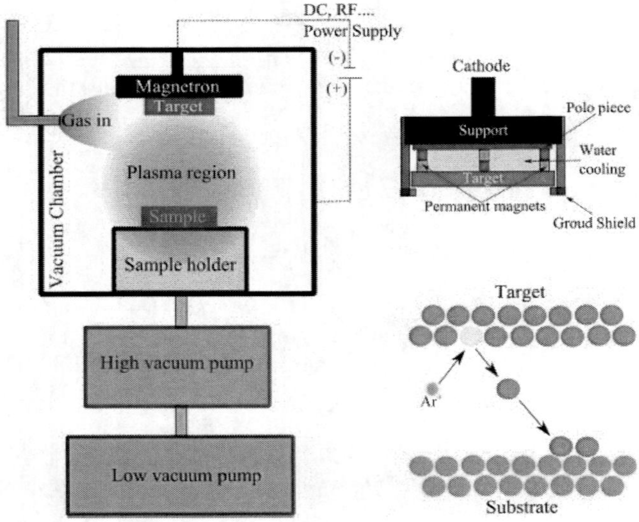

Figure 13. Schematic of sputtering process (Wender et al. 2013).

8.1.5. Laser Ablation

The process of laser ablation (LA) is intricate, as the laser may hit the sample surface depending on its wavelength and the refractive index of the target material (Figure 14). Strong electric fields generated by laser light are adequate to draw electrons out of bulk materials. When the newly formed free electron bumps into the atoms of the bulk sample, energy is transmitted. The surface heats up as a result, which leads to vaporisation. When the laser flux is sufficient, the material, including the atoms, molecules, ions, and clusters, will change into the plasma state.

Due to the difference in pressure between the seed plasma and its surroundings, it quickly expanded and cooled. The laser ablation may happen in a vacuum or a gaseous atmosphere. The laser ablation and a tube furnace are combined in a pulsed laser ablation process. This technique provides more precise control over the pressure, the type and quality of the gas used, and also the growth temperature (Haider, Haider, and Mehde 2018).

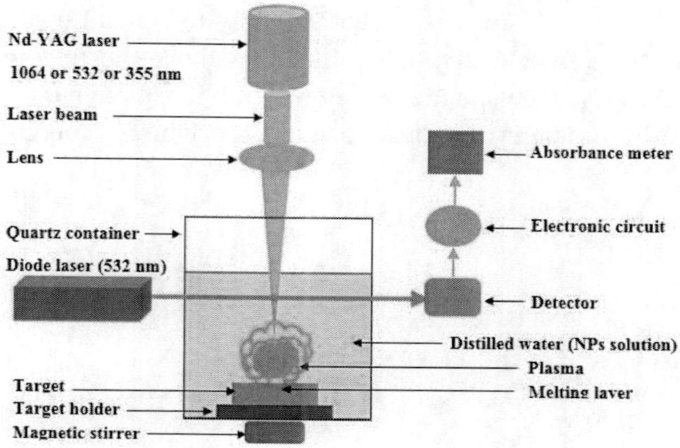

Figure 14. Schematic of laser ablation process.

8.1.6. Lithography

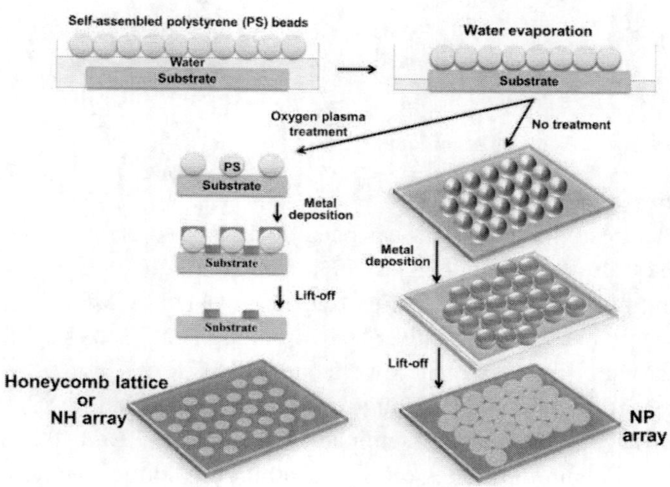

Figure 15. Micro and nanolithography technique.

Integrated circuits have been made for many years using micro- and nanolithography technologies. With this method, nm-sized patterns may be produced. High-resolution topography is often produced by combining lithography with deposition and etching. Masked and maskless lithography are two categories of lithography (Figure 15). In masked lithography, patterns will be created using masks or moulds. Photolithography, soft lithography, and nanoimprint lithography are a few examples of masked lithography types. Without the use of masks, lithography generates arbitrary patterns. For instance, scanning probe lithography, focused ion beam lithography, and electron beam lithography.

8.1.7. Aero-Sol Techniques

A combination of solid and liquid particles in a gaseous atmosphere called an aerosol. Aerosol technologies have an edge over other ones because they produce very pure products with lower environmental toxicity. The following categories apply to aerosol methods: Among the five methods, there are the furnace technique, flame method, electrospray, chemical vapour deposition (CVD), and physical vapour deposition method. Although the electrospray approach is said to be the best for producing nanoparticles, the procedure is sluggish (Figure 16).

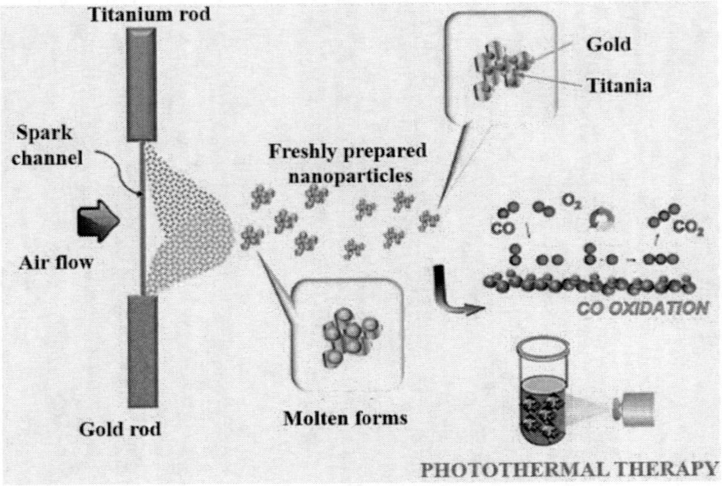

Figure 16. An overview of Aero-sol technique.

8.1.8. Electrospinning

Since 2012, the electrospinning process has been developed to create continuous fibres in the submicron to nanoscale size range (Figure 17). This technique is used to create composite, metal, ceramic, and polymer nanofibers. Scaffolds are created by electrospinning a mixture of nanoparticles and polymers. Additionally, electrospinning is utilised to assemble nanoparticles by aligning them with fibres, which lowers the Gibbs Free energy. Another benefit of electrospinning is that no functionalization procedure is needed; all that is needed is a solvent that can dissolve the polymer and disseminate nanoparticles. High electrostatic forces are essentially what electrospinning relies on. Concentration of polymers, solution viscosity and flow rate, electric field strength, and air humidity are all variables that affect the electrospinning process.

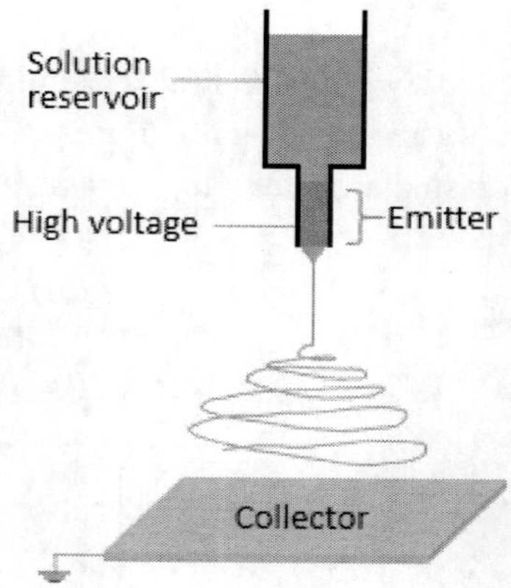

Figure 17. Schematic of Electrospinning ("Electrospinning - Nanoscience Instruments" n.d.).

8.2. Bottom-Up Approach

Atom by atom or molecule by molecule methods may be used to build nanostructures. Self-assembly is one method for doing this, in which the atoms or molecules organise themselves into a structure as a result of their inherent characteristics. Self-assembly may be seen in the growth of crystals for the semiconductor industry as well as in the chemical synthesis of big molecules.

8.2.1. Chemical Vapour Deposition

This technique uses a chemical process to deposit a sample material in a vapour medium (Figure 18). The substance is deposited as a thin layer, a powder, or a single crystal. The benefits of CVD include selective deposition on the desired pattern, thin films with consistent thickness and low porosity %, and great throwing power. Thin films made of dielectric, conductors, conductive oxides, tribological and corrosion-resistant coatings, heat-resistant coatings, etc. are all made using CVD. The creation of solar cells and high-temperature fibre composites are further uses.

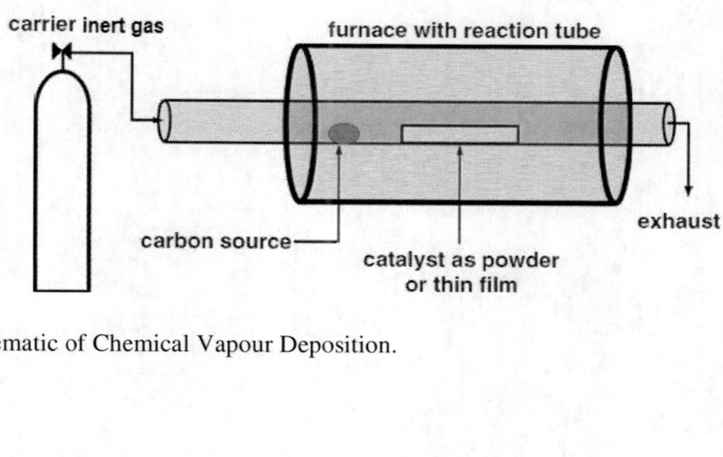

Figure 18. Schematic of Chemical Vapour Deposition.

8.2.2. Plasma Arcing

By conducting electricity through gas with the aid of a potential difference established between two electrodes, plasma is essentially an ionised gas (Figure 19). An inert gas and a plasma torch are used to create a constricted plasma arc (Shigeta, Tanaka, and Ghedini 2019).

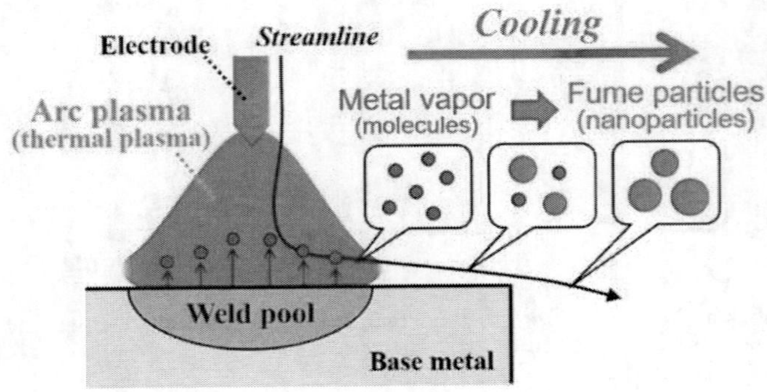

Figure 19. Schematic of plasma arcing process.

The most common technique for creating both single-walled and multi-walled nanotubes is the arc-discharge process. For this, the anode and cathode are often made of graphite electrodes. DC current causes the electrodes to melt (B100 A). Carbon rods are produced at the cathode side after the completion of the arc discharge for a length of time. MWCNTs are often produced by this approach, however single-walled CNTs may also be produced with the aid of metal catalysts including nickel, molybdenum, and iron.

8.2.3. Wet Chemical Methods

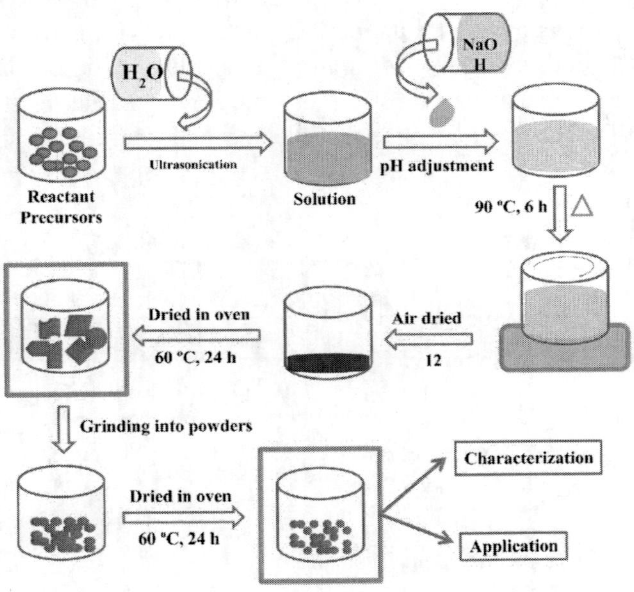

Figure 20. The process involved in wet chemical method.

The appropriate size of homogenous nanoparticles is produced using a wet chemical technique. Given that the size, shape, and crystallinity could all be controlled, this procedure was quite successful. Because it is inexpensive and simple to utilise, wet chemical synthesis is often employed to create inorganic nanomaterials (Figure 20). Due of the lengthy mixing time and unpredictable nucleation and development, this approach also has significant drawbacks on an industrial scale.

8.2.4. Hydrothermal Synthesis

The most used technique for creating nanomaterials is hydrothermal synthesis. The process essentially involves the nucleation of crystals and their development (Figure 21). Nanomaterials can be formed via hydrothermal synthesis over a broad range of temperatures, from very low to extremely high. Depending on the vapour pressure of the primary composition in the reaction, either low-pressure or high-pressure conditions may be utilised to regulate the morphology of the materials to be synthesised. This method has been used to successfully synthesis a wide variety of nanomaterials. The hydrothermal synthesis approach has a number of benefits over other ones. Nanomaterials produced by hydrothermal synthesis may not be stable at high temperatures. The hydrothermal approach allows for the production of nanomaterials with high vapour pressures while minimising material loss. Through liquid phase or multiphase chemical processes, the compositions of the nanomaterials to be produced may be tightly controlled during hydrothermal synthesis. By adjusting the temperature, pH, and reactant concentrations, particle shape may be managed.

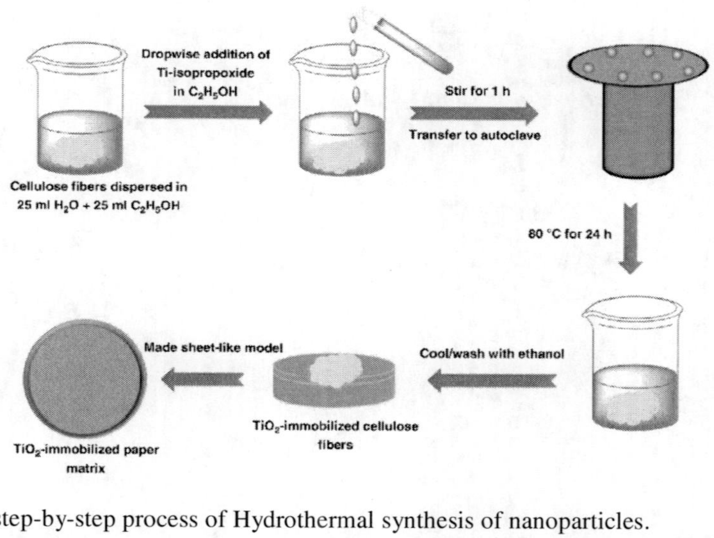

Figure 21. The step-by-step process of Hydrothermal synthesis of nanoparticles.

8.2.5. Reverse Micelle

A hydrophilic head and a hydrophobic chain make up surfactants, under certain circumstances, these amphiphilic molecules may self-assemble into a variety of forms. Polar cores generated with hydrophilic heads define the structure of reverse micelles. These solutions, known as microemulsions, are optically clear and thermodynamically stable (Figure 22). The size of the droplets in water-oil microemulsions ranges from 5 to 100 nm. By altering the parameters, such as the molar ratio of water to surfactant, the reverse micelles may be tailored to the nanoscale

scale. These micelles may be used to create nanostructures with a controlled shape by adsorbing surfactants to inorganic materials (Cheney, Lauterbach, and Chen 2011).

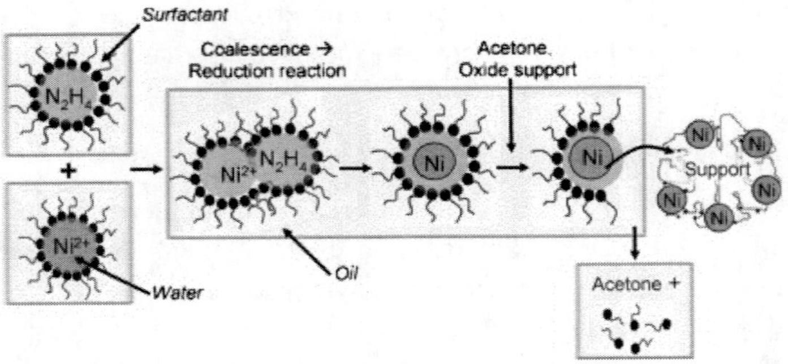

Figure 22. The reverse micelle process.

8.2.6. Sol-Gel Method

Sol-gel technique is mostly used to create mixed oxide composites and metal oxide NPs with desired nanostructures. The processes of a typical sol-gel technique include hydrolysis, condensation, and drying. The metal precursor first experiences hydrolysis to produce metal hydroxide, which is then followed by condensation to produce gels (Figure 23). After drying, the final gel is transformed into xerogel or aerogel (Tripathi et al. 2015).

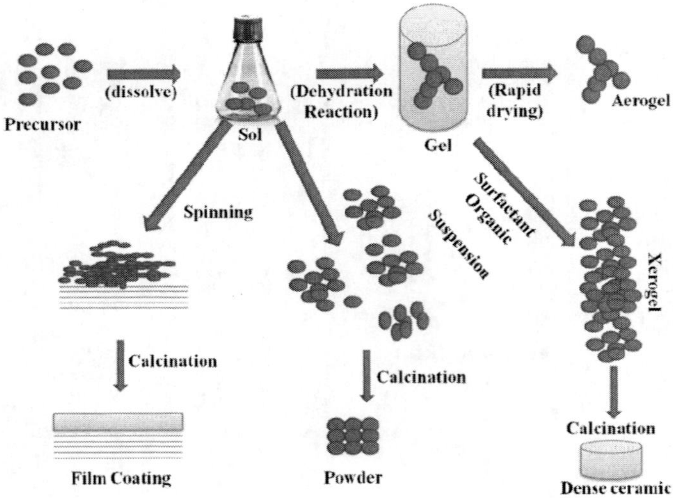

Figure 23. The steps involved in sol-gel method.

8.3. Magnetic Properties

The five fundamental types of magnetism in nanoparticles are diamagnetism, paramagnetism, ferromagnetism, antiferromagnetism, and ferrimagnetism based on the orientation behaviour of magnetic moments in regard with the applied electric and magnetic field (Figure 24).

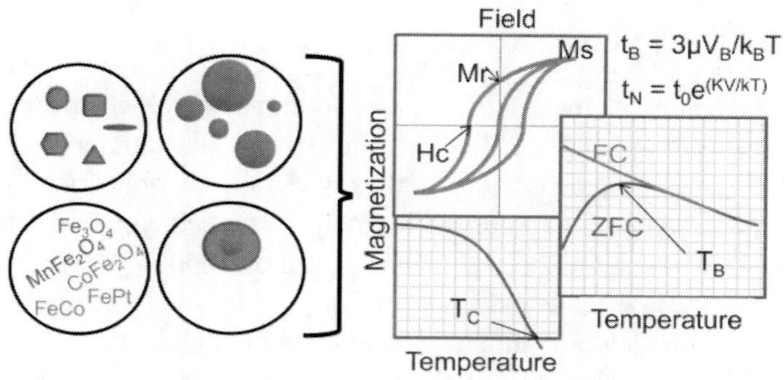

Figure 24. Change in magnetic properties of nanomaterials based on their size and shape (Kolhatkar et al. 2013).

Due to the orbital motion shown by the opposing/repulsive force of electrons, an atomic loop current will be generated on applying a magnetic field to the nanomaterial. Nearly all materials exhibit the repelling force that electrons exhibit; this behaviour is known as diamagnetism. The electronic subshells of the materials will neutralise this repulsive force, resulting in a negative susceptibility (example: quartz) (Jana, Chen, and Peng 2004; Hyeon et al. 2001; Park et al. 2004).

Although they are magnetic at the nanoscale, gold and platinum are not magnetic in bulk. In contrast to bulk atoms, surface atoms may be changed by interactions with other chemical species, such as capping by nanoparticles. By capping them with the suitable chemicals, this method makes it possible to change the physical characteristics of nanoparticles. In fact, when synthesised at the nanoscale, non-ferromagnetic bulk materials may show ferromagnetic-like properties. Magnetic Pd, Pt, and—in an unexpected turn—Au nanoparticles may be made from nonmagnetic bulk materials (which is diamagnetic in bulk). Because of size effects, structural changes in Pt and Pd result in ferromagnetism. The charge concentrated at the particle surface causes ferromagnetic-like behaviour, whereas the gold nanoparticles exhibit ferromagnetic behaviour only when the proper molecules cap their surface. Bulk Au becomes diamagnetic and has a very low density of states (Leslie-Pelecky and Rieke 1996).

All other magnetic properties of materials may be traced back to unpaired electrons in their atomic shells, which are frequently present in the 3d or 4f shells of each atom. Materials having uncoupled atomic magnetic moments exhibit paramagnetism, which is characterised by the absence of long-range order in the moments and a very low positive magnetic susceptibility (Chen, Liu, and Sun 2004; Sun et al. 2000).

Magnetic nanoparticles are used in a variety of industries, including disc recording and biomedical applications like magnetic resonance contrast media and cancer therapeutic agents. For each conceivable use, magnetic nanoparticles must have a particular set of characteristics. Particles need to be in a stable, switchable magnetic state in data storage applications, for instance, to represent information bits that are unaffected by temperature changes (Akbarzadeh, Samiei, and Davaran 2012; Niculescu, Chircov, and Grumezescu 2021).

Conclusion

Applied and device physics, materials science, supramolecular chemistry, mechanical and electronics engineering are just a few of the fields that fall under the umbrella of nanotechnology. Future equipment and products based on nanomaterials, whose quality has been approved by several regulatory authorities, could provide better comfort at reduced costs. With ramifications for biomedicine and industry, nanotechnology may be seen as either a nanoscale branch of more established scientific disciplines or as a new, modern concept of those disciplines. Future studies should concentrate on enhancing the properties of new nanomaterials so they may be used to a variety of applications.

Nanomaterials differ from its bulk counterparts in that they can't be studied in the same manner as bulk or tiny molecules since they have unique characteristics at the nanoscale. The characteristics of nanomaterials are influenced by their composition, chemistry, particle size, and interactions with other materials. They are being examined because of their larger surface area and smaller natural size. Nanoparticles may be produced using a variety of techniques depending on the material and composition. Despite their clear benefits, determining the toxicological and ecotoxicological characteristics of nanoparticles is a substantial problem.

Nanotechnological research should focus on the advancement of nanotechnology in the cosmetics, engineering, healthcare, and environmental sectors. Research should also be re-energized by academic publications, regulatory suggestions, and patent data on the use of nanotechnology in healthcare and personal care goods. Future study should concentrate on novel nanomaterials generated from diverse biopolymeric materials, as well as nanofibrous scaffolds for applications in effluent treatment, the food industry, cosmetics, and hard and soft tissue engineering.

The fields of nanoscience and nanotechnology are expanding rapidly; their results are already being put to use in fields as diverse as medicine (including pharmacology and biotechnology), consumer electronics, water treatment, and higher education. The Society of Toxicology said that the discoveries made in nanotechnology are among the most important in modern biological research, and that they are preparing for the next upheaval by developing a wide variety of novel materials. Researchers in the subject of nanotechnology are rediscovering ancient materials like carbon in order to advance the development of consumer and contemporary products, meet fundamental energy demands, enhance security frameworks, and advance medical practise. Nanoscale material production necessitates the introduction of novel, eminently primary and functional constituents in contrast to the original atoms or bulk material. A quick look into nanoparticles' physicochemical characteristics was included in this audit piece. There are a wide variety of nanomaterials, each with its own set of useful features. So, a deeper understanding of these characteristics will allow for the development of new and more efficient materials.

References

Akbarzadeh, Abolfazl, Mohamad Samiei, and Soodabeh Davaran. 2012. "Magnetic Nanoparticles: Preparation, Physical Properties, and Applications in Biomedicine." *Nanoscale Research Letters* 7 (1): 144. https://doi.org/10.1186/1556-276X-7-144.

Baalousha, M., J. R. Lead, and Y. Ju-Nam. 2011. "Natural Colloids and Manufactured Nanoparticles in Aquatic and Terrestrial Systems." *Treatise on Water Science* 3 (January): 89–129. https://doi.org/10.1016/B978-0-444-53199-5.00053-1.

Bai, Lina, Ting Xie, Qingqing Hu, Changyan Deng, Rong Zheng, and Wanping Chen. 2015. "Genome-Wide Comparison of Ferritin Family from Archaea, Bacteria, Eukarya, and Viruses: Its Distribution, Characteristic Motif, and Phylogenetic Relationship." *Die Naturwissenschaften* 102 (9–10). https://doi.org/10.1007/S00114-015-1314-3.

Barhoum, Ahmed, María Luisa García-Betancourt, Jaison Jeevanandam, Eman A. Hussien, Sara A. Mekkawy, Menna Mostafa, Mohamed M. Omran, Mohga S. Abdalla, and Mikhael Bechelany. 2022. "Review on Natural, Incidental, Bioinspired, and Engineered Nanomaterials: History, Definitions, Classifications, Synthesis, Properties, Market, Toxicities, Risks, and Regulations." *Nanomaterials 2022, Vol. 12, Page 177* 12 (2): 177. https://doi.org/10.3390/NANO12020177.

Besford, Quinn A, Francesca Cavalieri, Frank Caruso, Q A Besford, F Cavalieri, and F Caruso. n.d. "Glycogen as a Building Block for Advanced Biological Materials." https://doi.org/10.1002/adma.201904625.

Bishop, Kyle J. M., Christopher E. Wilmer, Siowling Soh, and Bartosz A. Grzybowski. 2009. "Nanoscale Forces and Their Uses in Self-Assembly." *Small* 5 (14): 1600–1630. https://doi.org/10.1002/SMLL.200900358.

Buchholz, Bruce A., Robert W. Dibble, and David Rich. 2003. "Quantifying the Contribution of Lubrication Oil Carbon to Particulate Emissions from a Diesel Engine." *SAE Technical Papers*, May. https://doi.org/10.4271/2003-01-1987.

"Chemical Properties of Nanomaterials | Winner Science." n.d. Accessed February 18, 2022. https://winnerscience.com/chemical-properties-of-nanomaterials/.

Chen, Min, J. P. Liu, and Shouheng Sun. 2004. "One-Step Synthesis of FePt Nanoparticles with Tunable Size." *Journal of the American Chemical Society* 126 (27): 8394–95. https://doi.org/10.1021/JA047648M/SUPPL_FILE/JA047648MSI20040601_120035.PDF.

Cheney, Beth A., Jochen A. Lauterbach, and Jingguang G. Chen. 2011. "Reverse Micelle Synthesis and Characterization of Supported Pt/Ni Bimetallic Catalysts on γ-Al_2O_3." *Applied Catalysis A: General* 394 (1–2): 41–47. https://doi.org/10.1016/J.APCATA.2010.12.021.

Chun, Youngsin, Hi Ku Cho, Hyo Sang Chung, and Meehye Lee. 2008. "Historical Records of Asian Dust Events (Hwangsa) in Korea." *Bulletin of the American Meteorological Society* 89 (6): 823–27. https://doi.org/10.1175/2008BAMS2159.1.

Czerwinski, J, P Comte, M Guedel, and P Bonsack. 2017. "Nanoparticle Emissions from Gasoline Vehicles DI & MPI." *Combustion Engines* 170 (3): 179–87. https://doi.org/10.19206/CE-2017-331.

Dai, Z. R., J. P. Bradley, D. J. Joswiak, D. E. Brownlee, H. G.M. Hill, and M. J. Genge. 2002. "Possible in Situ Formation of Meteoritic Nanodiamonds in the Early Solar System." *Nature 2002 418:6894* 418 (6894): 157–59. https://doi.org/10.1038/nature00897.

"Electrospinning - Nanoscience Instruments." n.d. Accessed February 18, 2022. https://www.nanoscience.com/techniques/electrospin/.

Ermolin, Mikhail S., Petr S. Fedotov, Natalia A. Malik, and Vasily K. Karandashev. 2018. "Nanoparticles of Volcanic Ash as a Carrier for Toxic Elements on the Global Scale." *Chemosphere* 200 (June): 16–22. https://doi.org/10.1016/J.CHEMOSPHERE.2018.02.089.

Filella, Montserrat. 2008. "NOM Site Binding Heterogeneity in Natural Waters: Discrete Approaches." *Journal of Molecular Liquids* 143 (1): 42–51. https://doi.org/10.1016/J.MOLLIQ.2008.04.018.

Freedy, Keren M., and Stephen J. McDonnell. 2020. "Contacts for Molybdenum Disulfide: Interface Chemistry and Thermal Stability." *Materials 2020, Vol. 13, Page 693* 13 (3): 693. https://doi.org/10.3390/MA13030693.

Gartia, Manas Ranjan, Austin Hsiao, Anusha Pokhriyal, Sujin Seo, Gulsim Kulsharova, Brian T. Cunningham, Tiziana C. Bond, and Gang Logan Liu. 2013. "Colorimetric Plasmon Resonance Imaging Using Nano Lycurgus Cup Arrays." *Advanced Optical Materials* 1 (1): 68–76. https://doi.org/10.1002/ADOM.201200040/FULL.

Geppert, Mark, and Martin Himly. 2021. "Iron Oxide Nanoparticles in Bioimaging – An Immune Perspective." *Frontiers in Immunology* 12 (June). https://doi.org/10.3389/FIMMU.2021.688927.

Griffin, Sharoon, Muhammad Irfan Masood, Muhammad Jawad Nasim, Muhammad Sarfraz, Azubuike Peter Ebokaiwe, Karl Herbert Schäfer, Cornelia M. Keck, and Claus Jacob. 2017. "Natural Nanoparticles: A Particular Matter Inspired by Nature." *Antioxidants 2018, Vol. 7, Page 3* 7 (1): 3. https://doi.org/10.3390/ANTIOX7010003.

Hadler, Jens, Christian Lensch-Franzen, Marcus Gohl, and Tobias Mink. 2015. "Emission Reduction A Solution of Lubricant Composition, Calibration and Mechanical Development." *MTZ Worldwide* 76 (9): 30–33. https://doi.org/10.1007/S38313-015-0040-8.

Haider, Adawiya J., Mohammad J. Haider, and Mohammad S. Mehde. 2018. "A Review on Preparation of Silver Nano-Particles." *AIP Conference Proceedings* 1968 (1): 030086. https://doi.org/10.1063/1.5039273.

Heiligtag, Florian J., and Markus Niederberger. 2013. "The Fascinating World of Nanoparticle Research." *Materials Today* 16 (7–8): 262–71. https://doi.org/10.1016/J.MATTOD.2013.07.004.

"High Energy Ball Milling Process for Nanomaterial Synthesis." n.d. Accessed February 18, 2022. https://www.understandingnano.com/nanomaterial-synthesis-ball-milling.html.

Hochella, Michael F., David W. Mogk, James Ranville, Irving C. Allen, George W. Luther, Linsey C. Marr, B. Peter McGrail, et al. 2019. "Natural, Incidental, and Engineered Nanomaterials and Their Impacts on the Earth System." *Science* 363 (6434). https://doi.org/10.1126/SCIENCE.AAU8299/ASSET/713F7638-CAC2-4D54-8D14-FDFC821E2E4B/ASSETS/GRAPHIC/363_AAU8299_FA.JPEG.

Hochella, Michael F., Michael G. Spencer, and Kimberly L. Jones. 2015. "Nanotechnology: Nature's Gift or Scientists' Brainchild?" *Environmental Science: Nano* 2 (2): 114–19. https://doi.org/10.1039/C4EN00145A.

Hyeon, T., Su Seong Lee, J. Park, Y. Chung, and Hyon Bin Na. 2001. "Synthesis of Highly Crystalline and Monodisperse Maghemite Nanocrystallites without a Size-Selection Process." *Journal of the American Chemical Society* 123 (51): 12798–801. https://doi.org/10.1021/JA016812S/SUPPL_FILE/JA016812S_S1.PDF.

"Image 129614 - UNESCO World Heritage Centre." n.d. Accessed February 25, 2022. https://whc.unesco.org/include/tool_image.cfm?id=129614.

Jana, Nikhil R., Yongfen Chen, and Xiaogang Peng. 2004. "Size- and Shape-Controlled Magnetic (Cr, Mn, Fe, Co, Ni) Oxide Nanocrystals via a Simple and General Approach." *Chemistry of Materials* 16 (20): 3931–35. https://doi.org/10.1021/CM049221K/SUPPL_FILE/CM049221KSI20040517_012350.PDF.

Kolhatkar, Arati G., Andrew C. Jamison, Dmitri Litvinov, Richard C. Willson, and T. Randall Lee. 2013. "Tuning the Magnetic Properties of Nanoparticles." *International Journal of Molecular Sciences 2013, Vol. 14, Pages 15977-16009* 14 (8): 15977–9. https://doi.org/10.3390/IJMS140815977.

Komeili, Arash. 2012. "Molecular Mechanisms of Compartmentalization and Biomineralization in Magnetotactic Bacteria." *FEMS Microbiology Reviews* 36 (1): 232–55. https://doi.org/10.1111/J.1574-6976.2011.00315.X.

Lerner, Marat I., Alexander V. Pervikov, Elena A. Glazkova, Natalya V. Svarovskaya, Aleksandr S. Lozhkomoev, and Sergey G. Psakhie. 2016. "Structures of Binary Metallic Nanoparticles Produced by Electrical Explosion of Two Wires from Immiscible Elements." *Powder Technology* 288 (January): 371–78. https://doi.org/10.1016/J.POWTEC.2015.11.037.

Leslie-Pelecky, Diandra L., and Reuben D. Rieke. 1996. "Magnetic Properties of Nanostructured Materials." *Chemistry of Materials* 8 (8): 1770–83. https://doi.org/10.1021/CM960077F.

"Lycurgus Cup: A Piece of Ancient Roman Nanotechnology | Amusing Planet." n.d. Accessed February 25, 2022. https://www.amusingplanet.com/2016/12/lycurgus-cup-piece-of-ancient-roman.html.

Moores, Audrey. 2018. "Bottom up, Solid-Phase Syntheses of Inorganic Nanomaterials by Mechanochemistry and Aging." *Current Opinion in Green and Sustainable Chemistry* 12 (August): 33–37. https://doi.org/10.1016/J.COGSC.2018.05.004.

Mubeen, Bismillah, Aunza Nayab Ansar, Rabia Rasool, Inam Ullah, Syed Sarim Imam, Sultan Alshehri, Mohammed M. Ghoneim, Sami I. Alzarea, Muhammad Shahid Nadeem, and Imran Kazmi. 2021. "Nanotechnology as a Novel Approach in Combating Microbes Providing an Alternative to Antibiotics." *Antibiotics 2021, Vol. 10, Page 1473* 10 (12): 1473. https://doi.org/10.3390/ANTIBIOTICS10121473.

Niculescu, Adelina Gabriela, Cristina Chircov, and Alexandru Mihai Grumezescu. 2021. "Magnetite Nanoparticles: Synthesis Methods – A Comparative Review." *Methods*, April. https://doi.org/10.1016/J.YMETH.2021.04.018.

"Northern Dust Brings Dirty Skies to Shanghai -- China.Org.Cn." n.d. Accessed February 14, 2022. http://english.china.org.cn/english/environment/205884.htm.

Park, Jongnam, Kwangjin An, Yosun Hwang, J. E.Geun Park, Han Jin Noh, Jae Young Kim, Jae Hoon Park, Nong Moon Hwang, and Taeghwan Hyeon. 2004. "Ultra-Large-Scale Syntheses of Monodisperse Nanocrystals." *Nature Materials 2004 3:12* 3 (12): 891–95. https://doi.org/10.1038/nmat1251.

Péwé, Troy Lewis, and American Association for the Advancement of Science. 1981. "Desert Dust: Origin, Characteristics, and Effect on Man." *Special Paper / The Geological Society of America (USA)*, 303. https://doi.org/10.3/JQUERY-UI.JS.

Poh, Tuang Yeow, Nur A. Tikah Binte Mohamed Ali, Micheál Mac Aogáin, Mustafa Hussain Kathawala, Magdiel Inggrid Setyawati, Kee Woei Ng, and Sanjay Haresh Chotirmall. 2018. "Inhaled Nanomaterials and the Respiratory Microbiome: Clinical, Immunological and Toxicological Perspectives." *Particle and Fibre Toxicology 2018 15:1* 15 (1): 1–16. https://doi.org/10.1186/S12989-018-0282-0.

Priyadarsini, Subhashree, Swaraj Mohanty, · Sumit Mukherjee, Srirupa Basu, and Monalisa Mishra. 2018. "Graphene and Graphene Oxide as Nanomaterials for Medicine and Biology Application." *Journal of Nanostructure in Chemistry 2018 8:2* 8 (2): 123–37. https://doi.org/10.1007/S40097-018-0265-6.

Sharma, Virender K., Jan Filip, Radek Zboril, and Rajender S. Varma. 2015. "Natural Inorganic Nanoparticles – Formation, Fate, and Toxicity in the Environment." *Chemical Society Reviews* 44 (23): 8410–23. https://doi.org/10.1039/C5CS00236B.

Shi, Zongbo, Longyi Shao, T. P. Jones, and Senlin Lu. 2005. "Microscopy and Mineralogy of Airborne Particles Collected during Severe Dust Storm Episodes in Beijing, China." *Journal of Geophysical Research: Atmospheres* 110 (D1): 1–10. https://doi.org/10.1029/2004JD005073.

Shigeta, Masaya, Manabu Tanaka, and Emanuele Ghedini. 2019. "Numerical Analysis of the Correlation between Arc Plasma Fluctuation and Nanoparticle Growth–Transport under Atmospheric Pressure." *Nanomaterials 2019, Vol. 9, Page 1736* 9 (12): 1736. https://doi.org/10.3390/NANO9121736.

Shih, Amy Y., Anton Arkhipov, Peter L. Freddolino, Stephen G. Sligar, and Klaus Schulten. 2007. "Assembly of Lipids and Proteins into Lipoprotein Particles." *Journal of Physical Chemistry B* 111 (38): 11095–104. https://doi.org/10.1021/JP072320B/SUPPL_FILE/JP072320BSI20070323_105921.MPG.

Sonntag, Darrell B., Chad R. Bailey, Carl R. Fulper, and Richard W. Baldauf. 2012. "Contribution of Lubricating Oil to Particulate Matter Emissions from Light-Duty Gasoline Vehicles in Kansas City." *Environmental Science and Technology* 46 (7): 4191–99. https://doi.org/10.1021/ES203747F/SUPPL_FILE/ES203747F_SI_001.PDF.

Stanley, Sarah. 2014. "Biological Nanoparticles and Their Influence on Organisms." *Current Opinion in Biotechnology* 28 (August): 69–74. https://doi.org/10.1016/J.COPBIO.2013.11.014.

Stolpe, Björn, and Martin Hassellöv. 2007. "Changes in Size Distribution of Fresh Water Nanoscale Colloidal Matter and Associated Elements on Mixing with Seawater." *Geochimica et Cosmochimica Acta* 13 (71): 3292–3301. https://doi.org/10.1016/J.GCA.2007.04.025.

Strambeanu, Nicolae, Laurentiu Demetrovici, and Dan Dragos. 2015. "Natural Sources of Nanoparticles." *Nanoparticles' Promises and Risks: Characterization, Manipulation, and Potential Hazards to Humanity and the Environment*, January, 9–19. https://doi.org/10.1007/978-3-319-11728-7_2.

Sun, Shouheng, C. B. Murray, Dieter Weller, Liesl Folks, and Andreas Moser. 2000. "Monodisperse FePt Nanoparticles and Ferromagnetic FePt Nanocrystal Superlattices." *Science* 287 (5460): 1989–92. https://doi.org/10.1126/SCIENCE.287.5460.1989.

Talapin, Dmitri V. 2008. "LEGO Materials." *ACS Nano* 2 (6): 1097–1100. https://doi.org/10.1021/NN8003179.

"The History Blog "Blog Archive" Lycurgus Cup Inspires Cool New Sensor Technology." n.d. Accessed February 25, 2022. http://www.thehistoryblog.com/archives/26724.

Tripathi, S. K., Ramneek Kaur, Harpreet Kaur, Mamta Rani, Jagdish Kaur, and Harmandeep Kaur. 2015. "Fabrication and Electrical Characterization of Memristor with TiO_2 as an Active Layer." *AIP Conference Proceedings* 1661 (1): 110027. https://doi.org/10.1063/1.4915472.

Wagner, Stephan, Andreas Gondikas, Elisabeth Neubauer, Thilo Hofmann, and Frank Von Der Kammer. 2014. "Spot the Difference: Engineered and Natural Nanoparticles in the Environment—Release, Behavior, and Fate." *Angewandte Chemie International Edition* 53 (46): 12398–419. https://doi.org/10.1002/ANIE.201405050.

Wender, Heberton, Pedro Migowski, Adriano F. Feil, Sérgio R. Teixeira, and Jairton Dupont. 2013. "Sputtering Deposition of Nanoparticles onto Liquid Substrates: Recent Advances and Future Trends." *Coordination Chemistry Reviews* 257 (17–18): 2468–83. https://doi.org/10.1016/J.CCR.2013.01.013.

"What Are the Different Types of Nanoparticles?" n.d. Accessed February 14, 2022. https://www.azonano.com/article.aspx?ArticleID=4938.

"What Nanomaterials Exist in Nature?" n.d. Accessed February 13, 2022. https://www.azonano.com/article.aspx?ArticleID=4837.

Xiong, Huiwen, Zhipeng Huang, Zhejun Yang, Qiuyuan Lin, Bin Yang, Xueen Fang, Baohong Liu, Hui Chen, and Jilie Kong. 2021. "Recent Progress in Detection and Profiling of Cancer Cell-Derived Exosomes." *Small (Weinheim an Der Bergstrasse, Germany)* 17 (35). https://doi.org/10.1002/SMLL.202007971.

Youssef, Ahmed M., Hesham Moustafa, Ahmed Barhoum, Abou El Fetouh A. Abdel Hakim, and Alain Dufresne. 2017. "Evaluation of the Morphological, Electrical and Antibacterial Properties of Polyaniline Nanocomposite Based on Zn/Al-Layered Double Hydroxides." *ChemistrySelect* 2 (27): 8553–66. https://doi.org/10.1002/SLCT.201701513.

Chapter 2

Potential Application of Nanomaterials

Learning Outcomes

- The reader will have a deep insight into the number of fields where nanotechnology could find a significant place.
- Various technologies deployed to synthesis the nanomaterials and functionalization of those to perfectly fit in the desired application can be identified.
- The reader will have a clear idea about the properties underlying in the nanomaterials that explicitly makes it an ideal choice of material for various applications.
- The reader will be able to independently choose a nanomaterial or nanocomposites with suitable synthesis and construction technology for their desired field of use.

Abstract

Nanotechnology over the last decade evolves to be the tiniest and invisible solution to bigger problems in the real day-to-day life. These tiny particles ranging in the order of 10^{-9} that also have a higher surface-to-volume ratio are potential candidates that find its usage in a lot of applications. Advances in nanotechnology have sparked the attention of both patients and physicians in tackling complex therapeutic concerns. Furthermore, depending on the application, the physicochemical and natural properties of nanoparticles can be modified. This chapter gives a deeper insight into the nanotechnology applications in various fields – electronics, healthcare, industry, environment, food and agriculture, biomedical, textiles, energy, transport, space science and sports. The advancement of steady and powerful nanoparticles requests a point by point handle of both the physicochemical qualities of nanomaterials and their expected applications. Besides, the wellbeing chances related with the utilization of adjusted nanoparticles should be respected cautiously.

Keywords: nanomaterial, industry application, electronics, food and agriculture, healthcare, textiles, space, sports

1. Introduction

Today, it's difficult to envisage a world without nanotechnology. Although the word "nanotechnology" is new, the principle behind it has been around for a long time. Nanotechnologies have already affected many facets of human existence. This chapter describes how nanomaterials and nanostructuring applications are employed in present and future commercial activities. After more than two decades of essential nanoscience research, nanotechnology applications are delivering on nanotechnology's promise to progress civilization in both expected and unforeseen ways. As a result of nanotechnology, several technologies and industrial areas have been significantly improved, if not completely transformed. The following is a sampling of the ever-expanding display of nanotechnology benefits and uses.

Many of the benefits of nanotechnology are based on the capacity to modify material designs at tiny sizes to obtain particular properties, considerably expanding the materials science toolset. Nanotechnology can make materials more grounded, lighter, stronger, more responsive, or better electrical transmitters. Many ordinary commercial devices that rely on nanoscale materials and cycles are now available and in use.

Nanoscale additives to texture surface medicines can give lightweight ballistic energy avoidance in the private body protective layer, as well as aid them resist wrinkling, staining, and bacterial growth. Nanoscale materials are being developed to create smart textiles with adaptable sensors and electronics in the nanoscale for healthcare, and solar energy collecting. Less-heavy vehicles, trucks, aeroplanes, boats, and spacecraft may compel essential fuel venture expenditures. Nanoscale added chemicals in polymer composite materials are used to manufacture lightweight, resilient, strong, and tough play clubs, tennis rackets, motorbikes, cruiser protective caps, vehicle components, gear, and power instrument housing. CNT sheets are currently being researched for use in cutting-edge execution planes. The combination of light weight and conductivity, for example, makes them ideal for applications such as electromagnetic shielding and heat control.

Surface treatments with nanoscale added compounds can resist wrinkles, stains, and bacterial development while also enabling lightweight ballistic energy evasion in the private body protective layer. Nanoscale coatings that are water and soil resistant, antireflective, self-cleaning, impermeable to brilliant or infrared light, antifog, antimicrobial, scratch-safe, or electrically conductive can be applied to eyeglasses, PC and camera presentations, windows, and other surfaces. Nano-designed materials in automobiles include high-power battery-powered battery frameworks, thermoelectric materials for temperature control, tyres with lower moving resistance, high-effective/low-cost sensors and gadgets, slim film brilliant sunlight based chargers, and fuel added substances for cleaner fumes and increased range. Traditional wear-resistant coatings for machine components are far weaker than nanostructured clay coatings. Nanotechnology-enabled greases and motor lubricants cut mileage as well, substantially prolonging the lifetime of moving components in everything from power equipment to modern gear.

To help complicated processes, nanoparticles are increasingly being employed in catalysis. This decreases the quantity of reactant materials needed to accomplish the desired results, hence conserving money and decreasing toxicity. Two important applications are oil refineries and automobile exhaust systems. Nano-engineered materials are employed to make exceptional domestic items such as degreasers and stain removers, as well as ecological sensors, air purifiers and channels, antibacterial cleaning agents, and speciality paints and fixes such as self-cleaning home paints that resist soil and stamps. To boost performance, nanoscale materials are also being included into a number of individual consideration items. Nanoscale titanium dioxide and zinc oxide have long been employed in sunscreen to give UV protection while being imperceptible to the human eye.

2. Electronics and Technology

Nanotechnology has facilitated tremendous progress in computers and electronics, resulting in fast, miniatured, and highly portable systems capable of managing and storing ever-increasing quantities of data. Nanotechnology has lowered the size of transistors, which are the switches that form the basis for modern computers. Around the turn of the century, a conventional transistor measured 130nm to 250nm in size. Because to smaller, faster, and better transistors, the computer's whole memory might one day be stored on a single little chip. Using magnetic random access memory, computers will be able to "boot" almost instantly (MRAM). MRAM (magnetic random access memory) allows the computers to boot instantly when they are used alongside nanoscale magnetic tunnel junctions which can store data at a faster rate and efficiently after a system shutdown or provide resume-play functionality. There are already ultra-high definition screens and televisions on the market that employ quantum dots to provide more bright colours while being more energy efficient.

Wearables, medicinal applications, aviation applications, and the Internet of Things are among the goods that adaptable, foldable, and stretchy technologies are being integrated into. Adaptable devices have been created, such as semiconductor nano films for usage in mobile phone and tablet displays. Other nanoparticles, such as graphene and cellulosic nanomaterials, are being employed to provide wearable and "tattoo" sensors, as well as sewable photovoltaics and moveable electronic paper. Other computing and electronic products include flash memory chips for smartphones and thumb drives, ultra-responsive hearing aids, antimicrobial/antibacterial coatings on keyboards and cell phone casings, conductive inks for printed electronics for RFID/smart cards/smart packaging, and flexible displays for e-book readers. Nanoparticle copper suspensions have been developed as a safer, less costly, and more trustworthy replacement for lead-based solder and other dangerous substances often used in the assembly process to fuse electronics.

2.1. Nanoscale Patterning of Electronic Circuits

Miniature/Nano designing is a one of the scaling down method for designs, particularly utilized for hardware. These days it turns into a norm in biomaterials designing and for central exploration on cell science by mean of delicate lithography. It for the most part utilizes photolithography strategies however numerous procedures have been created. The cluster creation of microstructures requires a minimal expense, high throughput surface designing innovation. The schematic of nano-device patterning procedures is given in Figure 25.

All living natural frameworks work because of sub-atomic communications of various subsystems. The sub-atomic structure blocks (proteins, DNA, nucleic acids, carbs, and lipids) can be seen as rousing conceivable technique on the most proficient method to plan elite execution MEMS and NEMS that have the qualities and properties required. Scientific and mathematical techniques are accessible to break down the elements and three-layered calculation, holding, and different highlights of particles and atoms. Thus, electromagnetic and mechanical, as well as other physical and synthetic properties can be considered. Nanostructures and nanosystems can be broadly utilized in medication and wellbeing. Among potential uses of nanotechnology are: drug amalgamation and medication conveyance (the remedial potential will be colossally improved because of direct powerful conveyance of new kinds of medications to the predefined body locales), nanosurgery and nanotherapy, genome combination and diagnostics, nanoscale actuators and sensors (infection finding and avoidance), nonrejectable nanoartificial organs plan and embed, and plan of elite execution nanomaterials.

These advancements actually should radically alter the creation and assembling of materials, gadgets, and frameworks through:

- improved safety
- environmental compatibility
- increased stability and durability
- better efficiency and capability,
- enhanced flexibility and integrated support

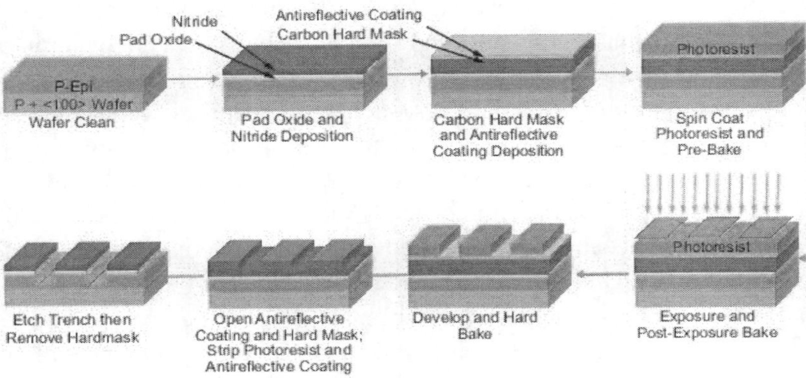

Figure 25. Schematic of semiconductor nano-device patterning ("Photolithography Overview" n.d.).

Nanolithography is a branch of nanotechnology that studies and applies nanofabrication of nanoscale structures, defined as nanopatterning with at least one dimension ranging from a single atom to about 100 nm. Nanolithography is employed in the nanofabrication of improved semiconductor integrated circuits (nano-circuitry), nano-electromechanical systems (NEMS), and nearly every other significant nano research application in a wide variety of scientific disciplines (Venugopal and Kim 2013). This technology might be used to nanofabricate different semiconducting integrated circuits, NEMS, and other research projects. Semiconductor device nanoscale changes are also feasible (in the range of 10^{-9} metres). Modern nano-lithographic technologies include photolithography, nano-imprint lithography (NIL), scanning probe lithography (SPL), AFM nanolithography, extreme ultraviolet lithography (EUVL), and X-ray lithography (Venugopal and Kim 2011; Kim, Latyshev, and Yamashita 1999).

2.2. Nano Conductive-inks

Nanomaterials are one of the latest dynamic materials utilised in ink definitions to make them conductive. There are various reasons for this, the most obvious being that some nanomaterials have exceptionally high electrically conductive properties that may be used in ink detailing. Graphene, silver nanoparticles, and silver nanowires are among the extremely conductive nanomaterials available. Furthermore, it is usually the case that a few wt percent of a design is all that is necessary to make it conductive, implying that they are not as expensive as many might believe. Because nanoparticles are being developed in place of mass materials, ink may be imprinted in an exceedingly thin layer on top of a substrate. The application cycle can be as simple or as complex as needed. In some cases, organisations use explicit added substance fabrication procedures (and accompanying programming) to supply conductive routes, but in other cases, organisations just print these conductive organisations using an inkjet printer. The conductive ink coating of silver nanoparticles and the NaCl over the glass or a silicon film is given in Figure 26 whereas the schematic of silver nanoparticle conductive ink coating over the cotton fibre textiles is given in Figure 27.

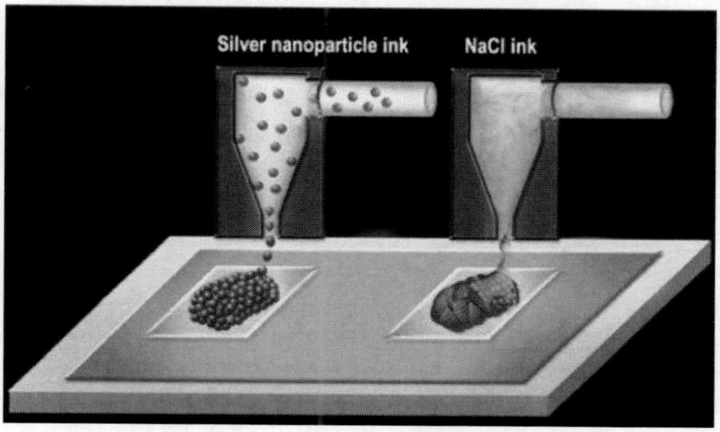

Figure 26. Film coated with silver nano conducting ink (Mo et al. 2019).

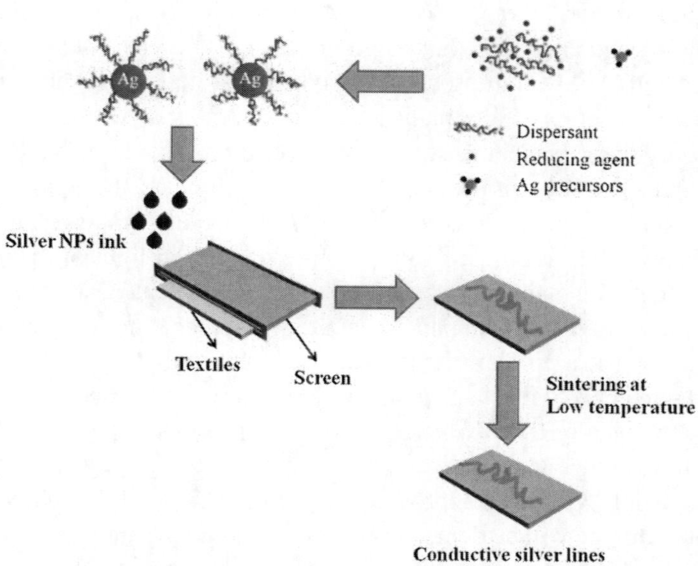

Figure 27. Nano-silver conductive ink coated on cotton fibre (Wang et al. 2016).

The key to the success of any nanoparticle ink is uniformity throughout each stage. Most importantly, the nanomaterials employed must be homogenous in terms of size, synthesis, surface functionalization, and dynamic properties (or as uniform as possible). Furthermore, the distribution of the nanomaterial inside the design is crucial, since it may be necessary to be as monodisperse as feasible. Otherwise, there will be areas of the ink with a dense cluster of nanoparticles and regions with none at all, leading in profoundly conductive and protecting zones, which will impair the ink's conductive presentation. Finally, the actual ink printing should be constant; otherwise, you may wind up with sections with no ink and areas with an excessive amount of ink, which may damage the printed circuit's display - not ideal for any electronic device application.

The ink details serve as a cover and may be employed similarly to protective materials in devices, with the exception that they can be printed. These criteria are frequently polymer-based inks that offer a protective barrier to the conductive medium while allowing the device to be tiny and multifunctional. Nanoparticle conductive inks are no longer isolated to the academic lab; firms are actively developing and combining these principles into real-world products. Since researchers develop new and more efficient techniques for scattering and determining nanoparticles into inks, an industry is certain to arise, as there is a need for the sorts of gadgets that these inks will aid in the creation of ("Nanoparticle Conductive Ink" n.d.).

2.3. High-Density Data Storage

Information capacity is one of the most significant components of data innovation. The constant desire for more capacity in a constantly shrinking structural factor, as well as the pressure to reduce the cost per capacity unit in $/Gbyte, have been significant driving forces for significant overall inventive work activities to increase capacity densities through various techniques. For a long time, silicon-based semiconductor memory chips and attractive hard drives (HDD) dominated the information storage business. Approaches that employ nanoscale — sharp methods for imaging and analysing the design of materials down to the nuclear scale — such as the nuclear power (AFM) and the checking burrowing magnifying instrument (STM) are well suited for the production of ultrahigh-thickness capacity gadgets (Eleftheriou et al. 2003).

In light of previous advancements, for example, hard disc drives (HDD), optical circle drives (ODD), and Flash Memories, which currently dominate the nonvolatile information storage market, nuclear power microscopy (AFM) as an empowering innovation for information capacity applications, as well as future gadgets in light of magnetoresistive and stage change impacts. The main concern is whether the intelligent AFM-based capacity, called "Test Storage," can provide a competitive approach for dealing with currently accessible innovations by serving as a problematic innovation. Test Storage will be separated from HDD and ODD, which are entirely mechanical and rely on a single test to address billions of bits of data, as well as nonvolatile RAM, which has no moving components. Limits, areal thickness, move rate, structural component, and cost of various data storage devices will be explored, as well as the incredible possibility provided by Probe Storage in exploiting monstrous parallelism. Probe Storage will be demonstrated to overcome any barrier between HDD, ODD, and other nonvolatile RAM by drawing on the benefits of each and adding a significant attribute that neither has: the capacity to tend to an extraordinarily high number of nanoscale pieces of information in equal time. This section varies from others in the book in that it covers the key question of whether a particular logical endeavour, in this instance Probe Storage, is developed enough to be commercially successful. The reply suggests that there is clearly a huge need in the information storage sector that such an invention can address, one large enough to justify a significant investment in fresh effort. To be sure, as various sections demonstrate, such work is moving quickly, with the goal of providing a usable product within a few years (Sarid, McCarthy, and Jabbour 2004).

2.4. Ferro Fluids

For biomedical applications like drug delivery and hyperthermia, the magnetic nanoparticles (MNP) with relative aqueous suspension of over 20 nm are required. But the MNPs ranging in this size spectrum have a tendency to aggregate forming a ferrofluidic sedimentation as shown in Figure 28. To overcome this, the multicore nanoparticles (MCNP) are used. Ferrofluids, which are composed of a stable colloidal solution containing permanently magnetised particles with diameters of 10 nanometers such as magnetite, are an excellent choice for biomedical applications. Brownian motion prevents the 10 nm particles from settling under gravity, and a surfactant is employed to induce short-range steric repulsion between particles in the presence of non-uniform magnetic fields, preventing particle agglomeration (Dutz et al. 2009; Zahn 2001).

Figure 28. Nano-magnetic Ferro fluids ("Ferrofluids, Nano Magnetic | Nanofluid | Nanotechnology Products | NPD" n.d.).

2.5. Single-Electron Transistors

The nanomaterials like carbon nanotubes, solid state quantum dots and molecular electronic devices form the best alternate for conventional transistors. The single electron transistor (SET) is a one-of-a-kind switching device that amplifies current using the controlled electron tunnelling phenomenon. A nano device can transport a single electron at a time. Electron transport is governed by Coulomb interaction, which occurs on a thin conductive layer known as an island. SETs are supposed to be the future's building blocks. SETs will be used to build very compact, low-power integrated circuits capable of sensing individual electron velocity.

SET is a three-terminal Single Electron Device (SED) that consumes little power and operates quickly ("Single-Electron Transistors – Physics World" n.d.; Krishnan, B-Tech 2014; Pal Singh and Raj 2013).

All of these gadgets share a little island that restricts the conducting charge in the form of electrons. This island corresponds to the channel of a MOSFET as shown in Figure 29. There are two types of electron confinement devices: single electron and quantum dots. The composition and device size of islands influence the gadget's specific features. Controlling these characteristics strategically enables the device designer to leverage quantum processes to influence electron passage on and off the islands in a variety of ways ("Analytical Discussion of Single Electron Transistor (SET)," n.d.; Khan 2013).

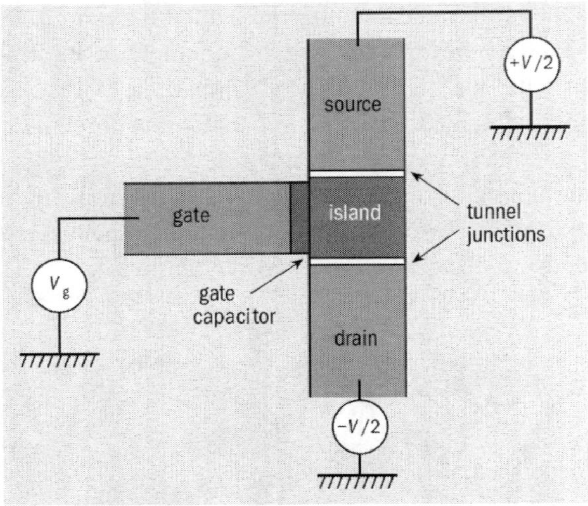

Figure 29. Schematic of single electron transistor ("Single-Electron Transistors – Physics World" n.d.).

2.6. Quantum Computers

The development of computer technology based on quantum theoretical notions is the focus of quantum computing. The quantum hypothesis is concerned with the nature and quantum behaviour of matter and energy. Quantum computers, like conventional computers, can handle data in more than two dimensions. Quantum computers can also deal with data that has a value of 0 or 1, enabling new types of simulations and calculations. A pulse originating from the Dutch FELIX laser may be used for depositing an electron in two different states simultaneously over the silicon chips and this state is called as quantum superposition. Once superposition is established, this situation might be managed to the point where electrons produce light at a predetermined time. This is known as photon reverberation, and it gives the iotas total control. As a result, advancements in silicon-based quantum computing may result in a shift in the skyline ("Nanotechnology in Quantum Computing" n.d.). The picture of first quantum computer with 100 qubits that has been developed in the year 2021 is given in Figure 30.

Figure 30. First Quantum Computer to pack 100 qubits (Ball 2021).

3. Healthcare

Currently, nanotechnology is broadening the breadth of clinical tools, data, and medicines accessible to clinicians. The use of nanotechnology to medicine, simply called as Nano-medicine, makes advantage of the usual scale of organic characteristics to provide precise solutions for preventing the infections, its diagnosis, and treatment. Business applications have changed gold nanoparticles into tests for determining assigned nucleic acid courses of action, and are found to be used for potential treatments against threatening development and other contaminations. Nanotechnology-enabled imaging and analytical technologies are opening the path for earlier diagnosis, more personalised treatment options, and improved success rates in supporting patients. Nanotechnology is being studied for its potential use in the prevention and treatment of atherosclerosis, as well as the advancement of plaque in supply lines. One method was utilised by researchers to manufacture a nanoparticle that resembled the body's "good" cholesterol, known as HDL (high-density lipoprotein), which assists in plaque contraction. The creation of revolutionary high-quality sequencing headways that engage single-molecule recognition at cheap cost and with little model preparation and equipment might be part of the design and planning of cutting-edge solid state nanopore materials.

Experts in nanotechnology are developing a number of therapies in which a nanoparticle can assist carry prescriptions directly to sick cells while minimising the potential of harm to solid tissue. This might change how doctors treat cancer and minimise the toxicity of chemotherapy. Nanotechnology is being used in regenerative medical research for a variety of applications, including bone and brain tissue creation. Novel materials can be developed to imitate the gem mineral structure of human bone, or they can be employed as a supporting pitch in dental applications. Scientists are seeking for novel approaches to manufacture complex tissues with the objective of one day producing transplantable human organs. Experts are also investigating ways to employ graphene nanoribbons to aid in the healing of spinal cord damage; preliminary research shows that neurons develop successfully on the conductive graphene surface. Experts in nanomedicine are investigating how nanotechnology might assist enhance vaccinations, such as antibody distribution without the need of needles. Analysts are also developing a wide vaccine platform for the yearly influenza antibody, which would cover more strains while requiring less resources to maintain each year.

3.1. Surgery

The capacity of nanoparticles to detect disease cells and destroy them with single-cell precision is a pipe dream for a 'great' therapeutic strategy to combat tumours. The principal use for nanoparticulate medication delivery may be the transportation of the medicine payload into the cerebrum. However, breaking through the blood-mind barrier, the cerebrum's protective barrier, is a significant problem. This is made feasible by the use of extraordinary nanoparticles. Specialists must be able to remove, control, and replace their innards in order to focus on particular parts of cells. Despite the restricted alternatives for this sort of procedure, scientists have devised a "nanoblade" that can cut through a cell's layer and insert mitochondria. This nanoblade might have broader applications in bioengineering. An illustration of a nanorobot performing the cell surgery is shown in Figure 31.

Rebuilding shattered parts of a human body without previous preparation has enthralled human creative minds for decades, from Mary Shelley's Doctor Frankenstein to cutting-edge professionals. A group of European academics working on the EUREKA project ModPolEUV have made a promising commitment to reconstructive medical operations owing to a unique interdisciplinary approach that blends cutting-edge medicine with the most recent breakthroughs in nanotechnology. Experts have created a biodegradable thin film with a thickness of around 20 nanometers that might be used to replace sensitive fasteners. The incredibly thin PLLA nanosheet gives superior fixing capability for stomach entry points as an original harm treatment that does not require cement professionals. In addition, the healing activity fully eradicated all scars and tissue grasp from the entry area. This approach would be the ideal option to typical stitch/ligation treatments from the standpoint of a less intrusive precise surgery as well as a decrease in activity times.

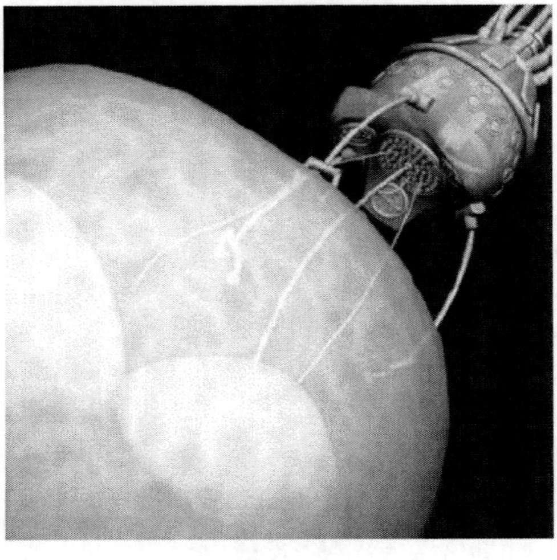

Figure 31. Nanorobot performing cell surgery (Murphy et al. 2006).

Neurosurgery, a delicate specialty that depends largely on mechanical innovations, is poised to profit considerably in the next years from a few graphene-based innovative improvements. Tattoos aren't simply for body adornment; they may also be utilised for medicinal reasons. Experts are applying them on patients to mark a region for future therapy, such as non-melanoma skin malignant growth, however the inks can cause certain complications. A better answer has been created by nanoscientists: a nanoparticle ink that lights only under specific light conditions and dissipates after a fixed length of time.

3.2. Dermatology/Cosmetics

Nanoparticles used in medication delivery systems, such as Nanoencapsulation vesicular delivery system, Nanoemulsion and nanocrystal, liposomes and Niosomes, Nicells, polymeric Nanocapsules, solid lipid nanoparticles, and so on, are of importance to the beauty industry. Concerns about the health and environmental risks of nanotechnology have increased considerably as a result of its widespread use in cosmetics. The nanomaterials find its application in various cosmetic fields as illustrated in Figure 32.

Titanium oxide nanoparticles in foundation will have a high iron concentration. The iron in the foundation darkens reversibly in reaction to brightness, and the aesthetic appearance stays consistent whether exposed to natural or artificial light. This behaviour is known as photochromic. An Australian company just produced ZO1, a sunscreen that contains microscopic zinc oxide particles. Estée Lauder and Johnson & Johnson are also developing products based on nanotechnology. Germatika Ageless Cream is an anti-aging cream based on nanotechnology. In Europe, an Eye Care treatment for crow's feet was produced, as well as a Hypoallergenic Wrinkle Nano-Remover cream that claims to smooth out wrinkles for up to eight hours (Hameed et al. 2019).

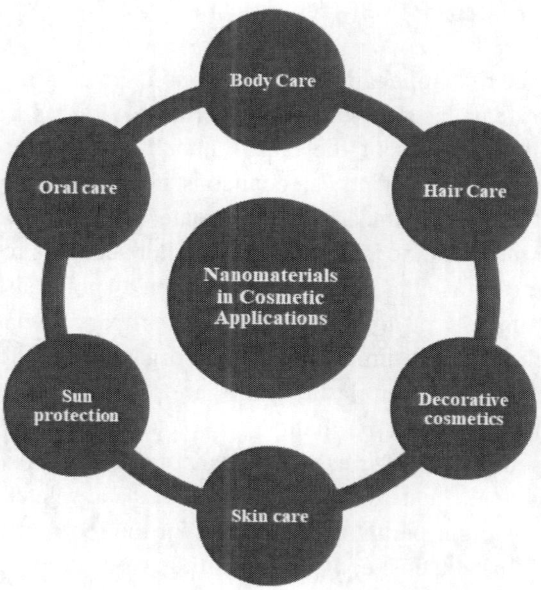

Figure 32. Nanomaterials in various cosmetic applications (Fytianos, Rahdar, and Kyzas 2020).

4. Industry

4.1. Automotive Industries

Although nanotechnology will play an important part in the vehicle industry, don't expect anything like the Volkswagen Nanospyder sophisticated concept car in the near future. In this entry to the 2006 Los Angeles Design Competition, hydrogen energy components, solar power, wheel-mounted electric engines, and inflated natural body boards combine to frame the odd state of the two-seater concept.

According to its creators at the Volkswagen Design Center in Santa Monica, the Nanospyder would be made up of a latticework of billions of small programmable micro gadgets spanning just a fraction of a millimetre in width. Each of these little gadgets may be customised to serve as safe havens for the really defenceless. The nano-cross section is covered with boards made of a variety of natural materials, some of which may expand to give additional cushioning as a consequence of an impact. Because polysynthesis produces modest amounts of electricity, the material can be employed as a power source. This, along with hydrogen energises, resulting in the ability to power the small electric engines located inside the four wheel hubs.

The fundamental patterns enabled by nanotechnology for autos are as follows:

- Materials that are lighter while remaining more grounded (for improved fuel use and expanded safety) for gas-powered autos, enhanced motor output, and fuel efficiency (impetuses; fuel added substances; greases)
- Cars fuelled by hydrogen and fuel cells have a lower environmental effect
- Electronic systems that have been enhanced and reduced in size
- Improved economic circumstances (longer assistance life; lower part disappointment rate; brilliant materials for self-fix)

Tyres were one of the first applications of nanostructured materials in cars. The automobile industry utilised carbon dark as the first nanomaterial in tyres as a colour and construction specialist. While the combination of the elastic and the filler material is the way to tyre execution, its streamlining requirements can be inconsistent (extremely complex substance and actual connections between the elastic and the filler material): while the tyre requires great hold, its moving obstruction must also be low. Supporting filler accounts for around 30% of the tyre cover and is in charge of attributes such as hold, scraped spot resistance, introduction wear resistance, and everlasting tear proliferation. The three primary components that influence the properties of regular elastic are sediment, silica, and organosilane. These particles, as well as their cross-connections with regular elastic atoms, are presently being produced in nanoscale structures and play a major role in tyre quality.

To build an electric vehicle, four essential components must be present: strong and safe energy storage to provide adequate driving range; motors and related electronic parts that use the stored energy; light-weight parts to compensate for the (currently still) additional load of the batteries; and all of this at a cost competitive with gas-powered vehicles.

Lithium particle batteries are now being extensively developed for use in electric vehicles. The majority of specialists assume that the battery will be built of lithium particles, but which of the lithium-particle sciences will be used remains a mystery. Hydrogen sensors will be a critical component of vehicle safety and will be widely used. They will find leaks before the gas becomes a serious concern. Small, tunable hydrogen sensors have been developed using nanostructured materials such as single-walled carbon nanotubes coated with palladium nanoparticles.

Another component of Ford's "Iotas to Engines" programme is the creation of a thermally splashed nanocoating that might someday replace the heavier cast iron liners that provide the necessary wear resistance of chamber exhausts in aluminium block motors. This thin wear-resistant coating saves weight and increases erosion performance while retaining the item's solidity and dependability.

Diesel engines, like their petroleum counterparts, utilise piezo fuel infusion technology. A syphon causes high strain due to direct infusion before delivering the precisely dosed fuel into the chamber's burning office via a spout. The accuracy with which this happens affects the ignition contact. The higher the tension and the more accurately the portion and season of infusion can be adjusted, the more successful fuel ignition is. These piezoelectric materials will be used in conjunction with nanocrystalline piezoelectric materials.

The bulk of the interior applications in the car will address comfort issues, such as dirt repellent and antimicrobial materials and surfaces, nanoparticulate air channels, and anti-glare mirror and instrument coatings. However, in light of thermoelectrics - materials that directly convert power into heat or cooling - how about environment-controlled automobile seats? Gadgets are a growth driver in the automobile industry as an increasing number of parts are operated electronically, electromechanically, or electromagnetically. Nanostructured entertainer components, for example, might be utilised to replace current microsystems innovation-based direct infusion frameworks.

Spintronics has the potential to revolutionise registration. While classic integrated metal-oxide semiconductors (CMOS), which are utilised in a wide range of hardware today, rely on electron charge to drive devices, the new area of spintronics utilises another characteristic of electrons: their twist, which may be manipulated by electric and attractive forces. When turned off, spintronics or electrical devices that employ nanoscaled attractive materials will not have a backup power scattering problem. This advantage enables the development of non-unpredictable hardware or gadgets that consume extremely little power.

4.2. Cement Industries

Despite their extensive usage, concrete-based materials have poor mechanical properties and are highly permeable to water and other powerful synthetic chemicals, lowering their strength. Furthermore, the concrete sector contributes significantly to CO_2 emissions, accounting for 5-6 percent of total man-made CO_2 emissions each year. Nonetheless, the increasing need for high-performance basic materials and components has accelerated the creation of new material classes.

Nanotechnology has the potential to play a key role in the development sector, ranking tenth among the most important nanotechnology applications. Nanoengineered concrete-based materials can offer exceptional or intelligent properties. Some of the concerns that may be addressed by the use of nanotechnology in the concrete industry include CO_2 emissions, unintended break blockage, long relieving periods, poor stiffness, excessive water input, low pliability, and a range of other mechanical issues. Nanoparticles such as nano-SiO_2, ZnO_2, Al_2O_3, TiO_2, carbon nanotubes, nano-muds, carbon nanofibers, and other nanomaterials significantly increase the mechanical properties and solidity of cementitious materials.

The purpose of the ongoing patent investigation is to build a picture of the patent landscape for nanotechnology applications in the concrete industry, with an emphasis on identifying nanomaterials for property enhancement, developing trends, key stakeholders, and applications in the field. All catchphrases relating to cementitious materials, as well as potential equivalents, were gathered by focusing on the whole text of chosen licences that directly pertain to the notion of water propelled concretes.

The patent evaluation is confined to nanotechnology uses in pressure-driven concrete (concrete that will set and solidify in the wake of being joined with water). Nanomaterials are more commonly utilised in the Portland concrete industry (made by baking earth and limestone and crushing the product) than in non-portland concretes, as evidenced by the number of patent applications in the two categories. The bulk of patent applications are for composite concrete (74 licences) and classic portland concrete (OPC) (71 licences), with gypsum-based concrete coming in second (27 patent filings).

4.3. Construction

According to a financial estimate, nanotechnology has a substantial influence on development. A few applications for this region have been created to work on the strength and enhanced execution of construction components, energy efficiency and security of structures, working with ease of maintenance, and giving increased living comfort. Although self-cleaning properties could be achieved using micron-sized coatings and surface medicines such as Teflon, polysilazane-based coatings, and so on, this feature has now become a marketing tool/witticism for nanotechnology applications, particularly in shopper markets such as construction and materials.

To transfer this property to the surfaces, nanoparticles of TiO_2, Al_2O_3, or ZnO are used as a final coating on progressing pottery manufacturing. TiO_2 is utilised on surfaces such as tiles, glass, and sanitaryware because of its capacity to separate soil or pollutants when exposed to UV light and then wash them away with water. ZnO provides UV resistance in both coatings and paints. Nanosized Al_2O_3 particles are used to create scratch-resistant surfaces. These surfaces also help to prevent/slow the production of bad odours, growth, and shape. Nanotechnology will also benefit important building materials such as concrete, cement, and steel. Nanoparticle expansion will result in more grounded, more substantial, self-mending, air sanitising, heat resistant, easier to clean, and rapidly compacting concrete. Some of the nanoparticles that might be utilised for these elements include nano silica (silica seethe), nanostructured metals, carbon nanotubes (CNTs), and carbon nanofibers (CNFs).

Researchers are investigating the use of nanotechnology to adjust the handling states of concrete, hence decreasing emissions, in order to minimise CO_2 emissions from concrete assembly. Significant designs benefit from nano-improved coatings that prevent spray painting and other undesirable stains from clinging to them. Regardless of these materials, nanotechnology is predicted to help in the development of new lightweight, fire-resistant, self-healing, and strong development materials, such as new nanocomposites.

The following is a list of nanomaterials that are utilised in construction.

- *Carbon nanotubes* – Mechanical toughness and break anticipation; enhanced mechanical and warm qualities; continuous primary wellness monitoring (NEMS/MEMS); and compelling electron intervention are just a few of the common advantages in solar cells.
- *Silicon dioxide nanoparticles* – Mechanical strength support (in concrete); cooling, light transmission, and fire resistance (in ceramics); and fire sealing and appearance resistance (in glass) are all fully anticipated benefits (in windows).
- *Titanium dioxide nanoparticles* – Expected benefits include quick hydration, higher hydration level, and self-cleaning (in concrete); superhydrophilicity, anti-hazing, and fouling-obstruction (in windows); and non-utility power age (in sun oriented cells).
- *Iron oxide nanoparticles* – Increased compressive strength and abrasion resistance in concrete are expected benefits..
- *Copper nanoparticles* – Steel features include weldability, corrosion resistance, and formability.
- *Silver nanoparticles* – Biocidal activity in coatings and paints is expected to be beneficial.
- *Quantum dots* – One of the anticipated benefits is effective electron mediation in solar cells.

Concrete is one specialised use of nanotechnology in the construction business, with researchers concentrating on the most effective method of building cement in order to increase its mechanical performance.

4.4. Nano Pigments

The benefits of employing nanoparticles as colours are vast and can be seen in a wide range of sectors. As fixatives, titania and silica nanoparticles are commonly used in paints. They operate as UV channels while also being antimicrobial and self-cleaning. Depending on the kind of paint and planned use, nanomaterials can be coordinated as loose powders or as settled particles in colloidal form, resulting in the nanomaterial being immovably implanted in the paint grid. ("Nanomaterials Colouring the World: (Nano)Pigments - NEcoTox | Ökotoxikologische Tests Und Untersuchungen" n.d.).

Nanoparticles used as shades can also give "enhancements." These are very useful in jewellery and beauty care products when surprising, gleaming, or dazzling effects are required. These effects are achieved by the exact placement of nanocrystals. Another sort of shadow is underlying variety, which is generated by regularly requesting nanoscale highlights. These create results akin to the opalescence of a butterfly wing, which is one of the most popular eye and nail make-up trends. Nanomaterials also cause "permanent" impacts on the hydrophobicity of beauty care products.

4.5. Transparent Conductive Coatings

Various electronic and optoelectronic devices have a similar structure, where the TCC is one of the upper layers. Nowadays, the most common materials used for the fabrication of TCCs are metal oxides, while the most widely used oxide being indium tin oxide (ITO), with a market share of more than 97% of transparent conductive coatings as shown in Figure 33. ITO has several disadvantages, such as high production cost, the need for special etching processes in order to achieve patterning, and the need for the relatively rare element, indium. Another disadvantage is the brittleness of ITO thin films which prevents its utilization for the fabrication of flexible, stretchable, and bendable devices. Therefore, much effort is being made to find alternatives for ITO, which are based on nanomaterials, and which can be printed directly on various substrates.

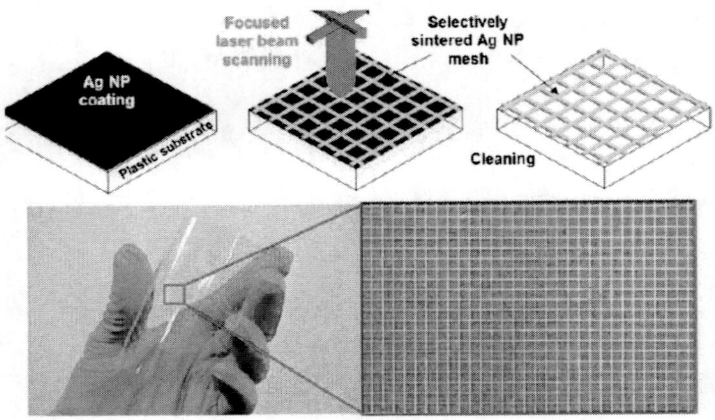

Figure 33. Transparent conductive coatings (Layani, Kamyshny, and Magdassi 2014).

The emerging materials for ITO replacement can be divided into three categories: carbon based materials, such as carbon nanotubes (CNTs) and graphenes; metallic nanoparticles (NPs) and nanowires (NWs); and conductive polymers, such as PEDOT:PSS. Depending on the application, the TCC replacement should have a sheet resistance in the range of 10 to 103 Ω cm^{-1} at a transparency ≥90% (Layani, Kamyshny, and Magdassi 2014). The typical sheet resistance of a graphene-based TCC is in the range of 200–1000 Ω ,cm^{-1} with a transparency lower than 90%. Improving the sheet resistance of graphene can be achieved by increasing the number of graphene sheet layers, but obviously, this would cause a decrease in transparency. The electrical resistivity of an individual CNT can be as low as 1 x 10-^{6} Ω cm for SWCNTs and 3 x 10-^{5} Ω cm for MWCNTs. Due to the superior mechanical properties of individual CNTs, thin films made of CNTs are promising candidates for the production of flexible electronic devices.

4.6. Superplastic Ceramics

Superplasticity is an intriguing property, not only because it, like metals, allows for the assembly of intricate artistic pieces for modern applications, but also because the combination of grain limit sliding (GBS) and diffusional processes makes superplasticity a fascinating instrument for joining ceramics pieces in shorter times and at lower temperatures than diffusional joining (Domínguez-Rodríguez and Gómez-García 2010). A few characteristics, such as the strain rate at which the material may be superplastically twisted, can have a significant impact on the superplastic conduct of pottery production. Grain size, second stages, pollutant isolation at grain limitations, and so on are commonly cited among them. Even ceramics, such as the high-temperature superconductors YBaCuO, exhibit superplastic behaviour.

4.7. Reinforced Plastics

The fiber-reinforced composites receive great importance due to their potential uses in many fields. In the last two decades with enormous developments in nanotechnology, the use of nanomodified matrix and nano-enhanced fiber-reinforced composites has gained an important attention. The use of nanotechnology such as graphene, carbon nanotube, POSS, and other nanostructures will significantly increase the interlaminar shear strength, tensile strength, impact strength, and other physical and chemical performance properties of the fiber-reinforced composites (Demiroglu et al. 2017).

The inclusion of these nanoparticles into various polymer networks has opened up a new field of study in composite materials. Furthermore, several inorganic nanoparticles, such as quantum dots, metal oxides, clay, and metallic nanoparticles, have been introduced into polymers for usage in a wide range of fields, including medicine, photovoltaics, bundling and underlying applications, and so on. One of the major objectives driving this method is to enhance the mechanical execution of the framework. Nanoparticles have a higher explicit surface region, surface energy, and thickness when compared to microparticles, so lower nanofiller fixations should achieve properties that are nearly identical to or far superior to those obtained by regular microfiller loadings, which aids in handling and limits composite weight expansion. In the wake of layered metal dichalcogenides such as WS_2 and MoS_2, inorganic fullerenes, nanotubes, and 2D nanomaterials have emerged as some of the most promising polymer fortifications (Díez-Pascual 2019).

4.8. Nano-phosphorous Displays

Any phosphor may convert a certain sort of external energy into visible light. This changing interaction may satisfy a range of criteria, and all common sense these applications may be classified into a few broad categories:

- Fluorescent lighting, liquid crystal display (LCD) backlighting, light-emitting diodes (LEDs), including phosphor-changed over white LEDs (pc-WLEDs), and so on.).
- Information displays (plasma display panels (PDPs), field emission displays (FEDs), cathode-ray tubes (CRTs), etc.).
- Radiation converters (X-beam escalating screens and other picture intensifiers, otherworldly converters for sun based cells, down-converters for excimer laser pillar

profilers and photolithography veil investigation instruments, seeing screens for electron microscopy, and so forth.).

- Fluorescent colours and tracers (non-horrendous testing, biolabeling, security marking, relaxation products, and so on.).

Brilliant substances, such as a continuous particulate sheet of different thickness that extends into the other two aspects, are widely used as a powder coating. Such a covering maintained on a basic substrate is typically referred to as a screen if it is utilised for the depiction of any visual data. Each application has its own set of requirements for phosphor covering aspects such as material arrangement, uniform or customised structure, and thickness. In most circumstances, a trade-off must be struck between the most extreme light outcome for the given excitation settings, the intended transitory features, the type of picture multiplication, mechanical and ecological resilience, and phosphor prices and handling (Kubrin 2014).

The particle size of phosphor screens is one of the most critical aspects influencing their appearance. Traditional phosphors frequently comprise particles with sizes ranging from a few micrometres to several micrometres, which are fundamentally bigger than the frequency of light they emit. If the chosen material's lighting attributes remain acceptable and the creative route stays conservative, the lowering of phosphor molecule size into the sub-wavelength system may be advantageous in a range of applications.

4.9. Functional Nanocomposites

Composites, which are materials that harden around a single section and are designed to disclose the best attributes of each part, are an important use for nanoparticles and nanotubes. In addition to mechanical properties, this multi-value connects optical, electrical, and appealing ones. Carbon fibres and multi-walled carbon nanotube storage are currently being used in polymers to regulate or update conductivity, with applications including antistatic packaging. Individual carbon nanotubes in composites are a popular long stretch application. One sort of nanocomposite is the use of nanoparticles as fillers in a system; for example, carbon dim is used as a filler in the production of vehicle tyres. In any case, because carbon dim particles range in size from tens to hundreds of nanometres, not all carbon dim fits inside our definition of nanoparticles.

4.10. Nano Clays

Nanoclays are mineral silicate nanoparticles. Depending on the material component and nanoparticle shape, nanoclays are included in a variety of social events, including montmorillonite, bentonite, kaolinite, hectorite, and halloysite. Figure 34 shows an example of a common nanoclay structure design. Muds containing naturally occurring nanoparticles have long been used as enhancement materials, and their popularity is growing. Mud particle-based composites with polymers and nano-sized dirt chips are also being used in applications such as vehicle monitoring (Murray 2006).

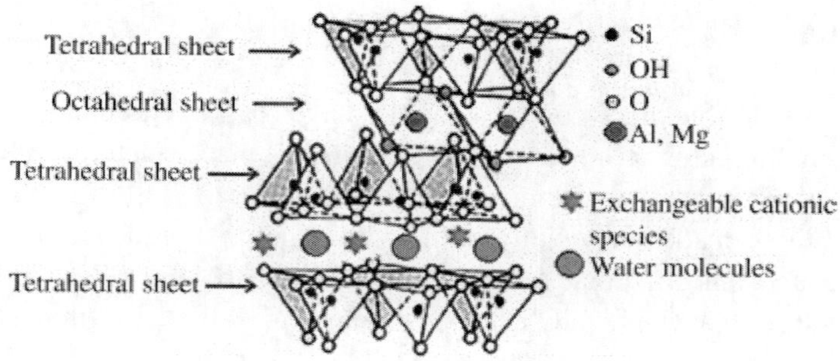

Figure 34. Schematic of nanoclay structure (Murray 2006).

Organoclays are a class of naturally modified inorganic nanomaterials that might be employed in polymer nanocomposites as rheological modifiers, gas sponges, and medicinal carriers. As nanoplate fillers, normal or manufactured muds, as well as changeable metal phosphates, can be employed. Mud-based nanocomposites have helped actual displays. The most often utilised are phyllosilicates. They have a glassy shell structure with a thickness of nanometers. Dirts are classified largely based on their glassy structures, as well as the number and arrangement of particles inside the rudimentary cross section.

4.11. Nanolubricants

By serving as nanosized 'metal rollers,' inorganic nanospheres might be employed as greases. Because of their controlled structure, they are anticipated to last longer than ordinary strong oils and require less additives. It would be interesting to observe if the higher financial and asset costs of delivering them are covered by the extended oil and component service life. These nanoparticles are also reported to minimise grinding between metal surfaces, particularly when average loads are high. If this is the case, they should focus on high-performance motors and drives, which might encompass both energy and transportation. Even if the metal surfaces aren't absolutely smooth, there's a case to be made for this type of ointment. The advantages of lower cost and asset input for machining should be balanced again against the production of nanolubricants. In this vast number of uses, the particles would be disseminated in a conventional fluid oil; as a result, the ointment structure should have mechanisms to retain and control waste. The usage of carbon nanomaterial as a lubricant and its internal structure is illustrated in Figure 35.

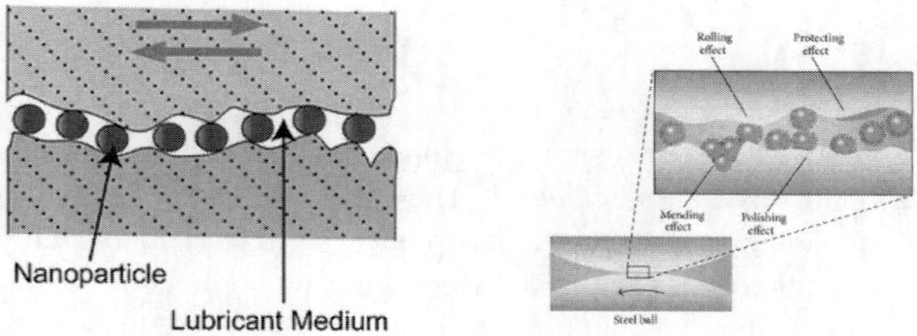

Figure 35. Carbon nanomaterial as lubricant (Ali et al. 2019).

4.12. Nanocoatings

Nano- or nuclear-scale thickness control coatings have long been employed in a number of applications, including atomic pillar epitaxy or metal oxide material fume affidavit for optoelectonic devices, as well as chemically dynamic and artificially functionalized surfaces. Self-cleaning windows made of specially formulated titanium dioxide that is both hydrophobic (water repellent) and antimicrobial (as shown in Figure 36), as well as coatings based on nanoparticulate oxides that chemically erase compound specialists, are among the most recently discovered applications.

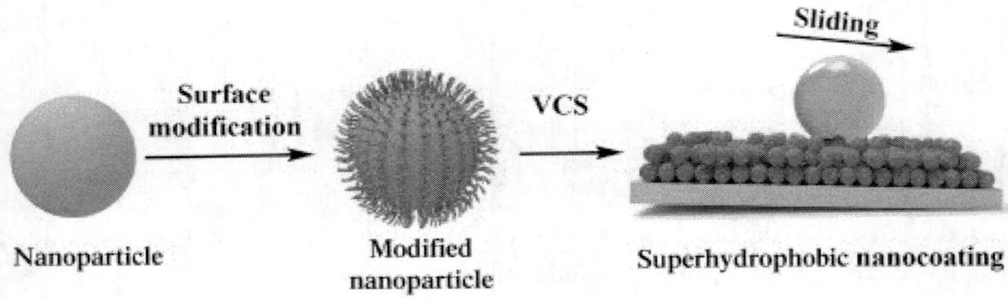

Figure 36. Schematic of nano coated surface (Ijaola, Farayibi, and Asmatulu 2020).

Hard coatings with nanoscale transitional layers (or multi-facets) between the hard outer layer and the substrate material provide higher wear and scratch resistance. The moderate layers enhance grip by providing great grip and a well-balanced blend of flexible and warm properties. Superior control of porosity and surface roughness at the nanoscale in a variety of polymers and inorganics has enabled a number of enhanced materials, including breathable, waterproof, and mess-proof textures. The many types of nano coatings available are described below:

- Gas-Barrier Coating
- Water-resistant Coating
- Antifouling Coating
- UV Blocking Coating
- Self-cleaning Coating
- Antimicrobial Coating

5. Environment

Nanotechnology products, cycles, and applications are projected to contribute to ecological and environmental protection by saving raw materials, energy, and water, as well as lowering ozone-depleting chemicals and toxic wastes. The use of nanoparticles in this manner assures distinct natural advantages as well as consequences for maintainability. Despite this, nanotechnology currently plays a little role in environmental protection, whether in research or in actual applications. Natural design groups in their various disciplines give nanotechnology barely a passing glance.

5.1. Environmental Catalysts

Organizations producing hydrogen-controlled inventions love being surrounded by the green brilliance of environmentally beneficial innovation that will save the world. While hydrogen fuel is undeniably a clean energy carrier, the source of that hydrogen is often dirty. The difficulty is that you can't access hydrogen by digging a well; instead, hydrogen must be created, which should be possible with a range of resources. Coal gasification is the dirtiest approach, at least until highly effective carbon capture and sequestration methods are discovered. Sustainable power electrolysis, by far the cleanest, employs ecologically friendly power enhancements such as wind, sun-focused, geo- and aqueous ability to split water into hydrogen and oxygen. Fake photosynthesis, which splits water into hydrogen and oxygen using solar energy, can provide a consistent and easy energy source equivalent to sunlight. A single water particle requires around 2.5 volts to be separated into oxygen with negatively charged electrons and positively charged protons. The extraction and division of these oppositely charged electrons and protons from water atoms generates electric power.

5.2. Waste Water Treatment

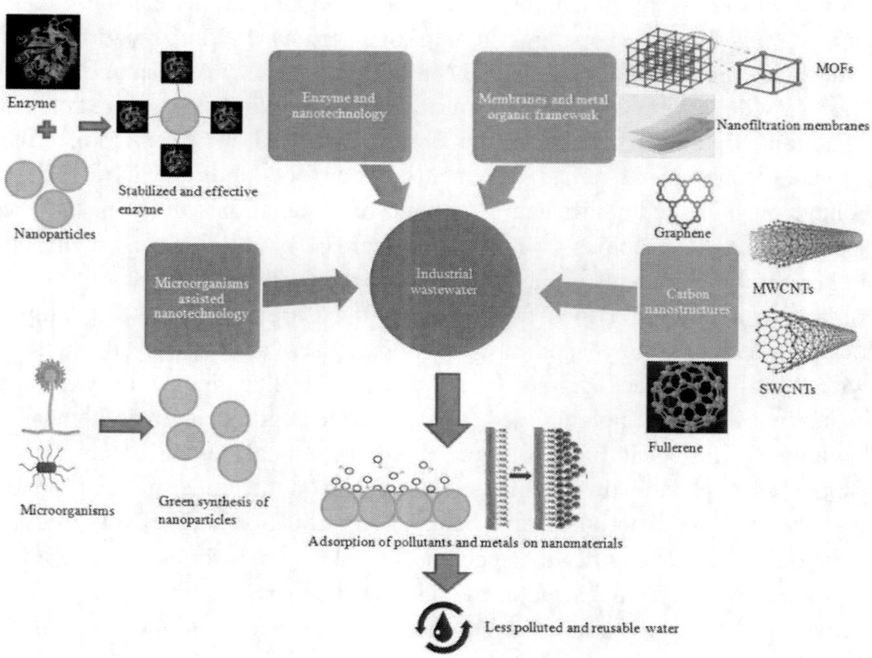

Figure 37. Industrial waste water treatment (Mandeep and Shukla 2020).

Researchers are developing a nanotechnology solution for radioactive waste remediation, notably the utilisation of titanate nanofibers as sponges to remove radioactive particles from water. According to scientists, the novel underlying properties of titanate nanotubes and nanofibers make them unrivalled materials for the evacuation of radioactive cesium and iodine particles in water. The three key effect areas for nanotechnology in water applications are treatment and remediation, detection and location, and contamination anticipation, with desalination developments being one of the most essential. Water purification systems based on nanotechnology have the potential to revolutionise the desalination business, for example, by harnessing the particle attachment polarisation peculiarity. The schematic of industrial waste water treatment based on various nanoparticels in shown in Figure 37.

5.3. Pollution Monitoring Sensors

Before CO_2 can be stored in carbon dioxide capture and storage (CCS) programmes, it must be separated from other waste gases created by burning or modern cycles. The most contemporary filtering processes are costly and include the use of synthetic chemicals. Nanotechnology methods for manufacturing nanoscale sparse films may result in the development of new layers, which might change this.

6. Food and Agriculture

If you've ever wondered where nanotechnology can take us, look no further than the possible applications in the realm of functional food, where organic atoms can be created to have capacities not found in nature, opening up an altogether new field of imaginative endeavour. Clearly, there are no boundaries to how food engineers will manage our food, and nanotechnology will provide them with an entirely new set of tools to explore new and extreme courses of action. Nonetheless, there are numerous positives. Let's take a look at the potential benefits of nanotechnology-driven advancements on our food kinds and, as a result, our health. Scientists have even created edible nanostructured representations. In the past, 3D images on food - especially, eatable graphics - were used to decrease bundling needs, such as by printing info about healthful components or names right onto the food item.

According to a definition in a document by the European Nanotechnology Gateway, nanoparticles, nanotechnology methods, or apparatuses are employed in the creation, manufacture, handling, or packaging of food. Food that has been molecularly changed or food created by nanomachines are not included. Analysts have begun to address the possible uses of nanotechnology for practical food sources and nutraceuticals by using new concepts and devising approaches related with nanomaterials to focus on the conveyance of bioactive mixes and micronutrients. Nanomaterials enable better exemplification and delivery of dynamic food fixings than traditional exemplifying specialists, and the development of nano-emulsions, liposomes, micelles, biopolymer structures, and cubosomes has resulted in superior properties for bioactive mixtures assurance, controlled conveyance frameworks, food network incorporation, and veiling undesirable substances.

Nanotechnology has the potential to revolutionise food processes that rely on proteins for nutrition and therapeutic purposes. Catalysts, for example, are commonly added to meals to hydrolyze antinutritive components, increasing the bioavailability of vital supplements like minerals and nutrients. Because of their enormous surface-to-volume proportions compared to standard macroscale support materials, nanomaterials may be employed to provide improved protein emotionally supporting networks, making these catalysts profoundly dynamic, comprehensive, and practical.

6.1. Food Quality and Safety

Nanotechnology allows for the selection of sensor stages for the confinement and localization of E. coli and other bacteria in a rapid, sensitive, solid, and fundamental manner. Some of the nanotechnology-enabled recognition methods include identification by glow utilising quantum specks, restricted surface plasmon reverberation of metallic nanoparticles, improved fluorescence, colour immobilised nanoparticles, and Raman columnist atom trapped metallic nanoparticles.

6.2. Plant Protectants

Nanostructured and nanoscale materials have a significant influence on the thermodynamics and energy of hydrogen intake and separation. By shortening the time required for dispersion and accelerating the dissemination interaction. At the nanoscale, material fitting variables' mass reciprocals can be arbitrarily adjusted. They also aided in the development of lighter hydrogen storage systems with increased hydrogen capacity.

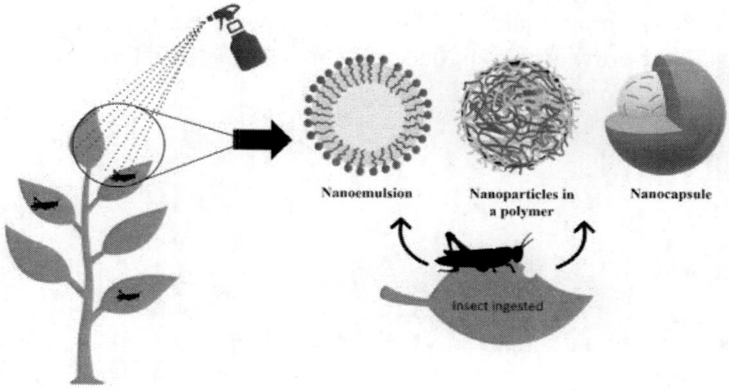

Figure 38. Nanotechnology in crop production (Medina-Pérez et al. 2019).

Copper and titanium dioxide nanoparticles are routinely utilised as manures, with CEOs conducting little study on plant disease. Chitosan is a notable nanoparticle because of its biodegradability, biocompatibility, non-allergenicity, antibacterial mobility, and low toxicity in both animals and humans. Chitosan nanoparticles aid in the prevention of viral infection in plant tissues by protecting them from diseases such as hay, snuff, nut, potato, and cucumber mosaic infections as shown in Figure 38 (Worrall et al. 2018; Medina-Pérez et al. 2019).

6.3. Plant Growth and Crop Production

In order to secure global food security, the rising need for global food production is now connected to the need to deploy innovative technologies to boost agricultural productivity. As a cutting-edge trend in this field, the utilisation of nanoscale arrangements can provide a response to key challenges such as iron, zinc, selenium, calcium, phosphorus, or vitamin A deficiencies in the human population. Through their usage in preparation, nanotechnology can give micronutrient arrangements in horticulture to ameliorate the deficit of these compounds in soil. Aside from the possible benefits, the use of nanoparticles in farming may provide unforeseen hazards owing to the intentional addition of nanomaterials to the climate, which may cause human sensitivity connected to bioaccumulation in yields and soil, and as a consequence, a shift in the pecking order (Predoi et al. 2020).

Plant feeding sculpting monitoring is a viable option for covering yield disease, but the main hurdle is that various harvests have varied supplement requirements, and the supplement interacts with the degree of plant infection in different ways. Micronutrients, for example, are important in the battle against crop disease, where tissue damage initiated responses that resulted in the creation of inhibitory auxiliary metabolites. These metabolites are typically formed by catalysts that require micronutrient cofactors like Mn, Cu, and Zn to activate, as well as host guard molecules like phenylalanine or alkali lyase. Micronutrient availability is connected to soil properties, such as the absence of Fe, Mn, and Zn in soluble soils, which limits root take-up and exposes roots to pollution. Another method for increasing disease resistance is to employ minor components such as Al or Si, which give protection from a variety of foliar and root microorganisms despite the fact that their availability in soil (for example, Si) is typically limited. Al's use has been restricted due to the idea that over-application may produce major harvest harm and output loss, and insufficient presence affects the sharpness of soil [7]. One of the most noteworthy aspects of nanoscale metals and metal oxides inside plants is their amazing accessibility and mobility.

7. Biomedical

7.1. Drug Delivery

The choice of system depends on the way they bind with the drug and the type of drug treatment as shown in Figure 39.

- *Nanocapsules:* these are pods that encapsulate drugs, which ensures the drugs are released more slowly and steadily in the body
- *Liposomes:* artificial vesicles made up of a lipid bilayer so they can fuse with and penetrate membranes easily. These have been used to treat diseases such cancer, fungal infections, hepatitis A, and influenza.
- *Dendrimers:* tree-shaped synthetic nanomolecules that carry drugs in the tips of the branches.
- *Buckyballs:* spherical nanoparticles can carry more than one drug at a time. They are useful in the treatment of diseases such as cancer and other diseases where monotherapy can lead to drug resistance
- *Nanobiomagnets* which carry drugs, for cancer for instance, into the body and are held at the target site by an external magnet. The purpose of this is to concentrate the drug at the tumour site for long enough for it to be absorbed.
- *Attapulgite* clays with nanometre-sized pores that are ideal for filtering out harmful bacteria from water

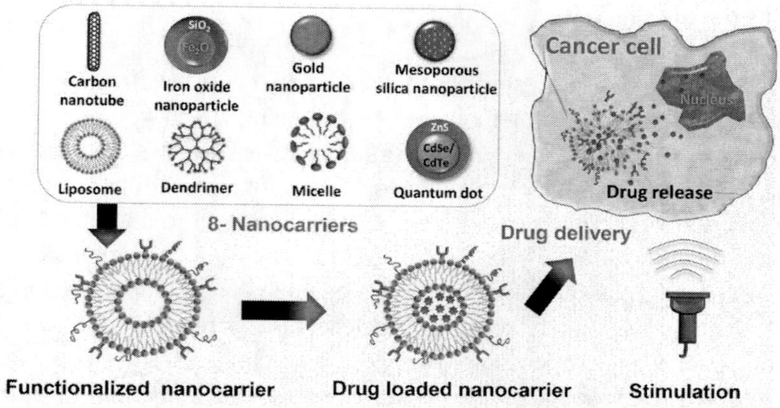

Figure 39. Schematic of nanoparticle drug delivery (Hossen et al. 2019).

Nanotechnology can also give alternatives to injectable vaccinations if the latent infection is hooked up with nanoparticles to induce the resistance response. Quantum dabs and large metal nanoclusters are extremely active and energising areas of bionanotechnology, with important advancements in the creation of new biosensors and bioelectronic devices occurring on a regular basis. The term "smart pills" refers to nano-scale electronic devices that are fashioned and constructed to resemble drug tablets but perform more sophisticated tasks such as detection, imaging, and medication distribution. Many smart pills, such as the PillCam, a shell with a small camcorder, and portion monitoring tablets, have recently been developed thanks to nanotechnology.

Meanwhile, MIT is working on 'Savvy Sensor Capsules,' which will eliminate the need for drugs to be administered into the stomach. Following the completion of appropriate documents for locating and checking therapy, it is provided orally and unfolds before being stored on the organ. It also has chambers that may be used to pre-load medications that will be administered to certain regions of the body. The 'Atmo Gas Capsule,' for example, is a sophisticated tablet that examines the gases in the human stomach and identifies any issues. Its sensors can detect oxygen and carbon dioxide levels in the body, as well as the presence of potentially hazardous chemicals. Its applications include detecting gastrointestinal problems, identifying potentially dangerous stomach organs, and tracking food aversions to enable customised diet and nutrition regimens.

7.2. Biomarkers

Combining DNA and other nucleic acids with nanoparticles in diverse physicochemical designs has resulted in a wide range of composite nanomaterials with unique or larger characteristics as a result of the two sections' synergistic development. Experts have created a DNA impedance biosensor, for example, to detect illness early or to swiftly identify a condition. As a result, these boundaries are attracting more attention from various research networks looking for new nanoscale instruments for grouped applications such as (bio)sensing, testing, assigned imaging, cell movement, diagnostics, therapeutics, theranostics, bioelectronics, and biocomputing, to name a few.

7.3. Cancer Therapy

Nanotechnology innovations have mostly focused on harming development, namely research and prescription movement. The chemotherapies paclitaxel, which slows cell proliferation, and lonidamine, which suffocates energy absorption in disease cells, have been used to treat multidrug-safe chest and ovarian cancer development with medications delivered by polymer-covered nanoparticles. The nanoparticles are designed to bind to an epidermal advancement factor receptor that is overexpressed in development cells. Early sickness discovery has a substantial influence on the rate of recovery. Using appealing nanoparticles in a miniaturised attractive resonation sensor, analysts may separate as little as two illness cells in one microlitre of a biosample, enhancing early area on a very basic level. Researchers at Stanford University in the United States have used nanotechnology to create a clear method for killing hazardous cells. They placed carbon nanotubes into disease cells before exposing the tissue to infrared laser light, which heated the nanotubes and destroyed the harmful development cells while leaving healthy cells alone. The nanomaterials can be used as cancer biomarkers as shown in Figure 40 and also in cancer diagnosis as illustrated in Figure 41 (Viswambari Devi, Doble, and Verma 2015; Y. Zhang et al. 2019).

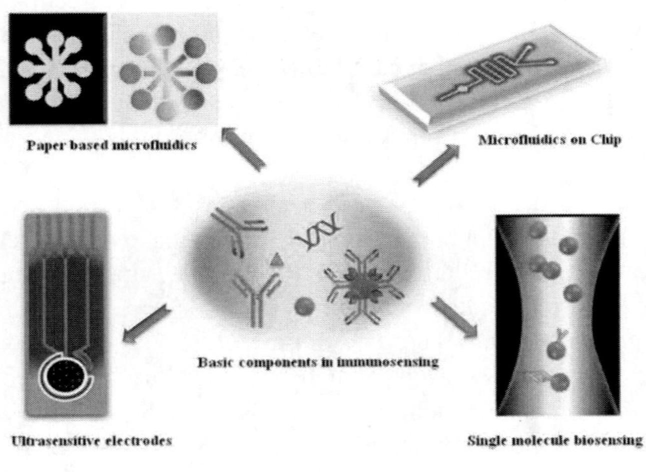

Figure 40. Nanomaterial as cancer biomarker (Viswambari Devi, Doble, and Verma 2015).

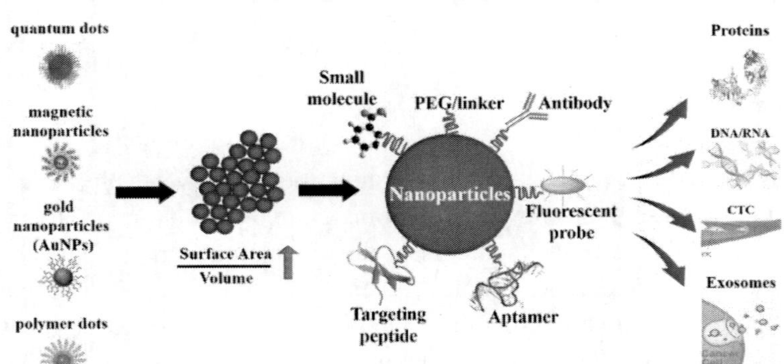

Figure 41. Nanotechnology in cancer diagnosis (Y. Zhang et al. 2019).

7.4. Biomedical Imaging

7.4.1. MRI Contrast Agents

Other imaging techniques, such as ionising radiation, are used to channel tissues, whereas attractive resonation imaging (MRI) makes use of protons' appealing properties. Frequently, protons are orientated for random reasons, reaching no generally appealing second, and then change to be equal to or against the fundamentally appealing field once the MRI device makes them. The inclination twists in the fundamental magnets change the fundamental alluring field, allowing MRI to picture in the x, z, or y axes.

Protons travel in and out of the lengthy centre of the basic appealing field at a pace proportionate to the attractive field's intensity. Using radio repetition pulses to induce a high-energy, in-work state in protons, the net magnetizable vector is then rotated 90 degrees towards the get over plane. When protons return to their amazing longitudinal out of stage condition via wind cross segment (T1) and turn (T2) loosening, they loosen up to their usual state. Because of attractive field in homogeneities, protons de-stage substantially quicker than T2; T2* is the amount of T2 loosening up as a result of these inhomogeneities. When protons move to the longitudinal plane, they relax up, creating changes in the net alluring vector, which may then be used to channel tissues as shown in Figure 42 (Bao, Sherwood, and Sun 2018). Contrast specialists have expertise in dealing with MRI images and are sought after for their ability to influence T1 or T2/T2 timings (Williams and Price 2017).

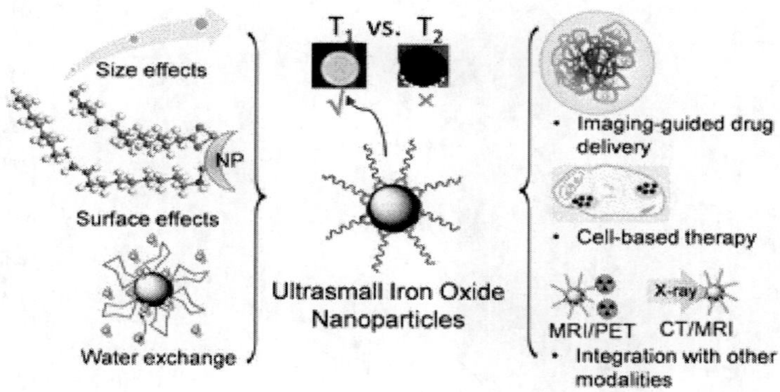

Figure 42. Magnetic iron oxide T1 contrast agent (Bao, Sherwood, and Sun 2018).

7.5. Bio-Composites

Biodegradable composites, also known as biocomposites, have grown in popularity across the world as a way to lessen the environmental effect of the petroleum industry. Nanotechnology has showed amazing potential as nanobiocomposites, which disclose critical qualities that are commonly not the same as those of standard polymers, working on mechanical demonstrations and so expanding contemporary and scholarly sectors of use. The most important characteristics to look for while fostering an item are unrivalled biocompatibility, low toxicity, security, antimicrobial insurance, heat resistance, and biodegradability. There are several forms of nanobiocomposites, each with its own set of benefits, that are suitable for each location and capacity (Ramos et al. 2022).

7.6. Dental Ceramics

Dentin, finish, cementum, mash, and periodontal tendon are all found in the oral pit. To protect the teeth, a variety of chemicals are used in dentistry. Nanoceramics are materials composed of ultrafine particles with dimensions less than 100 nm that are employed as a platform for the regeneration of bone design lost owing to illness, as well as flagging atoms in malignant development treatment. Because of their remineralizing properties, nanoceramics have been employed in a range of supportive and deterring dental biomaterials, including dental composites and glass ionomer concrete, as well as mouthwashes and toothpaste as shown in Figure 43 (Pasha et al. 2022; Priyadarsini, Mukherjee, and Mishra 2018).

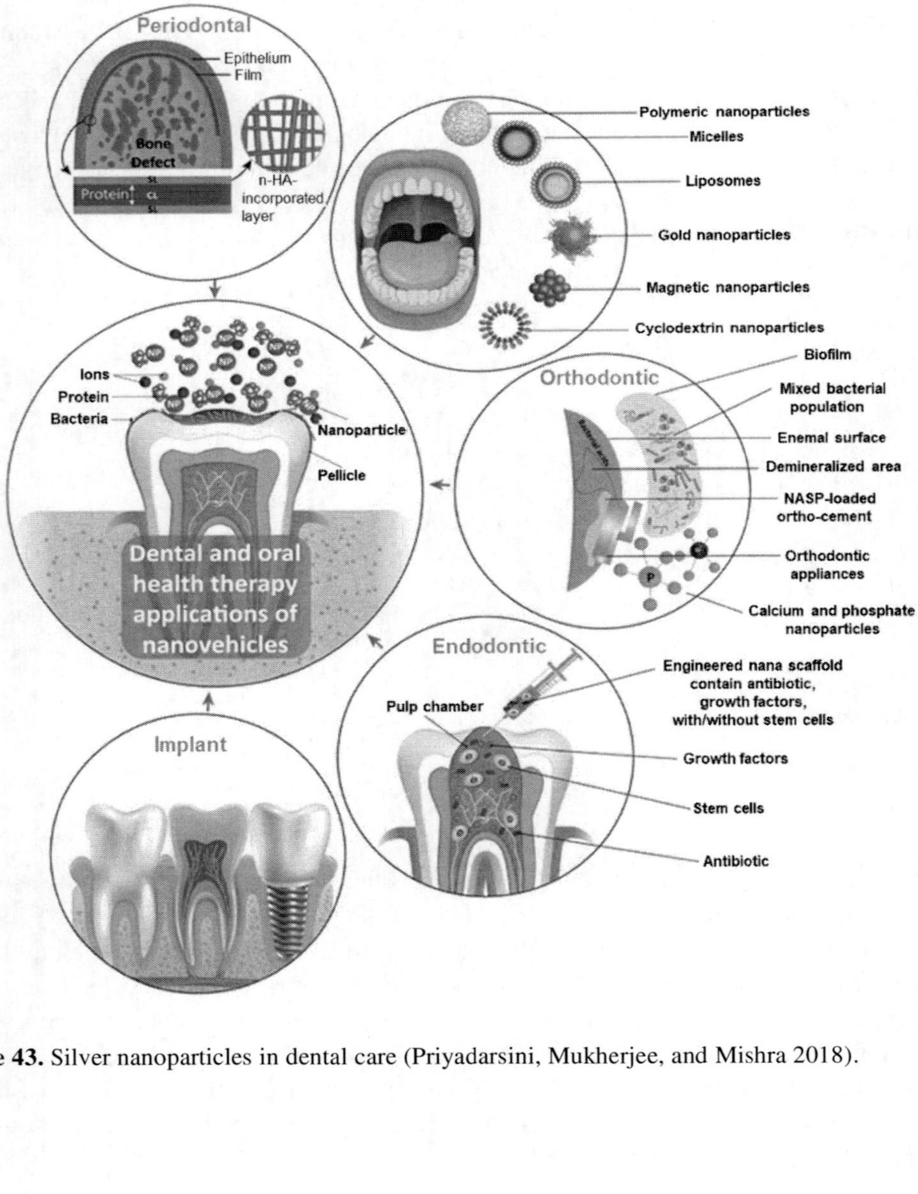

Figure 43. Silver nanoparticles in dental care (Priyadarsini, Mukherjee, and Mishra 2018).

7.7. Bone Growth

Bone insufficiency affects a vast number of people and is a leading cause of disability. Non-recovering breaks after a physical challenge, illnesses such as osteoporosis, and cancers all contribute to it. Metals, earthenware fabrication, polymers, and composites are currently used in the middle, and despite their long history and broad use, they have limitations such as limited biocompatibility, poor bone connection, and mechanical properties that differ from neighbouring bone. Because of the rising quest for novel therapeutics for basic tissue weakness, regenerative medicine has become a fascinating field of investigation. Biomaterials for bone regeneration should be osteoinductive as well as biocompatible and osteoconductive.

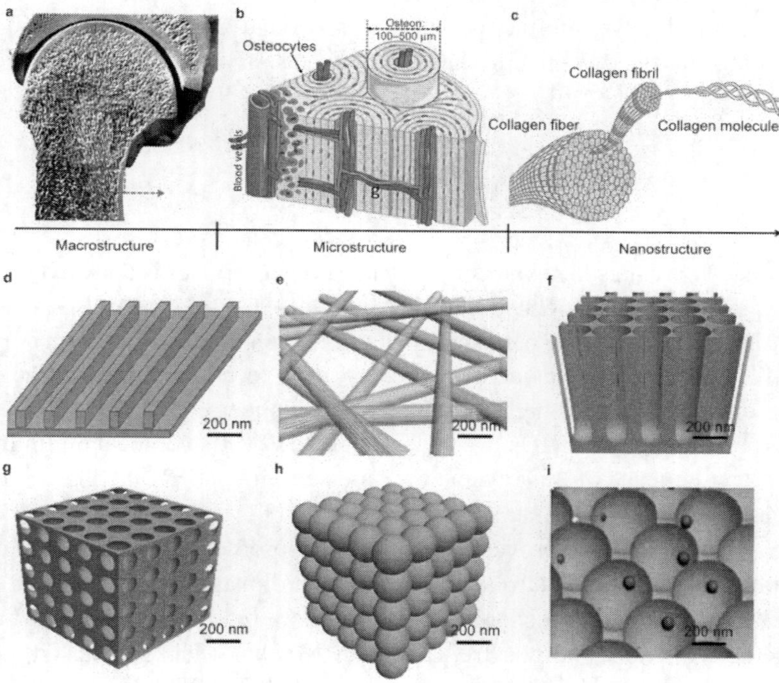

Figure 44. Nano materials in bone regeneration (Hajiali et al. 2021).

Nanostructured biomaterials have become unrivalled in supporting bone healing because to their planned and actual features (e.g., appealing, electrical) that are distinct to their mass reciprocals as shown in Figure 44 (Hajiali et al. 2021). The ability to perfectly mimic bone development and nanoarchitecture while exhibiting the nanoscale focus of critical biochemical milieu highlights is what distinguishes these technologies. This results in a greater ability to connect with the host's immune and progenitor cells at the nanoscale, leading in improved outcomes (Lyons et al. 2020).

7.8. Hyperthermic Treatment

Overheating (hyperthermia) can be used to treat a number of disorders, including cancer. Nearby warming of growth cells can be done by embedding attractive material (e.g., ferromagnetic poles) inside tumours and exposing them to AC attractive fields. To prevent a medical treatment and limit warming of solid tissue, it is better to disperse attractive nanoparticles in the target tissue, which is then warmed by the AC field (Pankhurst et al. 2003; S. Mornet et al. 2006; Stéphane Mornet et al. 2004; Xie et al. 2009; Bishop et al. 2009; Mørup, Hansen, and Frandsen 2011).

7.9. Wound Dressing

Nanofibers are used in injury dressings and protective materials, as well as implants, tissue design, and artificial organ components. Researchers are focusing on generating 'bright gauzes,' which will dissolve into the tissue after the wound has healed. Nanofibres contained in these ingenious gauzes can include thickening agents, antimicrobials, and even sensors to detect pollution.

8. Textiles

The application of fabric-based nanotechnology in medical services is a relatively new but well-established type of remote patient monitoring. Clinical data like as pulse, sweat parts, and circulatory strain are captured by nanosensors integrated in the fabric of such devices. It saves lives by alerting the wearer and medical professionals to any changes in the body's state. Nanowear, a company based in the United States, is a pioneer in this industry. NanoSENSE, a cardiovascular breakdown for CEOs, was released in 2019 with a clinical preliminary approval warning. As a consequence of this review, they created SimpleSENSE, a checking underwear and closed circular AI stage.

Using patented and FDA-approved fabric-based nanosensors, this piece of clothing collects and records phonocardiography, stroke volume, and cardiovascular outcomes. It also contains multi-channel ECG evaluating technology that monitors pulse abnormalities, breathing rate, thoracic impedance, action, and posture. In July 2020, Nanowear announced a collaboration with Hackensack Meridian Health Systems for extended COVID-19 remote analytic testing. Their objective is to employ Nanowear's fabric-based nanosensors to screen verified or suspected COVID-19 patients for physiological and biomarker changes that may indicate clinical weakness and necessitate therapeutic action.

8.1. Water-Repellent Textiles

Nano-Tex enhances the water-repellent property of texture by incorporating nano-stubbles, which are hydrocarbons 1/1000 the size of a regular cotton fibre, into the texture to create a peach fluff impression without losing cotton's strength. Water stays on the highest point of the hairs, or more particularly the texture's outer layer, since the intervals between the bristles on the texture are smaller than a typical drop of water but greater than water particles. Nonetheless, if tension is applied to the material, fluid can travel through it. The presentation lasts for an incredibly long time while retaining lung capacity. The most well-known use is the shark-skin suit worn during world-record-breaking Olympic swimming competitions. The suit, which includes a plasma layer strengthened by nanotechnology to resist water particles, is meant to let swimmers float through the water and has become a typical fixture of major swimming events as all participants want to improve their chances of winning (Soto et al. 2018).

8.2. Technical Textiles

The chips are built on nano strips, which are so adaptable that they can wrinkle over the edge of an amplifying focal point cover slide and are flexible enough to be curled into a wine gadget. Experts are focusing on application advancement in the clinical care industry, and they recognise that these small, versatile electronic sheets could one day be used to line the brain to distinguish development patients at risk for epilepsy, or be incorporated into cautious gloves to screen a patient's vital signs during an operation.

Nanotechnology has also revolutionised running shoes, tennis racquets, golf balls, skin treatments, and a variety of other outdoor products. In addition to developing materials that can withstand extreme conditions, scientists have sought for new building blocks for nanotechnology in viral micro particles that dwell in some of the most extreme environments on the planet. Clothing that recognises and interacts with the wearer's natural features is a popular issue. Such material-based nanosensors might offer a modified clinical benefits system, sensing significant physical processes as you run up an incline or recording environmental changes.

8.3. Medical Textiles

Antibacterial properties are provided by nano-sized silver, titanium dioxide, and zinc oxide. Metallic particles and metallic combinations are very sanitising. It is hypothesised that by catalysis with the metallic molecule, a portion of the oxygen in the air or water is changed to dynamic oxygen, causing the normal material to dissolve and have a cleaning effect. By employing nano-sized particles, the number of particles per unit region may be increased, and therefore the resistance to bacterial effects can be improved ("Application of Nanotechnology in Textile Industry, Nanotechnology Application in Textiles - Fibre2Fashion" n.d.).

8.4. Electro-Conducting Textiles

Intrinsically conductive polymers (ICPs) have limited incitation and detection properties. These polymers' electrical conductivity is owing to their peculiar chain structure, which includes the newly formed double bonds. Polyaniline, polyacetylene, polythiophene, and polypyrrol (PPy) are conductive polymers that are put into materials to function as detecting and impelling materials. Cycling strength was previously utilised to account for PPy-based supercapacitors. Electrical conductivity varies in polymers such as polypyrrol, polyacetylene, and poly 3, 4-ethylenedioxythiophene (PEDOT), among others, due to external deformities. Two elements that impact electrical behaviour are osmotic evacuation or the movement of dissolvable in/out, as well as a change in the polymer structural spine. Polypyrrol is extensively utilised as a conductive polymer due to its great flexibility and mechanical durability. The surface and state of the substance materials are never lost. As a result, natural piezo-resistive sensors were created using conductive polymers. They can be employed as imitation muscles due to their capacity to interact with physiological fluid by delivering mechanical force based on electrochemical responses.

8.5. Anti-wrinkle Textiles

Nano-Tex has introduced yet another nanotechnology-based anti-wrinkle treatment that is said to increase performance while keeping surface strength and authenticity, offering an alternative to harsh conventional cycles. Manufactured combinations and maintenance procedures decrease surface tear and unbending nature. This means that while some non-wrinkle surfaces and clothing are exceptional and useful for time-crunched clients, non-wrinkle development is unusual, such as lightweight surfaces or loose-fitting clothes. Surfaces may need to be over-planned or increased from time to time to survive the fibre damage produced by regular blemish-free operations. Without a doubt, current developments either do not work on all surfaces or demand the brand/retailer to invest more charges in order to adapt the damaging aspects of flaw-free science. The nano-scale sub-nuclear development of Nano-new Tex's Fortify DP innovation penetrates the fibre even deeper, resulting in wrinkle-free performance. Furthermore, it uses a longer and more adaptive cross-associating chain that minimises fibre stress under strain, hence lowering the huge strength disadvantage associated with conventional blemish-free technology.

8.6. Self-Cleaning Textiles

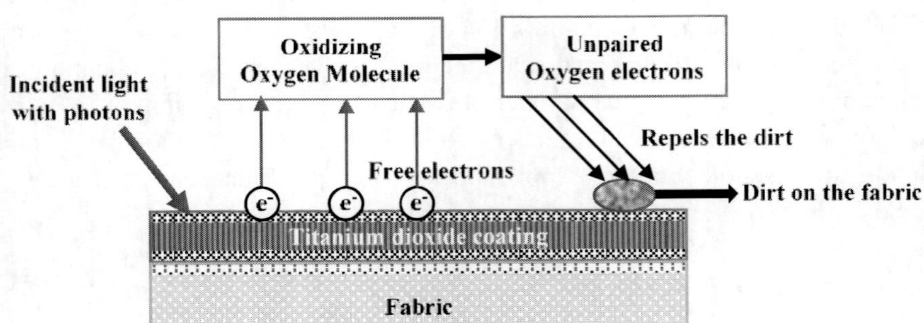

Figure 45. Schematic of self-cleaning fabric ("Self Cleaning Textile - an Overview - Free Technical Textile Industry Articles - Fibre2fashion.Com" n.d.).

Nanotex, an American business, developed and improved a self-cleaning cotton surface known as nano-care, and smirch-free jeans and khakis were debuted about 1990. Nanocare surfaces are created by changing the cylinder-shaped development of the cotton fibres that make up the surface. At the nanoscale, cotton threads resemble tree trunks. Thanks to nanotechnology, these tree trunks are coated with a puff of second hairs, which forms a stack of air around the fibre.

When water strikes the surface, it marks the qualities of the fibres, and the touches pack the air in the pits between the stubbles, making the stubbles extremely soft. The schematic of a self-cleaning fabric coated with titanium dioxide is given in Figure 45.

In technical terms, the fabric has been rendered super-non wettable or super-hydrophobic. The whiskers also create fewer points of contact for dirt. When water is applied to soiled fabric, the dirt adheres to the water far better than it adheres to the textile surface and is carried off with the water as it beads up and rolls off the surface of the fabric. Thus, the concept of Soil-cleaning is based on the leaves of the lotus plant.

8.7. UV Blocking Textiles

In specifically, the surface has been described as super-non-wettable or super-hydrophobic. Furthermore, the fibres deplete soil resources. As water is applied to a ruined surface, the dirt adheres to the water more than it does to the material surface and is trucked away with the water when it touches and rolls off the surface's outside layer. Thusly The lotus plant's leaves are used to clean the soil. Metal oxides, such as ZnO, are more consistent than other UV-blocking subject matter specialists. Nano ZnO's UV-deterring qualities will continue to improve as a result of their larger surface area and improved UV ingestion. In terms of cost-effectiveness, whiteness, and UV-blocking qualities for antibacterial finishing, ZnO nano particles exceed nano-silver.

UV-protected texture guarantees that the clothes deflects the sun's damaging dazzling beams, minimising UVR openness and protecting the skin from any injury. The degree of skin protection required by distinct types of human skin is determined by UV radiation production and transmission in relation to topographical area, season of day, and season. This protection is measured in SPF (Sun Protection Factor); the higher the SPF value, the better the UV ray protection as shown in Figure 46.

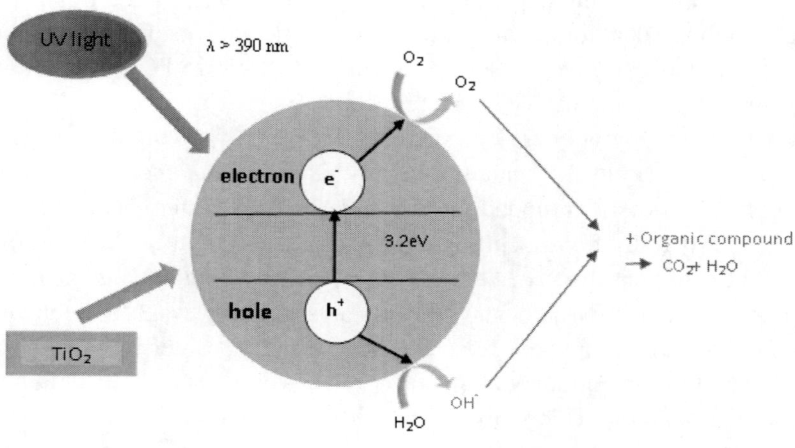

Figure 46. Schematic of UV blocking phenomenon in fabrics ("Impact of Nanotechnology in the Arena of Textile Apparel Finishing" n.d.).

9. Energy

Nanotechnology is being employed in traditional energy sources and is significantly enhancing alternative energy sources in order to assist satisfy the world's expanding energy demands. Many academics are researching tactics for developing ideal, inexpensive, and environmentally acceptable energy sources, as well as techniques for reducing energy use and alleviating climate-related concerns.

Nanotechnology is increasing the productivity of crude petroleum fuel production by enhancing catalysis. It also allows for less fuel use in autos and power plants due to more efficient combustion and decreased contact. Nanotechnology is also employed in oil and gas production, such as the use of nanotechnology-enabled gas lift valves in seaward activities or the use of nanoparticles to detect minor down-well oil pipeline cracks.

Carbon nanotube "scrubbers" and membranes are being investigated by researchers to recover carbon dioxide from power plant exhaust. Carbon nanotube-based wires with significantly lower resistance than the high-tension cables now used in the electric grid are being developed by researchers, resulting in lower transmission power loss. Nanotechnology can be included into solar panels to increase sunlight-to-energy conversion, resulting in more cheap solar electricity in the future. Nanostructured solar cells may be less expensive to make and install since they can be made using print-like technologies and are built in flexible rolls rather than discrete panels. Future solar converters may be "paintable," according to new research.

Nanotechnology is now being employed to create a range of new types of batteries that are quicker charging, more efficient, lighter, thicker, and can hold an electrical charge for extended periods of time. An epoxy containing carbon nanotubes is being utilised to make sharp edges that are longer, more grounded, and lighter-weight than conventional edges in order to maximise the amount of power that windmills can produce. Researchers are creating tiny film solar-powered electric boards that can be put inside PC casings, as well as adjustable piezoelectric nanowires woven into garments to generate useful energy fast from light, grinding, and even body heat to operate portable electronic gadgets. Other nanoscience-based approaches are also being researched to transform waste heat in PCs, autos, homes, power plants, and other locations into usable electrical power.

The number and variety of energy productivity and energy-saving devices is growing. In addition to the aforementioned, nanotechnology is allowing more productive lighting frameworks, lighter and more grounded vehicle undercarriage materials for the transportation industry, lower energy utilisation in cutting edge gadgets, and light-responsive smart coatings for glass. In addition to contributing in the growth of energy production, nanotechnology may help discover and clean up environmental toxins in a variety of ways. Nanotechnology may be able to help address the requirement for safe, clean drinking water by quickly and cheaply treating pollutants in water. Engineers have favoured a wobbly film layer with nanopores for energy-efficient desalination. This molybdenum disulphide (MoS2) film filtered two to four times as much water than conventional back-and-forth channels. Nanoparticles are being created to remediate existing ground water contamination by chemical reactions that render the toxins harmless. This link would be far less expensive than systems that involve syphoning groundwater for treatment.

Researchers have proposed for a nanofabric "paper towel" made of tiny potassium manganese oxide wires that can acclimatise to its weight in oil and be used for cleaning several times. Magnets have also been used by experts to accurately clean up oil spills by dissolving enticing water-repellent nanoparticles in the water. Various airline hotels and air channels use nanotechnology to produce "mechanical filtration," in which the fibre material forms small holes that trap particles greater than the pores' size. Furthermore, the tubes may have charcoal layers that absorb odours. Nanotechnology-enabled sensors and actions can now detect and identify material or natural experts in the air and soil with significantly higher sensitivity than previously. Researchers are investigating particles such as self-assembled monolayers on mesoporous supports (SAMMSTM), dendrimers, and carbon nanotubes to determine if they may be utilised to clean up hazardous waste sites. NASA has created a new wireless sensor that fire fighters may use to assess air quality around fires.

9.1. Lithium Ion Battery Electrodes

Electrons and lithium particles should explore more limited distances since nanoparticles and nanocrystallites are less intrusive than bulk materials. This allows for a higher rate limit (and hence greater power thickness) in typical cathode materials, as well as the use of protective materials that look insufficient. Another advantage is that nanoscale particles or precious stones may be able to survive the rigours of lithium implantation for a longer period of time (through intercalation or stage change). This suggests that lithium intercalation occurs throughout a broader spectrum of intercalation materials, resulting in a longer reversible cutoff. Because nanoparticles can sustain strain, they must be made of high-limit materials that go through stage changes when lithium is added (Hudak 2012).

9.2. Hydrogen Storage Materials

Hydrogen is a potentially new practical and clean energy option since it is a useful, safe, and long-lasting fuel source that can move from one form of energy to another in a matter of seconds without creating significant pollution. Because it minimises ozone-depleting material radiations, eliminates global reliance on oil-based things, and improves energy change viability, hydrogen is the ideal future fuel. Nanostructured materials may be acceptable under the hydrogen limit due to their unique features such as surface adsorption, inter and intragrain cutoff points, and mass absoprtion (Kutz 2008; "Carbon Nanotubes–A Successful Hydrogen Storage Medium" n.d.).

Nanostructured and nanoscale materials alter the thermodynamics and energy of hydrogen osmosis and detachment. By increasing the spread rate while decreasing the basic scattering length (Yumoto et al. 2003; "Nanocrystalline Materials for Hydrogen Storage (Book) | OSTI.GOV" n.d.). Material fitting variables at the nanoscale can be changed independently of their mass partners. They also backed the development of lighter hydrogen storage systems with greater hydrogen accumulation capacity (Niemann et al. 2008).

9.3. Paint-on Solar Cells

Combining nanoparticles in paints may limit their efficiency by making them lighter and offering other properties, for example. More lightweight paint coatings would lower plane weight, which might be beneficial to the environment. However, before all-around benefits can be demonstrated, the plane's whole life cycle must be analysed. It may also be feasible to drastically reduce the dissolvable substance in paints, which would result in the same benefits. Since the typical effects of tributyl tin (TBT) have been recognised, new forms of fouling-free marine paint may be produced and are desperately needed as replacements. Surface fouling prevention is especially crucial in process applications such as heat exchange, where it may induce energy hold reserves.

If fouling-resistant coatings can be created at a cheap enough cost, they may be used in common obligations such as pipes for local and current water infrastructure. It's still unclear if practical anti-fouling coatings may reduce the requirement for biocides like chlorine. Paints that change colour in reaction to changes in temperature or chemical environment, as well as paints with lower infrared absorptivity that minimise heat setback, are two more novel and far-fetched applications for nanoparticles. Concerns about nanoparticle success and standard impacts may need paying attention to the toughness and scratched district lead of nano-arranged paints and coatings, so scratched district items seem as coarse or tiny agglomerates rather than individual nanoparticles.

9.4. Dye-Sensitized Solar Cells

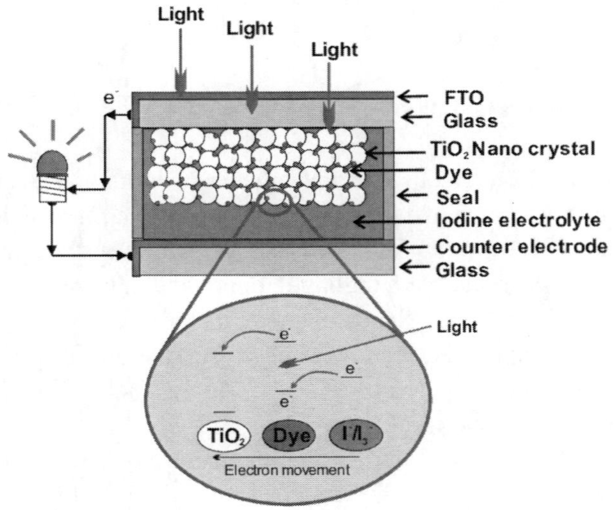

Figure 47. Typical structure of dye-sensitized solar cells (Rosa et al. 2017).

Dye-sensitized solar cells (DSSCs) are a promising third-generation photovoltaic (PV) breakthrough with cheap material costs and unrivalled performance. The flexibility of collecting cycles to function on a range of substrates and roll-to-move manufacturing considerably improves the capabilities of DSSCs in current applications. 1D nanostructured photoanodes, such as nanowires and nanotubes, can offer an electron transport channel from the site of electron implantation to the event substrate, which can be single-glasslike or polycrystalline, increasing power change efficiency (PCE) in DSSC applications. Because of their tiny surface area, 1D nanostructures have a poor light-social occasion efficiency. Because of their huge surface area, 3D moderate nanostructures may gather a considerable amount of light and charge carriers (W. Zhang 2019). The schematic of a typical dy-sensitized solar cell structure is shown in Figure 47 (Rosa et al. 2017).

10. Transports

Nanotechnology has the potential to create multifunctional materials that can assist in the construction and upkeep of lighter, safer, more attractive, and more usable vehicles, planes, transporters, and ships. Nanotechnology can also help with transportation infrastructure in a variety of ways as illustrated in Figure 48.

Polymer nanocomposites essential parts; high-power battery-controlled battery structures; thermoelectric materials for temperature control; lower rolling-resistance tyres; high-capability/low-cost sensors and devices; simple film brilliant daylight-based chargers; and fuel added substances and chipped away at exhaust frameworks for cleaner vapour and expanded territory are all examples of nano-planned materials in vehicle things. Nano-planning of aluminium, steel, dark top, concrete, and other cementitious materials, as well as their reused structures, provides exceptional assurance in terms of reducing the appearance, strength, and life expectancy of roadway and transportation infrastructure parts while lowering their life cycle costs. Innovative limitations, such as self-repairing structures or the ability to generate or transmit energy, might be included into traditional establishment materials utilising novel techniques.

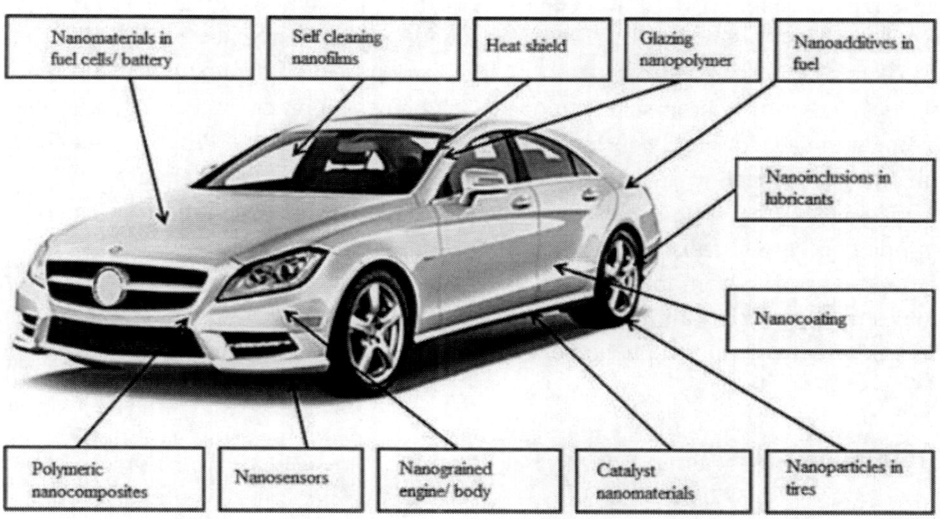

Figure 48. Various parts of automobile in which nanotechnology is applied (Mathew, Joy, and George 2019).

Nanoscale sensors and devices may provide long-term monitoring of bridges, tunnels, railways, parking structures, and pavements at a low cost. Nanoscale sensors, communications devices, and other nanoelectronics-based advancements can also aid drivers in maintaining lane position, avoiding collisions, changing travel routes to avoid congestion, and improving drivers' interfaces with onboard electronics. The advantages of adopting lightweight, high-strength materials provided by nanotechnology in virtually any transportation vehicle would be "game changing." According to an early NASA research, creating and deploying advanced nanomaterials with double the strength of traditional composites might cut the gross weight of a launch vehicle by up to 63%. This would not only save a considerable amount of energy necessary to put spacecraft into orbit, but it would also allow for the construction of single-stage to orbit launch vehicles, cutting launch costs and enhancing mission dependability.

11. Space Science

11.1. Propulsion Systems

The bulk of today's rocket engines employ chemical propulsion. Chemical rockets are used to launch all contemporary spacecraft, and most employ them to regulate attitude (the angular position and rotation of the spacecraft, either relative to the object it is circling or relative to the celestial sphere). Real rocket scientists, on the other hand, are focused on inventing new sorts of space propulsion technology. (Yan et al. 2018; Levchenko et al. 2018; "Nanomaterials | Special Issue : Advanced Materials and Nanomaterials for Space Technology" n.d.). Electric impetus (EP) frameworks significantly reduce the required force mass when compared to conventional material rockets, allowing for increased payload capacity or reduced launch mass. EP frameworks include field discharge electric drive (FEEP), colloid engines, and various field outflow engine types (FETs). EP has been successfully shown as a crucial drive architecture by the ESA's SMART-1, Japan's HAYABUSA, and NASA's Deep Space 1 missions. An EP idea for nanotechnology uses electrostatically charged and propelled nanoparticles as a force.

A vast number of micron-sized nanoparticle engines might be packed into a single square centimetre, allowing for highly customizable engine demonstrations. Proposals that harnessing Casimir forces may lead to an interplanetary spaceship propulsion system are a long way off. The primary idea is that future space explorers will have access to a limitless energy source due to zero-point energy by utilising the fact that vacuum is an energy reservoir. Of course, all they seek is a propulsion system that takes the necessary energy from the vacuum. According to a 1984 newspaper, this isn't altogether ridiculous. Several laboratories are attempting to harness the Casimir effect and related phenomena for vacuum energy conversion.

11.2. Radiation Shielding

In the subject of radiation shielding, nanotechnology has the potential to make a substantial contribution to human space travel. The likelihood of room radiation exposure, according to NASA, is the most major worry restricting people's ability to engage in long-duration space missions. As a result, substantial research is underway to create remedies to safeguard space travellers from these dangers. Designers of spacecraft are seeking for materials that will allow them to build multifunctional spaceship bodies that meet radiation security standards as well as other criteria such as low weight and basic durability (Sttar, Mkhaiber, and Majeed 2019).

Enhanced nanomaterials, such as recently developed isotopically upgraded boron nanotubes, might pave the way for future spacecraft with nanosensor-coordinated architectures that provide excellent radiation shielding and energy storage. Another area where radiation protection is critical is the security of installed hardware. It has already been demonstrated that as electronic device dimensions are lowered, they become more radiation friendly. For example, multi-quantum well or quantum speck electronics can be tens or hundreds of times less radiation resistant than normal mass devices. It was even proved that photovoltaic devices based on quantum specks/CNTs were five times safer than conventional mass-produced solar cells (Cheraghi, Chen, and Yeow 2021). A few recent studies on the influence of high-energy particles like protons, electrons, and heavy ions on nanomaterials like carbon nanotubes and nanowires have focused on the structural properties of the nanomaterials after irradiation. (Thibeault et al. 2015).

11.3. Anti-Satellite Weapon Countermeasure

In January 2007, China successfully tested an Anti-satellite (ASAT) rocket framework by destroying their own ageing LEO spacecraft, resulting in large quantities of room waste. This ASAT test fuelled considerable worry about the vulnerability of satellites and other space resources, as well as the prospect of a space arms race (Pillsbury, n.d.). In response to the problems posed by ASAT rocket frameworks, military planners and analysts are proposing new ideas to secure their space resources. As a result, Raytheon has created a countermeasure architecture that employs quantum dots to safeguard space assets such as satellites against rocket strikes. They've created a bait comprised of quantum dots of various sizes and shapes that are designed to transmit radiation in accordance with the resource's radiation profile.

11.4. Space Elevator

After joining the outermost lengths of a piece of ribbon, turn a stone over your head. The stone will be drawn tight as it turns. Consider a 62,000-mile-long strip of land sheltered at the equator, with a mound on one end. The diffusive force of the world's metamorphosis will compel it to behave similarly. You'll receive the world's most renowned nunchuck, as well as a type of space elevator. A space elevator that considers carbon nanotube joins is one of those crazy 1950s-style futuristic concepts that could just work. Every year, a Space Elevator Conference is conducted to review progress, and a Japanese technology company has expressed aspirations to create a space lift by 2050 (Pugno 2006; "Space Elevator: A Futuristic Application of Carbon Nanotubes - Nanografi Nano Technology" n.d.; "Nanotechnology Application: Space Elevator – MRSEC Education Group – UW–Madison" n.d.; "Using Nanotechnology to Build a Space Elevator" n.d.).

11.5. Protecting Satellites

Earth sensing satellites will mostly be found in LEO and MEO orbits, where they will be utilised for environmental sensing, earth imagery for both vegetation and minerals, and other applications. Because they move slowly in a highly predictable orbit, these satellites are vulnerable to physical harm and can be destroyed by directed energy weapons. The solar panels, which have a large surface area, and the optical sensors, which may be blinded by adjusting heat controls, are easily targetable and vulnerable satellite components. Hard, solid surfaces with coatings that can withstand extreme temperatures, scraped areas, and wear are necessary for spacecraft. The thermal or electromagnetic effects of a coordinated energy weapon may be felt first on the surface of the satellite. Depending on the design and pattern, the surface covering would either reflect, retain, or transmit the incoming energy, or a combination of the three. According to a recent study report issued by the Air Force's Air War College, several countries, including the United States, Russia, and China, have either produced or are developing the ability to build ground-based directed energy weapons. (Huntington 2007; Sharma 2012).

11.6. Space Instrumentation

The nanomaterial being developed by a team of 10 technologists at the NASA Goddard Space Flight Center in Greenbelt, Md., is a thin coating of multi-walled carbon nanotubes. While carbon nanotubes would find use in the Space Elevator thanks to their extraordinary strength, in this application, NASA is interested in using the technology to help suppress errant light that has a funny way of ricocheting off instrument components and contaminating measurements ("Nanotechnology in Space" n.d.).

MIT hosts the Space Nanotechnology Laboratory whose primary mission is to develop nano-fabrication, advanced lithography and precision engineering technology for building high performance space instrumentation, including x-ray telescopes and high resolution x-ray spectrometers, magnetospheric imagers and solar physics instrumentation ("Nanomaterials | Special Issue : Advanced Materials and Nanomaterials for Space Technology" n.d.).

12. Sports

Developments complicate a game, but they also increase a competitor's performance and lessen the chance of injury, making the sport more pleasurable for both spectators and competitors. Even little changes in sports equipment might be the difference between winning and losing in the realm of competitive gaming ("Sport and Nanotechnology: Are the Big Sports Looking to Go Small?" n.d.). The athletic gear sector has recently emerged as a sophisticated yet financially viable hello there tech industry, with technological advancements upending sports. Furthermore, big global championships such as the Olympics and World Cups have bolstered the athletic gear industry as a whole, making it a rich economic prospect for both public and private investors (Aoki and Nakamura 2018).

Nanotechnology has had a substantial impact on sportsmanship, as have a few other cutting-edge materials science concepts. In the world of sporting products, nanotechnology offers a wide range of benefits and has enormous promise, making competitors safer, more pleasurable, and more agile than at any other time in recent memory. Nanotechnology has improved the look and strength of products such as tennis and badminton racquets, hockey sticks, running bicycles, golf balls/clubs, skis, fly-projecting shafts, toxophilism bolts, and other sporting equipment ("Nanotechnology in Sports Equipment: The Game Changer" n.d.).

Carbon nanotubes (CNTs), silica nanoparticles (SNPs), nanoclays fullerenes, and other nanomaterials are employed in sports equipment to enhance competitor and hardware performance. Each of these nanomaterials provides a unique benefit in sports equipment, such as improved strength and solidity, toughness, weight reduction, scraped area blockage, and so on.

Some of the progressive advancements/items include Wilson's Double CoreTM and BLXTM (tennis), InMat's Air D-FenceTM and NanolokTM (tennis), Holmenkol's nano-CFC® (skiing), Easton's N-FUSEDTM (bows and arrows), St. Croix's NSiTM (fly-fishing), and Yonex's NanopremeTM (golf). The weight of athletic equipment has dropped but its strength has almost grown as materials in sporting gear have advanced. CNTs are the most often utilised material in nano-improved sports equipment because they have better explicit strength and stiffness than other materials. They are ideal for sporting equipment where light weight and incredible strength are required. They are many times more grounded than steel, yet many times lighter and as rigid as precious stone (Harifi and Montazer 2015).

Furthermore, nanocomposites containing CNTs are far stronger and more durable than other popular materials used in sporting gear. CNTs are being utilised to produce strong zones for trail-blazing bicycles, long-lasting tennis racquets, and lightweight bike designs. Cadel Lee Evans and Floyd Landis, for example, have won multiple races, including the Tour de France, on bikes with CNT integrated plastic casings. These cyclists rode on BMC SLC01 Pro Machine bikes from the Swiss business Bicycle Manufacturing Co. (BMC), which weighed only 1055 gms yet had many times the rounded strength of steel ("Nanotechnology in Sports," n.d.).

Footballs and tennis balls with nano-mud linings as a hindrance material sustain strain for longer game play. The consolidation of nanomaterials (such as fullerene) makes golf clubs simpler to wield by lowering the centre of gravity, resulting in increased power and precision during game play. In the body and poles of kayaks and race boats, graphene oxide and buckypaper (a sheet of carbon nanotubes) are utilised to boost coast while making them more grounded and lighter. Although self-collecting bionic nanoparticles are employed in land speed vehicles to boost economy and speed even with half the nitro methane fuel combination, extremely hard nanoceramics make ice skate edges ultra-sharp.

Nanocomposite goods that are lighter and more durable are being employed in Formula One racing, where the engine body weight and type of tyres play a significant influence in the race. Nanoparticles of ZrO2, ZnO, CuO, and other oxides are also employed to minimise wear and friction in greases, carbon nanofibers are cemented in cracks, and nano-based paints are utilised for heat resistance and drag reduction. Furthermore, KERS employ nano-phosphate lithium particles batteries, which are lighter, have a larger charge/release limit, and are more effective. The intriguing discovery of nanotechnology has had an influence on every part of our everyday existence, including games. In terms of strength, solidity, and sturdiness, nano-improved sports gear is far more performance-enhancing and widespread than standard athletic gear. Several organisations are investigating new and novel methods to employ nanotechnology to further the development of athletic equipment.

To improve their performance, several well-known sportsmen are turning to nano-integrated sporting equipment. Despite the fact that nano-improved sporting equipment is an expensive innovation, it cannot be managed by every single sportsperson, and it would be unjustifiable and difficult to pass judgement on the regular brandishing capacity and human soul unless each capable competitor is provided with a similar sporting equipment. As a result, rather than increasing the expense of ultramodern sporting equipment, a specific objective should be to push the boundaries of execution levels. Similarly, appropriate monitoring of their production methods, duration of use, and removal is intended to prevent nanoparticles from arriving owing to debasement or other reasons, addressing the possibility for nano-improved sports gear to be harmful.

Conclusion

The future of nanotechnology may have both bright and gloomy phases. The technological advancements, the government and private backup and investments, and ever increasing need for the miniaturization of devices are projected to propel the industry forward internationally. On the other hand, the hazards of nanotechnology to the environment, health, and safety, as well as concerns about its commercialization, may limit market expansion. As nanotechnology steps forward, devices or material based on this technology become smaller and more sensitive.

Wearable wellness innovation implies that we can track our health by strapping sophisticated technologies to our bodies. Electronic tattoos in the shape of models that can monitor our critical bodily functions are also available. By scaling down this technology, we may be able to go much further by embedding or injecting miniscule sensors into our bodies. This would allow clinicians to adjust their medication by capturing far more specific data while causing less danger to the patient. The original is about material science and improving attributes through the use of "inactive nanostructures." This might take the form of coatings or the use of carbon nanotubes to strengthen polymers. Because they are bioactive, dynamic nanostructures, for example, will be employed in the future to deliver medication to a specific target cell or organ. If the nanoparticle is coated with certain proteins, this should be possible. The intricacy rises in the third and fourth centuries. In the fourth era of nanomaterials, beginning with a developing nanosystem, such as nanorobotics, and proceeding to a sub-atomic nanosystem to manage the growth of phoney organs.

Researchers are now investigating ways to improve the 'Protected by-plan' idea for nanomaterials. The key argument is that security evaluations should be integrated into the planning and development stages of a nanomaterial's life cycle, rather than after they are launched.

References

Ali, Imran, Al Arsh Basheer, Anastasia Kucherova, Nariman Memetov, Tatiana Pasko, Kirill Ovchinnikov, Vladimir Pershin, et al. 2019. "Advances in Carbon Nanomaterials as Lubricants Modifiers." *Journal of Molecular Liquids* 279 (April): 251–66. https://doi.org/10.1016/J.MOLLIQ.2019.01.113.

"Analytical Discussion of Single Electron Transistor (SET)." n.d.

Aoki, Hirooki, and Hidetoshi Nakamura. 2018. "Non-Contact Respiration Measurement during Exercise Tolerance Test by Using Kinect Sensor." *Sports* 6 (1). https://doi.org/10.3390/SPORTS6010023.

"Application of Nanotechnology in Textile Industry, Nanotechnology Application in Textiles - Fibre2Fashion." n.d. Accessed May 2, 2022. https://www.fibre2fashion.com/industry-article/7135/application-of-nanotechnology-in-textile-industry.

Ball, Philip. 2021. "First Quantum Computer to Pack 100 Qubits Enters Crowded Race." *Nature* 599 (7886): 542. https://doi.org/10.1038/D41586-021-03476-5.

Bao, Y., J. A. Sherwood, and Z. Sun. 2018. "Magnetic Iron Oxide Nanoparticles as T1 Contrast Agents for Magnetic Resonance Imaging." *Journal of Materials Chemistry C* 6 (6): 1280–90. https://doi.org/10.1039/C7TC05854C.

Bishop, Kyle J. M., Christopher E. Wilmer, Siowling Soh, and Bartosz A. Grzybowski. 2009. "Nanoscale Forces and Their Uses in Self-Assembly." *Small* 5 (14): 1600–1630. https://doi.org/10.1002/SMLL.200900358.

"Carbon Nanotubes–A Successful Hydrogen Storage Medium." n.d. Accessed May 2, 2022. https://publications.waset.org/16692/carbon-nanotubes-a-successful-hydrogen-storage-medium.

Cheraghi, Elahe, Siyuan Chen, and John T.W. Yeow. 2021. "Boron Nitride-Based Nanomaterials for Radiation Shielding: A Review." *IEEE Nanotechnology Magazine* 15 (3): 8–17. https://doi.org/10.1109/MNANO.2021.3066390.

Demiroglu, Sibel, Vivekanandhan Singaravelu, M. Özgür Seydibeyoğlu, Manjusri Misra, and Amar K. Mohanty. 2017. "The Use of Nanotechnology for Fibre-Reinforced Polymer Composites." *Fiber Technology for Fiber-Reinforced Composites*, January, 277–97. https://doi.org/10.1016/B978-0-08-101871-2.00013-8.

Díez-Pascual, Ana María. 2019. "Nanoparticle Reinforced Polymers." *Polymers* 11 (4): 625. https://doi.org/10.3390/POLYM11040625.

Domínguez-Rodríguez, A., and D. Gómez-García. 2010. "Superplasticity in Ceramics: Applications and New Trends." *Key Engineering Materials* 423: 3–13. https://doi.org/10.4028/WWW.SCIENTIFIC.NET/KEM.423.3.

Dutz, Silvio, Joachim H. Clement, Dietmar Eberbeck, Thorsten Gelbrich, Rudolf Hergt, Robert Müller, Jana Wotschadlo, and Matthias Zeisberger. 2009. "Ferrofluids of Magnetic Multicore Nanoparticles for Biomedical Applications." *Journal of Magnetism and Magnetic Materials* 321 (10): 1501–4. https://doi.org/10.1016/J.JMMM.2009.02.073.

Eleftheriou, E., P. Bächtold, G. Cherubini, A. Dholakia, C. Hagleitner, T. Loeliger, A. Pantazi, et al. 2003. "A Nanotechnology-Based Approach to Data Storage." *Proceedings - 29th International Conference on Very Large Data Bases, VLDB 2003*, January, 3–7. https://doi.org/10.1016/B978-012722442-8/50008-2.

"Ferrofluids, Nano Magnetic | Nanofluid | Nanotechnology Products | NPD." n.d. Accessed April 27, 2022. https://product.statnano.com/product/8392/ferrofluids-nano-magnetic.

Fytianos, Georgios, Abbas Rahdar, and George Z. Kyzas. 2020. "Nanomaterials in Cosmetics: Recent Updates." *Nanomaterials 2020, Vol. 10, Page 979* 10 (5): 979. https://doi.org/10.3390/NANO10050979.

Hajiali, Hadi, Liliang Ouyang, Virginia Llopis-Hernandez, Oana Dobre, and Felicity R.A.J. Rose. 2021. "Review of Emerging Nanotechnology in Bone Regeneration: Progress, Challenges, and Perspectives." *Nanoscale* 13 (23): 10266–80. https://doi.org/10.1039/D1NR01371H.

Hameed, Awais, Gull Rida Fatima, Kainat Malik, Ayesha Muqadas, and M. Fazal-ur-Rehman. 2019. "Scope of Nanotechnology in Cosmetics: Dermatology and Skin Care Products." *Journal of Medicinal and Chemical Sciences* 2 (1): 9–16. https://doi.org/10.26655/JMCHEMSCI.2019.6.2.

Harifi, Tina, and Majid Montazer. 2015. "Application of Nanotechnology in Sports Clothing and Flooring for Enhanced Sport Activities, Performance, Efficiency and Comfort: A Review:" *Http://Dx.Doi.Org/10.1177/1528083715601512* 46 (5): 1147–69. https://doi.org/10.1177/1528083715601512.

Hossen, Sarwar, M. Khalid Hossain, M. K. Basher, M. N.H. Mia, M. T. Rahman, and M. Jalal Uddin. 2019. "Smart Nanocarrier-Based Drug Delivery Systems for Cancer Therapy and Toxicity Studies: A Review." *Journal of Advanced Research* 15 (January): 1–18. https://doi.org/10.1016/J.JARE.2018.06.005.

Hudak, Nicholas S. 2012. "Nanostructured Electrode Materials for Lithium-Ion Batteries." *Lithium-Ion Batteries: Advances and Applications*, September, 57–82. https://doi.org/10.1016/B978-0-444-59513-3.00004-2.

Huntington, Joseph. 2007. "IMPROVING SATELLITE PROTECTION WITH NANOTECHNOLOGY."

Ijaola, Ahmed Olanrewaju, Peter Kayode Farayibi, and Eylem Asmatulu. 2020. "Superhydrophobic Coatings for Steel Pipeline Protection in Oil and Gas Industries: A Comprehensive Review." *Journal of Natural Gas Science and Engineering* 83 (November): 103544. https://doi.org/10.1016/J.JNGSE.2020.103544.

"Impact of Nanotechnology in the Arena of Textile Apparel Finishing." n.d. Accessed May 2, 2022. https://www.textiletoday.com.bd/impact-of-nanotechnology-in-the-arena-of-textile-apparel-finishing/.

Khan, Fozia Haque. 2013. "Chemical Hazards of Nanoparticles to Human and Environment (A Review)." *Oriental Journal of Chemistry* 29 (4): 1399–1408. https://doi.org/10.13005/OJC/290415.

Kim, S. J., Yu I. Latyshev, and T. Yamashita. 1999. "Submicron Stacked-Junction Fabrication from Bi2Sr2CaCu2O8+δ Whiskers by Focused-Ion-Beam Etching." *Applied Physics Letters* 74 (8): 1156. https://doi.org/10.1063/1.123472.

Krishnan, B-Tech, Radha. 2014. "Single Electron Transistors." *International Journal of Scientific & Engineering Research* 5.

Kubrin, Roman. 2014. "Nanophosphor Coatings: Technology and Applications, Opportunities and Challenges." *KONA Powder and Particle Journal* 31 (1): 22–52. https://doi.org/10.14356/KONA.2014006.

Kutz, Myer. 2008. "Environmentally Conscious Alternative Energy Production." *Environmentally Conscious Alternative Energy Production*, January, 1–308. https://doi.org/10.1002/9780470209738.

Layani, Michael, Alexander Kamyshny, and Shlomo Magdassi. 2014. "Transparent Conductors Composed of Nanomaterials." *Nanoscale* 6 (11): 5581–91. https://doi.org/10.1039/C4NR00102H.

Levchenko, I., S. Xu, G. Teel, D. Mariotti, M. L. R. Walker, and M. Keidar. 2018. "Recent Progress and Perspectives of Space Electric Propulsion Systems Based on Smart Nanomaterials." *Nature Communications 2018 9:1* 9 (1): 1–19. https://doi.org/10.1038/s41467-017-02269-7.

Lyons, Joseph G., Mark A. Plantz, Wellington K. Hsu, Erin L. Hsu, and Silvia Minardi. 2020. "Nanostructured Biomaterials for Bone Regeneration." *Frontiers in Bioengineering and Biotechnology* 8 (August): 922. https://doi.org/10.3389/FBIOE.2020.00922/BIBTEX.

Mandeep, and Pratyoosh Shukla. 2020. "Microbial Nanotechnology for Bioremediation of Industrial Wastewater." *Frontiers in Microbiology* 11 (November): 2411. https://doi.org/10.3389/FMICB.2020.590631/BIBTEX.

Mathew, Jinu, Josny Joy, and Soney C. George. 2019. "Potential Applications of Nanotechnology in Transportation: A Review." *Journal of King Saud University - Science* 31 (4): 586–94. https://doi.org/10.1016/J.JKSUS.2018.03.015.

Medina-Pérez, Gabriela, Fabián Fernández-Luqueño, Rafael G. Campos-Montiel, Katia B. Sánchez-López, Laura N. Afanador-Barajas, and L. Prince. 2019. "Nanotechnology in Crop Protection: Status and Future Trends." *Nano-Biopesticides Today and Future Perspectives*, January, 17–45. https://doi.org/10.1016/B978-0-12-815829-6.00002-4.

Mo, Lixin, Zhenxin Guo, Li Yang, Qingqing Zhang, Yi Fang, Zhiqing Xin, Zheng Chen, Kun Hu, Lu Han, and Luhai Li. 2019. "Silver Nanoparticles Based Ink with Moderate Sintering in Flexible and Printed Electronics." *International Journal of Molecular Sciences 2019, Vol. 20, Page 2124* 20 (9): 2124. https://doi.org/10.3390/IJMS20092124.

Mornet, S., S. Vasseur, F. Grasset, P. Veverka, G. Goglio, A. Demourgues, J. Portier, E. Pollert, and E. Duguet. 2006. "Magnetic Nanoparticle Design for Medical Applications." *Progress in Solid State Chemistry* 34 (2–4): 237–47. https://doi.org/10.1016/J.PROGSOLIDSTCHEM.2005.11.010.

Mornet, Stéphane, Sébastien Vasseur, Fabien Grasset, and Etienne Duguet. 2004. "Magnetic Nanoparticle Design for Medical Diagnosis and Therapy." *Journal of Materials Chemistry* 14 (14): 2161–75. https://doi.org/10.1039/B402025A.

Mørup, S., M. F. Hansen, and C. Frandsen. 2011. "Magnetic Nanoparticles." *Comprehensive Nanoscience and Technology* 1–5 (January): 437–91. https://doi.org/10.1016/B978-0-12-374396-1.00036-2.

Murphy, D., B. Challacombe, M. S. Khan, and P. Dasgupta. 2006. "Robotic Technology in Urology." *Postgraduate Medical Journal* 82 (973): 743–47. https://doi.org/10.1136/PGMJ.2006.048140.

Murray, Haydn H. 2006. "Chapter 2 Structure and Composition of the Clay Minerals and Their Physical and Chemical Properties." *Developments in Clay Science* 2 (C): 7–31. https://doi.org/10.1016/S1572-4352(06)02002-2.

"Nanocrystalline Materials for Hydrogen Storage (Book) | OSTI.GOV." n.d. Accessed May 2, 2022. https://www.osti.gov/biblio/122756.

"Nanomaterials | Special Issue : Advanced Materials and Nanomaterials for Space Technology." n.d. Accessed May 2, 2022. https://www.mdpi.com/journal/nanomaterials/special_issues/Nanomaterials_Space_Technology.

"Nanomaterials Colouring the World: (Nano)Pigments-NEcoTox | Ökotoxikologische Tests Und Untersuchungen." n.d. Accessed May 2, 2022. https://necotox.de/en/nanopigments-background-and-regulatory-context/.

"Nanoparticle Conductive Ink." n.d. Accessed April 27, 2022. https://www.azonano.com/article.aspx?ArticleID=5248.

"Nanotechnology Application: Space Elevator – MRSEC Education Group – UW–Madison." n.d. Accessed May 2, 2022. https://education.mrsec.wisc.edu/nanotechnology-application-space-elevator/.

"Nanotechnology in Quantum Computing." n.d. Accessed May 2, 2022. https://www.azonano.com/article.aspx?ArticleID=3251.

"Nanotechnology in Space." n.d. Accessed May 2, 2022. https://www.understandingnano.com/space.html.

"Nanotechnology in Sports." n.d.

"Nanotechnology in Sports Equipment: The Game Changer." n.d. Accessed May 2, 2022. https://www.nanowerk.com/spotlight/spotid=30661.php.

Niemann, Michael U., Sesha S. Srinivasan, Ayala R. Phani, Ashok Kumar, D. Yogi Goswami, and Elias K. Stefanakos. 2008. "Nanomaterials for Hydrogen Storage Applications: A Review." *Journal of Nanomaterials* 2008 (1). https://doi.org/10.1155/2008/950967.

Pal Singh, Gurinder, and Balwinder Raj. 2013. "Single Electron Transistor Theory: A Review." *2013 IEEE International Conference on Emerging Trends in Computing, Communication and Nanotechnology, ICE-CCN 2013*, 95–99. https://doi.org/10.1109/ICE-CCN.2013.6528471.

Pankhurst, Q. A., J. Connolly, S. K. Jones, and J. Dobson. 2003. "Applications of Magnetic Nanoparticles in Biomedicine." *Journal of Physics D: Applied Physics* 36 (13): R167. https://doi.org/10.1088/0022-3727/36/13/201.

Pasha, Mehwish, Nawshad Muhammad, Saad Shahnawaz, Yasna Najmi, Nabiya Shahroz, and Saad Liaqat. 2022. "Ceramic Nanomaterials in Dental Applications." *Nanoengineering of Biomaterials*, February, 123–44. https://doi.org/10.1002/9783527832095.CH22.

"Photolithography Overview." n.d. Accessed May 2, 2022. https://www.newport.com/n/photolithography-overview.

Pillsbury, Michael P. n.d. "AN ASSESSMENT OF CHINA'S ANTI-SATELLITE AND SPACE WARFARE PROGRAMS, POLICIES AND DOCTRINES."

Predoi, Daniela, Rodica V. Ghita, Simona Liliana Iconaru, Carmen Laura Cimpeanu, and Stefania Mariana Raita. 2020. "Application of Nanotechnology Solutions in Plants Fertilization." *Urban Horticulture - Necessity of the Future*, March. https://doi.org/10.5772/INTECHOPEN.91240.

Priyadarsini, Subhashree, Sumit Mukherjee, and Monalisa Mishra. 2018. "Nanoparticles Used in Dentistry: A Review." *Journal of Oral Biology and Craniofacial Research* 8 (1): 58. https://doi.org/10.1016/J.JOBCR.2017.12.004.

Pugno, Nicola M. 2006. "On the Strength of the Carbon Nanotube-Based Space Elevator Cable: From Nanomechanics To." *Journal of Physics: Condensed Matter* 18 (33): S1971. https://doi.org/10.1088/0953-8984/18/33/S14.

Ramos, Sergiana dos Passos, Michele Giaconia, Monize Burck, Daniella Carisa Murador, Maria Carolina Bezerra Di-Medeiros Leal, Ailton Cesar Lemes, Mariana Buranelo Egea, and Anna Rafaela Cavalcante Braga. 2022. "Biocomposites Potential for Nanotechnology." *Biodegradable Polymers, Blends and Composites*, January, 489–510. https://doi.org/10.1016/B978-0-12-823791-5.00001-6.

Rosa, Erlyta Septa, Shobih, Lilis Retnaningsih, Lia Muliani, and Jojo Hidayat. 2017. "The Influence of the Addition of Dye Surface Modifier on the Performance of Transparent Dye Sensitized Solar Cells." *AIP Conference Proceedings* 1904 (1): 020050. https://doi.org/10.1063/1.5011907.

Sarid, Dror, Brendan McCarthy, and Ghassan E. Jabbour. 2004. "Nanotechnology for Data Storage Applications." *Springer Handbook of Nanotechnology*, 899–920. https://doi.org/10.1007/3-540-29838-X_30.

"Self Cleaning Textile - an Overview - Free Technical Textile Industry Articles - Fibre2fashion.Com." n.d. Accessed May 2, 2022. https://www.technicaltextile.net/articles/self-cleaning-textile-an-overview-2646.

Sharma, Ravi. 2012. "Nanotechnology in Space Exploration: Needs and Applications." *Journal of Pure Applied and Industrial Physics* 2 (3A): 286–402.

"Single-Electron Transistors – Physics World." n.d. Accessed April 27, 2022. https://physicsworld.com/a/single-electron-transistors/.

Soto, Dan, Asli Ugur, Taylor A. Farnham, Karen K. Gleason, and Kripa K. Varanasi. 2018. "Short-Fluorinated ICVD Coatings for Nonwetting Fabrics." *Advanced Functional Materials* 28 (33). https://doi.org/10.1002/ADFM.201707355.

"Space Elevator: A Futuristic Application of Carbon Nanotubes - Nanografi Nano Technology." n.d. Accessed May 2, 2022. https://nanografi.com/blog/space-elevator-a-futuristic-application-of-carbon-nanotubes/.

"Sport and Nanotechnology: Are the Big Sports Looking to Go Small?" n.d. Accessed May 2, 2022. https://www.azonano.com/article.aspx?ArticleID=4859.

Sttar, Muhammed Abdul Kadar Abdul, Ahmed Fadhil Mkhaiber, and Aseel Mustafa Abdul Majeed. 2019. "Study of the Effect of Using Nanomaterial in Radiological Shielding." *AIP Conference Proceedings* 2190 (1): 020077. https://doi.org/10.1063/1.5138563.

Thibeault, Sheila A., Jin Ho Kang, Godfrey Sauti, Cheol Park, Catharine C. Fay, and Glen C. King. 2015. "Nanomaterials for Radiation Shielding." *MRS Bulletin* 40 (10): 836–41. https://doi.org/10.1557/MRS.2015.225.

"Using Nanotechnology to Build a Space Elevator." n.d. Accessed May 2, 2022. https://www.azonano.com/article.aspx?ArticleID=4011.

Venugopal, Gunasekaran, and S. J. Kim. 2011. "Temperature Dependent Transfer Characteristics of Graphene Field Effect Transistors Fabricated Using Photolithography." *Current Applied Physics* 11 (3 SUPPL.). https://doi.org/10.1016/J.CAP.2011.03.030.

Venugopal, Gunasekaran, and Sang-Jae Kim. 2013. "Nanolithography." *Advances in Micro/Nano Electromechanical Systems and Fabrication Technologies*, May. https://doi.org/10.5772/55527.

Viswambari Devi, R., Mukesh Doble, and Rama S. Verma. 2015. "Nanomaterials for Early Detection of Cancer Biomarker with Special Emphasis on Gold Nanoparticles in Immunoassays/Sensors." *Biosensors and Bioelectronics* 68 (June): 688–98. https://doi.org/10.1016/J.BIOS.2015.01.066.

Wang, Zehong, Wei Wang, Zhikang Jiang, and Dan Yu. 2016. "Low Temperature Sintering Nano-Silver Conductive Ink Printed on Cotton Fabric as Printed Electronics." *Progress in Organic Coatings* 101 (December): 604–11. https://doi.org/10.1016/J.PORGCOAT.2016.08.019.

Williams, Harry M, and Helen Price. 2017. "The Application of Magnetic Nanoparticles in the Treatment and Monitoring of Cancer and Infectious Diseases." *Bioscience Horizons: The International Journal of Student Research* 10 (January): 2017. https://doi.org/10.1093/BIOHORIZONS/HZX009.

Worrall, Elizabeth A., Aflaq Hamid, Karishma T. Mody, Neena Mitter, and Hanu R. Pappu. 2018. "Nanotechnology for Plant Disease Management." *Agronomy 2018, Vol. 8, Page 285* 8 (12): 285. https://doi.org/10.3390/AGRONOMY8120285.

Xie, J., J. Huang, X. Li, S. Sun, and X. Chen. 2009. "Iron Oxide Nanoparticle Platform for Biomedical Applications." *Current Medicinal Chemistry* 16 (10): 1278–94. https://doi.org/10.2174/092986709787846604.

Yan, Qi Long, Guo Qiang He, Pei Jin Liu, and Michael Gozin. 2018. "Nanomaterials in Rocket Propulsion Systems." *Nanomaterials in Rocket Propulsion Systems*, October, 1–592. https://doi.org/10.1016/C2017-0-00465-2.

Yumoto, A., F. Hiroki, I. Shiota, and N. Niwa. 2003. "In Situ Synthesis of Titanium-Aluminides in Coating with Supersonic Free-Jet PVD Using Ti and Al Nanoparticles." *Surface and Coatings Technology* 169–170 (June): 499–503. https://doi.org/10.1016/S0257-8972(03)00152-X.

Zahn, Markus. 2001. "Magnetic Fluid and Nanoparticle Applications to Nanotechnology." *Journal of Nanoparticle Research* 3: 73–78.

Zhang, Wei. 2019. "Nanomaterials in Dye-Sensitized Solar Cells." *Advanced Nanomaterials for Solar Cells and Light Emitting Diodes*, January, 69–95. https://doi.org/10.1016/B978-0-12-813647-8.00003-5.

Zhang, Ye, Maoyu Li, Xiaomei Gao, Yongheng Chen, and Ting Liu. 2019. "Nanotechnology in Cancer Diagnosis: Progress, Challenges and Opportunities." *Journal of Hematology and Oncology* 12 (1): 1–13. https://doi.org/10.1186/S13045-019-0833-3/FIGURES/2.

Chapter 3

Lab-on-a-Chip Devices: Needs and Trends

Learning Outcomes

- The reader will develop a deep understanding about the lab-on-a-chip (LoC) technology.
- The reader will be able to identify the ardent need of LoC based devices or systems in the modern technological era for various fields.
- The reader could be able to independently identify the material and technology required for developing a particular LoC system to meet the desired application.
- The reader will have a deep insight into the current trends and existing technologies of LoC systems that are being used in numerous sectors in medicinal and day-to-day life.
- The reader will be able to clearly distinguish between the advantages and the limitations of using or developing the LoC devices.

Abstract

In recent years, the development of systems based on microfluidic channels has resulted in substantial breakthroughs in lab-on-a-chip technologies that integrate numerous functions onto a single platform. The miniaturised devices have evolved over time into a potentially useful mechanism to rapidly analyse various tests and display them with exceptional precision and accuracy. However, the vast majority of lab-on-a-chip devices are not made commercially available in the market for usage by the common population. These devices can be extensively used in the biological, healthcare, food safety and environmental sectors. Rapid diagnostic testing is one of the most successfully commercialised lab-on-a-chip technologies, which uses the lateral flow or spot testing concepts in the basic diagnostic instruments. In this chapter, a detailed discussion about the lab-on-a-chip technology, their need, advantages limitations and the current trends in various applications is carried out.

Keywords: lab-on-a-chip, biopolymers, proteomics, biosensors, drug production, DNA analysis, microarray, food analysis

1. Introduction to Lab-on-a-Chip

The preceding chapters provided an in-depth look at the realm of nanomaterials, nanotechnology, and its potential real-time applications. The Lab-on-a-chip (LoC) technology will be explained in this chapter, which allows nano devices to be combined with other electronic devices to suit their application requirements. The LoC is a miniaturized device that have the ability to combine one or several laboratory-based experiments like DNA sequencing or biological analyte detection, onto a single chip. Numerous LoC applications are being researched in recent years, including DNA analysis and human diagnostics to a greater extent. Miniaturizing biochemical activities that are traditionally performed in a lab has various advantages, including cost effectiveness, parallelization, ergonomics, diagnostic speed, and sensitivity. The two major technologies behind the growth of the LoC sector are microfluidics and molecular biology.

The LoC device integrates many laboratory procedures onto a small, sometimes mm or cm sized component. Working on this scale has various advantages. Instead of being transferred to a large laboratory facility, samples may be analysed immediately where they are created, in situ. Because of changes in fluid dynamics at extremely small scales, the movement and interaction of samples may be more readily controlled, increasing reaction efficiency and lowering chemical waste. The major barrier to the development of LoC technologies is the design, synthesis and fabrication of functional devices at reduced cost on a very small scale. Several LoC-style devices are recently being designed and tested as a result of advances in materials and micro and nanofabrication processes. The components of a LoC system is mentioned in Figure 49.

The purpose of the LoC devices is to combine hundreds of biochemical processes which are performed by splitting a few μl drop of the patient's blood sample acquired for accurate sickness diagnosis onto a single device. We will demonstrate that, while we are still a long way from this, existing technology is already capable of performing a number of single tests utilising specialised lab-on-a-chip, such as HIV or glucose detection, which brings us closer to realising this goal. Diagnostic processes will alter over the next few decades as lab-on-a-chip technology progresses.

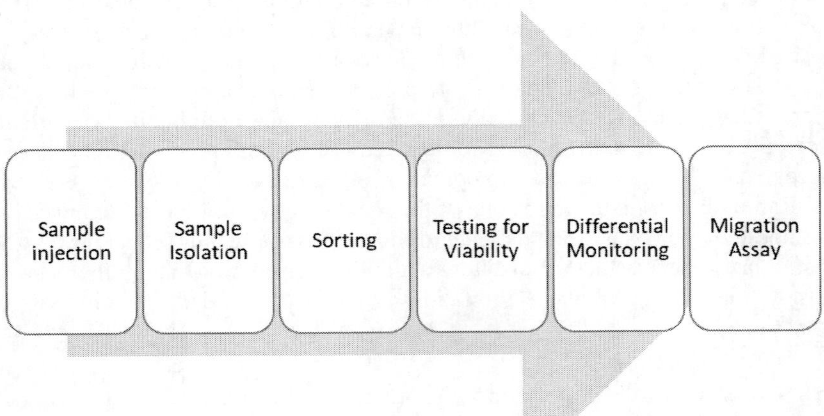

Figure 49. Components of a LoC system.

Specific breakthroughs in the field of nanotechnology have enabled the creation of lab-on-a-chip technologies. Lithography has been utilised in particular to create tiny, micro-scale pumps, valves, and other elements for regulating flow from polydimethylsiloxane. Previously, lithography was employed to create nano-scale features on metal and semiconductor surfaces (PDMS). PDMS is a transparent, flexible elastomer that is more suited for quick prototyping and facilitating optical investigations in microfluidic systems than traditional lithography materials. Nanosensor technology is an important component of many lab-on-a-chip systems.

Carbon nanotubes and other nanomaterials have been utilised to construct sensors capable of detecting very low concentrations—even single molecules in some circumstances. These are extremely useful in giving a LoC system with an increased level of analytical adaptability without escalating the overall physical size of the device.

With the ability to perform a patient's whole diagnosis during a consultation, LoC technology might change the conventional way of medicinal practise. People with less expertise will do the diagnosis, allowing specialists to focus solely on therapy. Real-time diagnosis increases patients' chances of survival in emergency services and allows each patient to receive the appropriate care. A comprehensive diagnosis can considerably reduce antibiotic resistance, which is currently one of the decade's major issues. The ability to diagnose at a cheap cost will frequently change the way we perceive medicine, allowing us to recognise issues sooner and begin treating them immediately. The use of lab-on-a-chip technology will enable healthcare practitioners in developing countries to spread diagnoses to a broader population and provide appropriate care to those who actually require it without the use of rare and expensive drugs.

2. History

Microfluidics is intrinsically linked to lab-on-a-chip and has a lengthy history. Furthermore, the evolution of microfluidics is inherently associated with the growth of semiconductor technology. As part of the Apollo programme, the United States invested tens of billions of dollars to miniaturise calculators that can be launched to space. In the early 1950s, researchers adapted photolithography techniques to build μm ranged transistors, as a result of which the vast field of micro technologies and microfabrication were evolved. These discoveries triggered the transistor/electronic technological revolution that in turn has given rise to the current information and telecommunications sectors. Figure 50 shows the historical growth of the LoC technology.

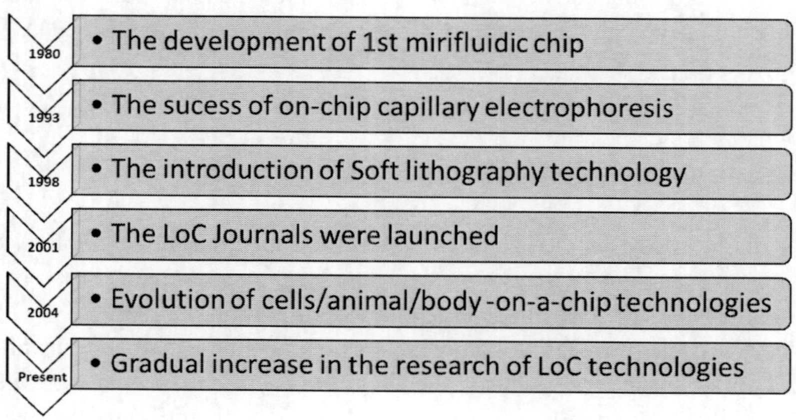

Figure 50. Historical growth of the LoC technology.

Scientists adopted similar techniques ten years later, in the 1960s, to create MEMS, or micromechanical electronic systems, which enabled the fabrication of smaller accelerometers for application that can be used in everyday things such as airbags and telephones. These manufacturing processes were used by the Stanford University created the first true LoC for gas chromatography in 1979. However, the advent of microfluidics and microfabrication techniques for fabricating the polymer chips in the late 1980s marked the beginning of serious LoC research evolution. Soft-lithography refers to the use of this microfabrication process to polymers.

Although silicon production for microelectronics was efficient, it was costly and required specialised expertise. Because of the ease with which polymer microchips may be manufactured, several research laboratories have initiated their own LoC systems research and development. Owing to the freestanding LoC manufacturing capabilities in the present day, a LoC device can be fabricated in any lab even without a clean room. In the 1990s, various research in the field of microfluidics in order to miniaturise biological processes like PCR have been instigated. Earlier the LoC researchers were focusing more on cell biology, owing to the fact that the microchannels and human blood cells were of similar sized. The research has been carried out extensively which led the scientists for the first time to accomplish operations at a single cell level. Extensive study have been further conducted to miniaturize the genomic biochemical processes that involves electrophoresis, cell lysis, DNA microarray, and so on. After which the scientists were able to integrate all of the procedures, from collecting the sample to analysing them, on a single chip, showcasing the entire potential of LoC technology. The phrase "micro total analysis system" (TAS) refers to a type of LoC technology that allows researchers to do all tasks as mentioned.

Military organisations like DARPA and DGA grew interests in LoC technology rapidly owing to the fact that it could be able to swiftly identify the biological/pathogenic threats to the population. These organisations extensively financed the growth of lab-on-a-chip research around 30 years ago through their work with semiconductors and space exploration programmes. In certain circumstances, LoC not only integrates and parallelizes well, but it also exceeds more traditional technologies in terms of performance. For example, the incorporation of PCR onto a lab-on-a-chip, known as microPCR, in which the DNA can be amplified 10 times faster than that of a typical PCR equipment (a method used to multiply DNA for disease identification).

3. Need for Lab-on-a-Chip Devices

The medical diagnostics is a vast field in which the modern LoC technology constitutes a paradigm shift. LoC technology may be used by healthcare practitioners to assess patients using the point-of-care (POC) facilities in remote and resource-limited areas. This will extensively reduce the check-in time from days to minutes. It is no longer necessary to transmit test findings to other laboratories for analysis. This rapid diagnosis is critical for time-consuming medical procedures like the examination of exposed soldiers and identifying the warfare agents, detecting a viral infection in in the elder population with a compromised immune system, detecting an infection or disease at the early stage which helps to proceed with the treatment that avoids further life-threatening complications.

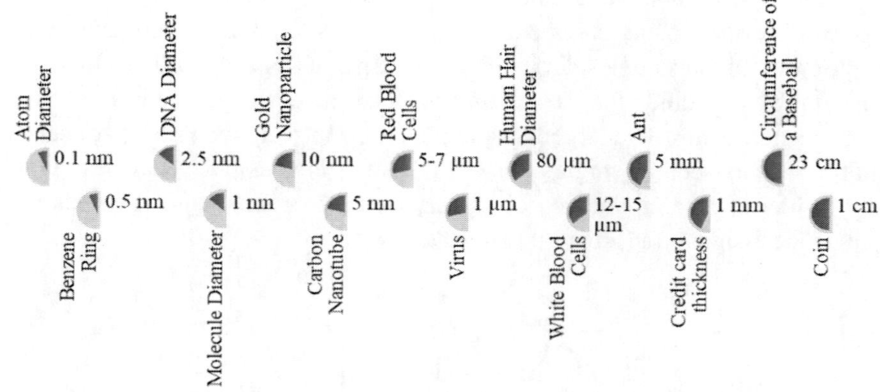

Figure 51. Micro and Nano fluidic scale.

The disease-causing pathogens like bacteria and viruses can be detected using LoC devices that can even be made wearable. LoC is used in nucleic acid immunoassays and analysis from which the antibody binding towards the particular antigen that occurs due to the presence of the associated disease can be identified and measured. LoCs have also been used in clinical diagnostics (disease detection), biochemical analysis, and disease monitoring by performing quantitative and qualitative tests on proteins or chemical compounds from the analyte of interest (e.g., cardiac arrest biomarkers that releases in the blood prior to a heart attack). LoC can be used for protein, DNA, and other sorts of analysis, such as in DNA and protein microarrays. The microarrays are nothing but an assemblage of scaled-down tests that are built on a simplified membrane to allow several concurrent tests to increase throughput. The gene expression microarray technique has been used to diagnose malignant tumours (growths), predict patient outcomes, and discover new genes related with certain disorders.

3.1. Microfluidics

Microfluidic technologies are used to generate thousands of microchannels, each ranging between a few μms that are integrated onto a single chip the size of a human hand. The real-time examples and the micro and nano fluidic scale of various nanomaterials are mentioned in Figure 51. These microchannels present on a microchip is capable of controlling the biological processes and the management of fluids in amounts as little as a few picoliters. To make a comprehensive LoC diagnostic system, the devices must require various processing components integrated together such as pumps to enable microfluidic mixing and centrifuging of the sample as demanded by the requisite, the conducting and reference electrodes, the fluidic channel valves to control the fluid flow, the electrical fields to facilitate the electrical connections between the components, and electronics. Elveflow's flow control equipment may be used to create a LoC system.

To detect target analytes from biological serum, it basically consists of a network of wells and channels scraped onto polymer or silicon substrates and linked with nanostructured, biofunctionalized surfaces. LOC is more often referred to as microfluidics. Small amounts of liquid are transported through the wells and channels of microfluidic devices such arrays of microcantilevers, carbon nanotubes, or graphene petals by electrokinetic or pressure forces moving over a defined capture/detection zone as illustrated in Figure 52. Bacterial cell suspensions, entire blood samples, antibody or protein solutions, as well as other reagents and buffers, are fluids and liquids that are often used. The same procedures were used to develop or construct the microchip. The target analyte inside the moving medium is connected with by biorecognition agents such as single-stranded DNA, antibodies, and enzymes that have been chemically or physically immobilised on the sensor sensing area using a detector, triggering electrical or optical signal transduction using a transducer.

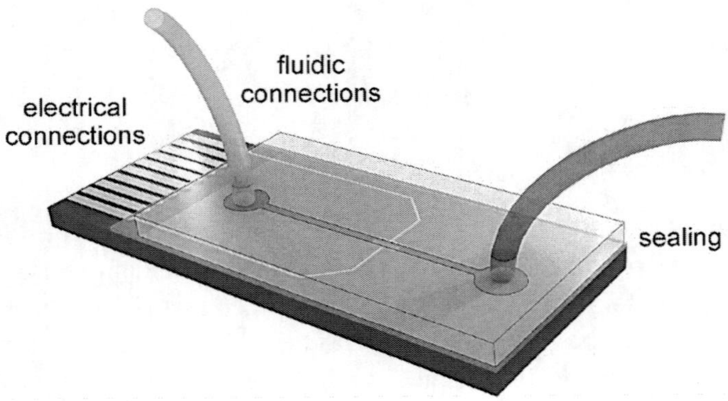

Figure 52. Schematic of a Microfluidic sensor (Temiz et al. 2015).

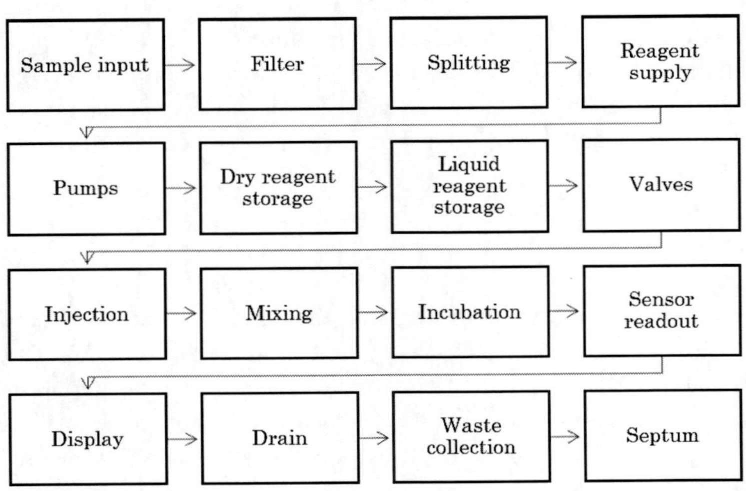

Figure 53. Building blocks of a Microfluidic system (Neumann and Rapp 2015).

On a single, integrated system, the LOC allows for sample dilution, mixing, handling, electrophoresis, chromatographic detection, staining, and separation. Short analytical timeframes, ease of use, minimal reagent and sample consumption, high throughput processing, and higher repeatability because of automation and standardisation are the LOC's main benefits.

Liquid-conducting parts make up the overwhelming majority of "lab-on-a-chip" systems, most often in the form of microfluidic channel networks. Sample input, reagent supply, pumps, sensor readout, drain, waste output, septum, (8) mixing channels, and (9) liquid reagent storage were the components of the LOC (Figure 53). While the LOC system is being developed, it is not always necessary for the architecture of these structures to be modified.

Synthetic nanomotors give a new approach for organising and evaluating biological samples utilising LOC devices. The capacity of receptor-functionalized nanomotors, which move by catalysing the breakdown of hydrogen peroxide, to capture and isolate biological targets from untreated biological samples, including E. coli and pancreatic cancer cells, has been demonstrated. To cut off the target analyte or break up cellular tissue from unprocessed biological media, such as a piece of tissue biopsy, these micro/nanomotors may be attached to micro fluidic-based equipment. There may be less of a requirement for discrete nanostructured, biodetection zones to be generated or immobilised within the structure if Forster Resonance Energy Transfer (FRET)-based nanosensors are used. Rapid interactions between enzyme and substrate, antibody-antigen binding, and DNA hybridization may lead to diffusion-limited kinetics, bound/free reagent separation, and washing processes. Luminous semiconductor Quantum Dots (QDs), which are very sensitive and capable of multiplexed bioanalytical examination, might be used to build an optofluidic system (FRET-based biodetection technique). The intriguing characteristics of these systems include high quantum yields, size-tunable photoluminescence (PL), increased avidity/sensitivity to biomolecular probes, biosensing, and photobleaching resistance (nanocrystalline materials).

3.2. Nanofluidics

Nanofluidics is the study, manipulation and control of fluids that include nanoparticles enclosed in structures with a diameter of a nanometer or nm. The fluid dynamics of substances at the nanoscale differ significantly from those at higher length scales. Nanofluidics and nanofluids have the potential to be useful in a broad variety of applications because they may display distinct physical behaviours not present in larger structures, such as increased viscosity, improved heat conductivity, and unique rheological and acoustical features. Advances in nanotechnology have recently inspired greater interest in researching nanofluidics and nanofluids in depth.

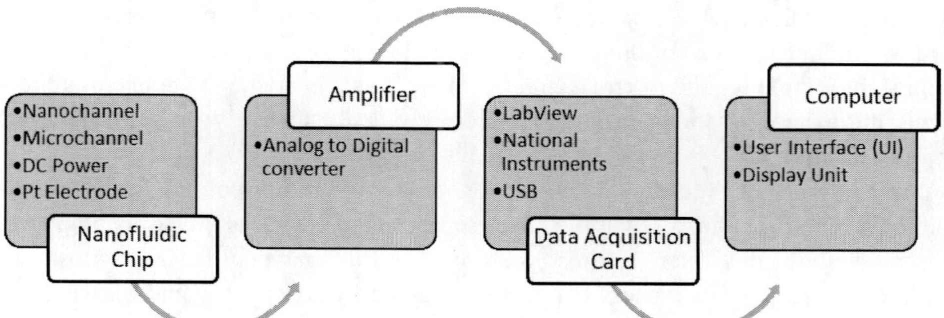

Figure 54. General schematic of a Nanofluidic chip used to detect the nanoparticles and DNA (Peng and Li 2017).

Size (or scale) matters in nanofluidics, maybe much more than in microfluidics. For instance, several of the nanoscale molecule diameters are equal in size to the channels that confine them. Nano electro mechanical systems (NEMS), which are often built utilising complementary metal oxide semiconductor (CMOS) technology, frequently incorporate nanofluidics. There have been several descriptions of nanofabrication technology. Some of them, such focused ion beam (FIB) etching, scanning probe lithography (SPL), and nanoimprinting, are designed specifically for the nanoscale. The cross-sectional area of a nanochannel, such as 10 nm by 10 nm, is often preferred over the aspect ratio. Either from the bottom up or from the top down, nanofluidic channels may be made.

On addition to enabling basic research into fluid-flow and transport phenomena, numerous applications are concentrated in the bio-sciences, bio-nanotechnology, and bio-analytical chemistry fields (with many studies aimed at the study of naturally occurring processes in biological nanochannels). Nano-pores (used for biological purposes and DNA sequencing, for example) and even the study of fluid-flow in nano-porous medium are examples of applications outside of standard nano-fluidics. The general schematic of nanofluidic chip used to detect the nanoparticles and DNA is given in Figure 54. NEMS had been developed for the chemical analysis of single proteins and the detection of airborne nanoparticles. This condensed list indicates that nanofluidics has the potential to develop into a transformative technology deserving of more study.

The use of nanofluidic devices for ultrasensitive electroanalytical applications is at the forefront of innovative detection methods. They are ideal for performing extremely effective redox cycling (repeated oxidation and reduction) to magnify electrochemical signals and enable detection and quantification of analyte molecules. Two independently addressable electrodes are often inserted within a nanochannel in these devices. Because of their low capacitance current and steady-state voltametric response, these devices are ideal for single molecule electrochemical detection.

4. Advantages of LoC Over Traditional Technologies

The same manner that microtechnologies decreased the cost of calculations, they will do the same for analysis. The cost of each individual investigation will be drastically lowered because to integration, which enables several tests to run concurrently on a single chip. Lab-on-a-chip technology will enable tens or hundreds of tests to be done simultaneously on the same device due to its ability to integrate microchannels. This will enable doctors to concentrate on certain ailments during consultations in order to promptly and effectively provide the appropriate antibiotic or antiviral.

A vast number of procedures may be condensed into a tiny volume using lab-on-a-chip technology. Finally, a chip a few millimetres square and a device as tiny as a computer will be able to do analyses equivalent to those performed in full analytical labs. Diagnostics using a lab-on-a-chip will involve far less handling and complex procedures, and they may often be performed on-site by a nurse. Automated diagnoses utilising lab-on-a-chip would greatly lower the likelihood of human mistake as compared to traditional laboratory analytical processes.

Chemical, flow switch, and heat diffusion are all quicker at the micrometric scale. You can rapidly alter the temperature (allowing for, for instance, quicker DNA amplification using PCR) or quickly combine substances using diffusion (to enable faster biochemical reactions, for example). Lab-on-a-chip devices will lower the cost of analysis by using cheaper chemicals since they only need a little volume of blood for each experiment. Last but not least, it will enable early disease detection over a broad range without needing a lot of patient blood. Fast microscale reactivity enables the lab-on-a-chip to govern the surroundings of a chemical reaction in real time, producing more precise findings. Lab-on-a-Chip devices may be utilised in outdoor settings for air and water monitoring without the need for human involvement because of its cheap cost, automation, and low energy consumption. The expense of infrastructure, infrastructure training, and diagnostic costs will be decreased through the implementation of lab-on-a-chip technology. As a result, lab-on-a-chip technology will lower the cost of sophisticated medicine for underdeveloped nations. It seems that lab-on-a-chip will save a great deal of lives.

5. Nanomaterials Used in LoC Devices

The development of modern medicine has been significantly impacted by the usage of polymers as biomaterials. Initially, silicon-based materials like glass and quartz were used because charged DNA could be moved within a microscale channel with ease. Microdevices made of silicon are, however, delicate and costly. They also need sophisticated manufacturing methods and facilities. Numerous researchers are looking into new materials that have less drawbacks in terms of ease of fabrication, low cost of manufacturing, and device resilience as a result of these shortcomings. As a consequence, a variety of plastic materials have been promoted as plastic material options, including PDMS, PMMA, PC, PI, and PET.

5.1. Biopolymers

Biopolymers or the traditional source of polymers, are those produced or derived by living organisms such as plants and bacteria. Biopolymers are mostly derived from renewable resources. The properties of biopolymers can be enhanced by adding nanoscale supports or creating a composite of biopolymers, which will increase their utility. By considering the comparability, strength, and viability of the coupling of new materials with biopolymers, it is possible to create new materials with unparalleled electrical, mechanical, thermal, and optical capabilities. There are two major categories under which the biopolymers can be classified based on their utility in the LoC devices: Shape-Memory Polymers and Biodegradable Polymers.

5.1.1. Shape-Memory Polymers

Mechanically, shape-memory polymers react to external stimuli, most notably changes in temperature. These materials undergo a considerable transformation from a rigid polymer to an elastic state, allowing for elastic deformation of up to 400%, when heated over their "activation" temperature. The imposed form is said to be "frozen," returning the structure to a hard but "non-equilibrium" condition if the material is cooled after manipulation. The material returns to its initial non-deformed condition when heated above the vitreous transition temperature, also known as the "activation temperature." Most manufacturers can provide a range of polymers with activation temperatures ranging from -300C to 2600C, depending on the application. As a result, a variety of medical devices may make use of these thermomechanically linked active materials with an excellent deformation recovery rate. Shape memory polymers' low density and cheap production cost are additional benefits that may be used for LoC applications. The illustration of the shape memory hydrogels responding to external stimuli is shown in Figure 55. Drug supply reservoirs may be incorporated directly into the implanted medical device if biodegradable shape memory polymers are employed. After implantation, the organism starts to take the polymer in and releases the medication. For the aim of storing pharmaceuticals, it is feasible to create temporary geometries with micro-reservoirs that may release the medications when activated by body heat.

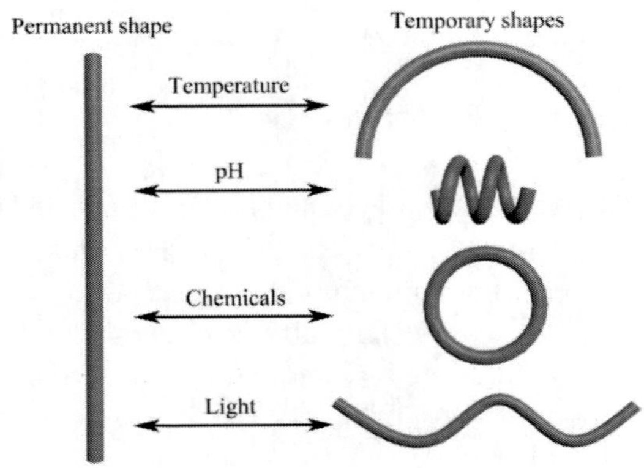

Figure 55. Shape memory hydrogels responding external stimuli (Liang et al. 2019).

5.1.2. Biodegradable Polymers

In 24 hours to several months, biodegradable polymers utilised in tissue engineering and drug administration break down. They show promise for microstructured tissue scaffolds in a microfluidic manner. The capacity to separate and discard polymeric biomaterials once they have served their purpose is afforded by their biodegradability. Degradable polymers have several medical applications, including precise sutures, scaffolds and implants. Materials having appropriate physical, chemical, biomechanical, and biocompatible qualities should be selected in order to satisfy benefit interest. Thankfully, a wide range of organic and synthetic degradable polymers have been investigated for biomedical applications, and new materials are consistently developed to deal with brand-new problems. For thousands of years, collagen and other natural polymers are often employed in biomedicine. The biodegradable biopolymers include PGA, PLA, PCL, polyanhydrides, PLGA starch plastic and cellulose.

5.1.2.1. Polyglycolic Acid (PGA)

The most basic linear, aliphatic polyester is PGA, a biodegradable thermoplastic polymer and the chemical structure is given in Figure 56. To create glycolide, a cyclic lactone, ring-opening polymerization is used. Because of its high crystallinity — which ranges between 45 and 55% — it isn't dissolvable in natural solvents. It has a glass progress temperature between 35 and 40 °C and a high softening point between 225 and 230 °C. The mechanical properties of PGA are unsurpassed. Nonetheless, due to its terrible dissolvability and rapid rate of dissolution, which leads in acidic consequences, its natural uses are restricted. Trimethylene carbonate, caprolactone, and glycolide copolymers have therefore been developed for application in therapeutic devices.

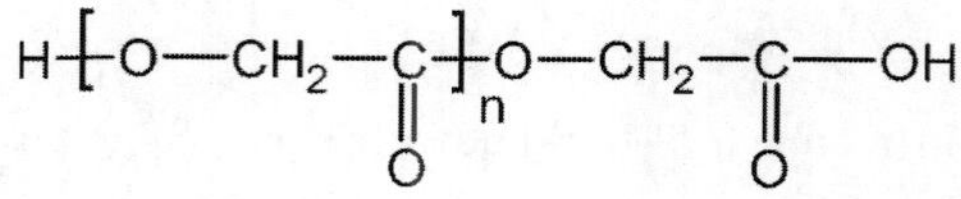

Figure 56. Chemical structure of PGA ("PGA" n.d.).

5.1.2.2. Polylactic Acid (PLA)

Polylactic acid or polylactide (PLA, Poly) is a biodegradable thermoplastic aliphatic polyester produced from renewable resources such as maize starch, tapioca roots, chips, starch, or sugarcane. Figure 57. The primary process for making PLA is the ionic polymerization of lactide, a ring closure of two lactic acid molecules. When catalytic tin compounds, such as tin oxide, are present, a ring-opening polymerization takes place at temperatures between 140 and 180 °C. Lactic acid may be fermented to produce lactide utilising a variety of microorganisms and renewable resources like starch. Polycondensation is a further technique for producing PLA directly from lactic acid. Low molecular weight polymers are produced by this technique; however industrial manufacturing has a solvent disposal issue. The molecular mass, crystallinity level, and maybe the percentage of co-monomers of PLA all affect its characteristics. Tg, Tm, tensile strength, elastic modulus, and strain following fracture all increase with molecular mass. The substance is hydrophobic or water-repellent because of the CH3 side group. Numerous organic solvents, such as dichloromethane and similar substances, are soluble in PLA. Compared to other biodegradable polymers, PLA is more transparent and has improved weather resistance and workability.

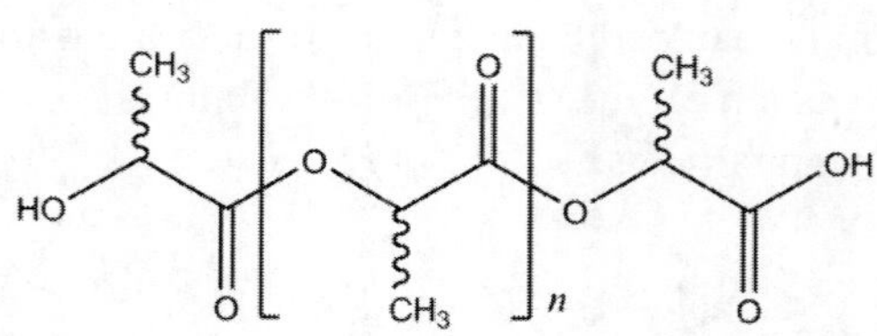

Figure 57. Chemical structure of PLA ("Poly Lactic Acid (PLA) Is Gaining Traction in the Market - Polymer Innovation Blog" n.d.).

5.1.2.3. Polycaprolactone (PCL)

Scaffolds are produced using a biodegradable polyester called Poly(e-caprolactone) (PCL) for a variety of tissue engineering applications. With a glass transition temperature of around 60°C, it has a low melting point of roughly 60°C. It is used to create polyurethanes that are resistant to solvents, oil, and chlorinated water as well as chlorine. PCL may be added to resins to enhance their impact resistance and processing capabilities. PCL may be combined with other biodegradable polymers to reduce costs and boost biodegradability since it is compatible with a variety of other materials. It serves as a polymeric plasticizer for other polymers, such as PVC. Polycaprolactone is further used in modelling, splinting, and as a feedstock for prototype equipment like 3D printers. The chemical structure of PCL is given in Figure 58.

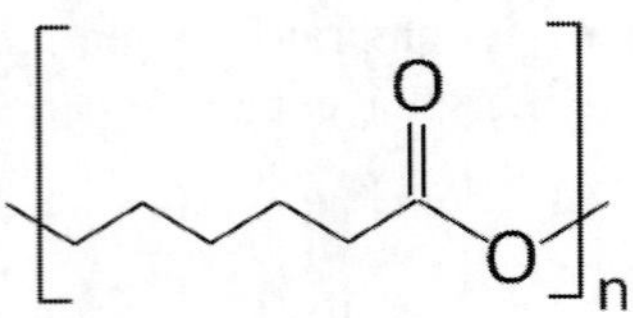

Figure 58. General chemical structure of PCL ("File:Polycaprolactone Structure.Png - Wikipedia" n.d.).

A number of cell types have been shown to be supported by PCL, a biocompatible and biodegradable polymer. PCL is another commonly used FDA-approved biomaterial in the biomedical industry. Water is naturally resistant to the PCL. Its low surface moisture and interactions with biological fluids prevent cell attachment and growth. Ring opening polymerization of -caprolactone with a catalyst results in PCL. Under physiological settings, PCL is eliminated as a result of the hydrolysis of its ester bonds, which has improved its applicability in several tissue engineering applications. This feature allows it to be used for a variety of implantable biomaterials. This material is suggested for long-term implanted devices since it degrades more slowly than polylactide.

5.1.2.4. Polyanhydrides

A class of surface-eroding polymers called polyanhydrides, which consist of two carbonyl groups connected by an ether bond (as shown in Figure 59), are mainly researched for use in biological applications. They couldn't be extensively employed due to their hydrolytic instability. One of the peculiar characteristics of polyanhydrides is that the chemistry of the polymer backbone heavily influences how easily the anhydride bond may break down. The chemistry of the monomer may cause the breakdown rate to differ by more than six orders of magnitude. Polyanhydrides have gained popularity in drug delivery applications because they allow precise regulation of the payload release rate via surface erosion and control over the rate of degradation. To make injectable, oral, or aerosol distribution easier, polyanhydrides are often synthesised as micro- or nanoparticles.

```
[ – C – R – C – O – ]n
    ||      ||
    O       O
```

Figure 59. General chemical structure of polyanhydrides ("Polyanhydrides" n.d.).

The use of polyanhydrides in tissue engineering is not possible due to their low molecular weights and poor mechanical qualities, despite substantial research into these compounds for drug delivery applications. Methacrylated polyanhydrides have been investigated as a way to strengthen injectable, cross-linkable biomaterials. When methacrylic groups are introduced, acids and methacryloyl chloride generally react to generate dimethacrylate monomers. For tissue engineering applications, dimethacrylate monomers are available as liquids or soft solids that may be injected into solid scaffolds and cross-linked. Due to the flexibility of the monomer, cross-linked polyanydrides have the potential to be exploited in a range of tissue engineering applications, such as drug delivery and structural support for the development of bone tissue.

5.1.2.5. Polylactic-Co-Glycolic Acid (PLGA)

The most extensively researched degradable polymer for biomedical applications, poly(lactide-co-glycolide, or PLGA, was created by random copolymerization of PLA (both L- and D,L-lactide forms) and PGA as shown in Figure 60. It has been used in tissue engineering scaffolds, drug delivery systems, and sutures. Researchers do not need to be experts in polymer synthesis to utilise PLGA in their study because of the broad selection of commercial suppliers and ease of processing available for polymers. Because PLA and PGA have quite varied characteristics, PLGA may be tailored for particular purposes by carefully choosing the copolymer mix. Compared to the more stable homopolymers, PLGA copolymers need substantially more property modification because they form amorphous polymers with a 25–75% lactide content that are significantly more hydrolytically unstable.

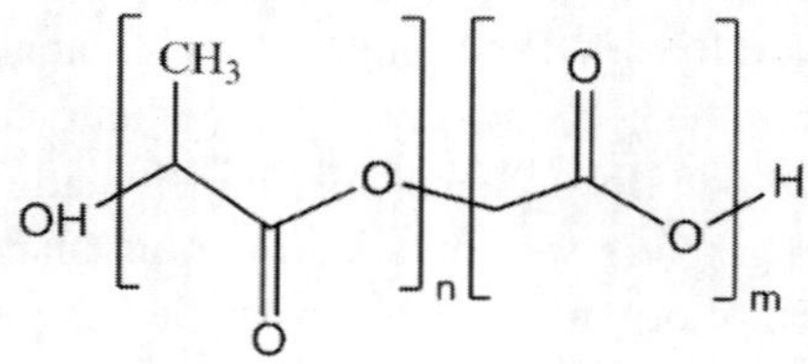

Figure 60. Chemical structure of PLGA ("PLGA Polymer: The Ultimate Buying Guide | Polylactide" n.d.).

The use of PLGA in drug delivery applications has been particularly common since it degrades more quickly than other polyesters. PLGA has been used to deliver proteins, antibiotics, anti-inflammatory drugs, and vaccinations. To permit controlled dispersion of payloads that have been encapsulated or adsorbed, PLGA is typically manufactured into microcapsules, micro or nano spheres, or nanofibers. Drug or protein release patterns may vary depending on the PLGA composition and how the payload and polymer interact. The better cell attachment and proliferation capabilities of PLGA make it a fantastic material for tissue creation. Several methods, including gas foaming, electrospinning, microsphere sintering, and polymer printing, have been used to create PLGA scaffolds.

5.1.3. Starch Plastic

The flexible and disposable polymer known as starch is made up of amylose and amylopectin particles. The most well-known kind of carbohydrate used by plants to store energy is starch, which resembles animal glycogen. Both starch and glycogen are glucose polymers that differ in the extent to which the glucose monomers are stretched and linked together by glycosidic bonds as shown in Figure 61. The most well-known common polymer is likely starch, which is also the least costly and most biodegradable. With properly handled starch, several commercial products may be created. They have only been used in a limited number of particular applications due to their moisture sensitivity and brittle mechanical characteristics. Research found that adding strands might affect the mechanical and obstructive properties of starch products. During composite construction, nano and small structured strands are dispersed throughout the polymer lattice, which affects the material's elasticity and flexibility.

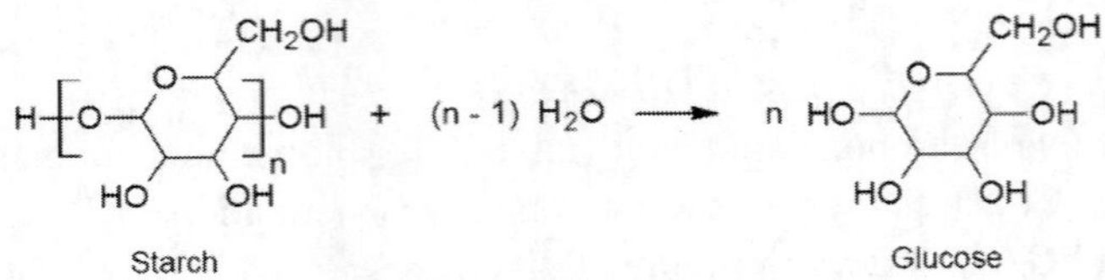

Figure 61. Chemical structure of modified starch plastic ("Starch" n.d.).

Because they can be mass-produced utilising simple manufacturing techniques in resource-constrained settings, flexible microfluidic devices are now in great demand. This industry has become more complex due to the use of specialised detection methods for various lab-on-chip applications. Recently, a wide array of plastic polymers have become popular for use in several lab-on-chip applications. Future disposable organic electronics now have access to biodegradable starch paper that is environmentally friendly and doesn't need a planarizing layer. By gelatinizing the starch with PVA and a cross-linker, which improves the PVA's mechanical characteristics, starch sheets are created. Synthetic components are employed less often as a result of this method. Because of this, paper made with starch has exceptional mechanical stability and strength. We find outstanding nonpolar solvent stabilities. They play a crucial role in the production of organic electronics made using nonpolar solution techniques. Starch paper's potential as a flexible substrate is assessed using the creation of flexible organic transistors.

5.1.3.1. Cellulose

Despite being a biodegradable linear polysaccharide, cellulose cannot be melted or dissolved. Cellulose may be directly dissolved or chemically modified to create a liquid that can be processed. Adsorbents such as beads, films, hydrogels, microfibers, and microcrystals have been created from cellulose and its derivatives. More research is being done on cellulose nanocrystals (CNCs), which have a greater surface-to-volume ratio. Applications for CNCs and nano fibrillated cellulose (CNF) might include anything from small-scale medicinal products to massive sorbent goods. For applications that need flexibility, like wearable electrochemical ones, CNF, for instance, offers promise. CNF-based aerogels are intriguing since they are simple to make using freeze drying or critical point drying. CNF-based nanocomposite hydrogels may be used as sophisticated origami actuators to study material/cell interactions. Among other uses, these actuators have been employed in biosensors, medical equipment, artificial tissue, and diagnostics. The chemical structure of the cellulose is given in Figure 62.

Paper-based microfluidics is a term used to describe the integration of cellulose with microfluidics. Paper-based microfluidics, sometimes known as "lab on paper," is a cutting-edge method of fluid management and analysis. The method is touted as being disposable, inexpensive, easy to use, and equipment-free. Paper is a desirable substrate for these devices because it is widely available, affordable, and accessible. The substance may thus be used for a variety of further chemical, biological, biochemical, and medical purposes. Liquid is moved by capillary forces alone, without the aid of external forces. To measure the concentration of different analytes in a solution, point-of-care diagnostics and LoC devices, for example, might be built on top of microfluidic paper-based analytical devices.

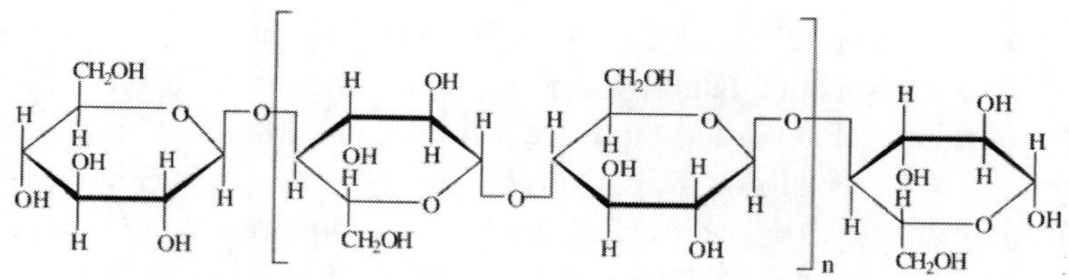

Figure 62. Chemical structure of Cellulose ("3: Chemical Structure of Cellulose | Download Scientific Diagram" n.d.).

5.2. Hydrogel

Water may absorb up to 80% of a hydrogel's bulk, and its molecular structure is comparable to that of extracellular matrix. Due to their great porosity, hydrogels are an excellent matrix for cellular biology research. Given that nutrient diffusion can only occur a few hundred micrometres below the surface of bulky hydrogels, direct tissue formation from these materials is challenging (Choi et al. 2007). The capacity of microfluidic technology to deal with this important problem has been proved using both top-down and bottom-up methods. Inside the hydrogel, microchannels are created top-down, whilst hydrogels are created bottom-up to fill the spaces left by the microchannels. The two animal-derived hydrogels that are most often employed are MatrigelTM and collagen (Bang et al. 2016). Excellent cellular adhesion and proliferation are fostered by them. Synthetic hydrogels provide more flexibility and formulation adaptability for fine-tuning goals, but having less cellular adherence than animal-derived hydrogels. Droplet-based microfluidics has employed hydrogel formulation, morphology, topology, and flexibility in high-throughput applications (Chau et al. 2014). The schematic of hydrogel-based blood-brain barrier built on a chip is shown in Figure 63 (Sood et al. 2022).

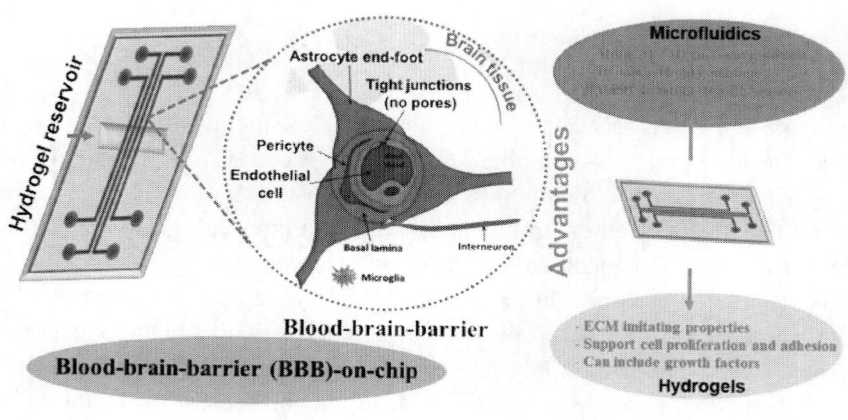

Figure 63. Schematic of Hydrogel-based Blood-Brain Barrier (BBB) on chip (Sood et al. 2022).

5.3. PDMS

For lab-on-a-chip prototypes, research labs often employ PDMS. Due to the ease and low cost of producing PDMS labs-on-a-chip, polydimethylsiloxane, or PDMS, is a popular and transparent elastomer. The general fabrication steps involved in developing a PDMS based LoC device is given in Figure 64 (Fujii 2002). Air permeability for cell culture and research, as well as the simple integration of shock microvalves for quick flow switching, are further advantages of PDMS labs-on-a-chip. PDMS has significant limitations in industrial manufacturing while being routinely utilised for lab-on-a-chip prototyping. Due to PDMS's ageing susceptibility and propensity to absorb hydrophobic chemicals, it is difficult to include electrodes onto PDMS chips. Last but not least, PDMS cannot be used with high-throughput chip production techniques like hot embossing or injection moulding.

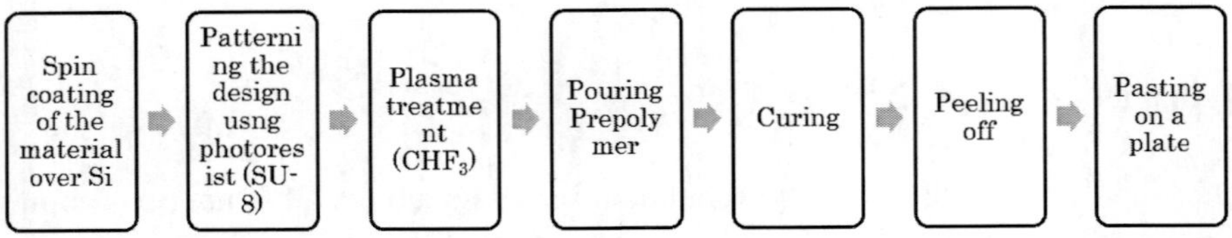

Figure 64. Fabrication steps of PDMS LoC device (Fujii 2002).

5.4. Thermopolymers

The construction of labs-on-a-chip is often done by researchers using thermoplastic polymers. Despite being more difficult and costly to implement than PDMS, thermoplastics are excellent candidates for lab-on-a-chip construction because they are transparent, compatible with micrometer-sized lithography, and more chemically inert than PDMS. The industrialization of various labs-on-a-chip may be a good fit for thermoplastic materials since microelectrodes may be placed into them. Thermoplastic labs-on-a-chip were employed by many research teams to get very good results in specialised applications.

5.5. Silicon

Because microtechnologies rely on the microfabrication of silicon chips, silicon was chosen as the material for the first lab-on-a-chip. Currently, silicon is not commonly utilised by researchers for labs-on-a-chip, owing to its expensive cost, lack of optical transparency (apart from IR), and need for a clean environment, as well as a thorough grasp of microfabrication. Furthermore, due to the high electrical conductivity of silicon, lab-on-a-chip activities needing high voltage cannot be done on it (like electrophoresis). Nonetheless, we believe silicon is still a feasible alternative for the industrialization of certain difficult LoC applications, even if it appears to be a dated candidate for lab-on-a-chip industrialization. This assumption is founded on the fact that silicon microfabrication is very accurate, the process is mature, investments in the silicon industry have been made, and any form of microelectrode or even electronics may be combined on the same chip.

5.6. Glass

Glass is a great choice for the industrialization of labs-on-a-chip because it is transparent, suitable for micrometer-sized machining, chemically inert, provides a variety of well-known chemical surface treatments, and has repeatable electrode integration. Research-wise, creating glass labs-on-a-chip need clean rooms and microfabrication specialists. As a consequence, glass lab-on-a-chip is not available for usage in all research institutes.

5.7. Paper

Devices using paper-based labs-on-a-chip should provide accurate results for uses needing very cheap prices. Paper labs-on-a-chip may soon find a market thanks to G. Whiteside, one of the most well-known microfluidic researchers. We truly hope it succeeds, because the innovative idea has the potential to broaden diagnostics and make them more accessible to populations with lower incomes and less resources (Suresh et al. 2018) developed a paper based LoC device on a screen printing electrode to detect the urea present in the given analyte through the enzyme catalysis method.

6. Fabrication Techniques used in LoC Devices

6.1. Inkjet Printing

Microfluidics that can be produced cheaply and quickly are more important than ever with the rise of LoC and dispersed healthcare. Microfluidics inkjet printing, which has a number of potential electrical and sensor applications, could be able to resolve this issue. The inkjet printing process (shown in Figure 65), which is low-cost in terms of equipment and creates no waste, places individual ink droplets, known as functional materials, with diameters of several m on a substrate. It is also simple to understand, has a rapid manufacturing time, and a low planar resolution in the μm range. The produced dots may be combined to create pictures with a resolution of up to several thousand dots per inch that are of photo-quality. Inkjet printing is especially suitable for the scalable creation of minute features of materials such as polymers, bioparticles, and even ceramics because of its extraordinary accuracy and flexibility.

Using hydrophobic materials like silicon with dimer conductive inks, hydrophobic barriers may be selectively printed onto a paper substrate without the need for heating. Metallic inks are examples of printable conductive materials for inkjet printers used to create biosensors. These materials include graphene, silver nanoparticles, carbon nanotubes, and quantum dots.

On the PDMS substrate, inkjet printing produces metal nanoparticle microelectrodes. This method has two significant drawbacks: the hydrophobic properties of PDMS, which restrict printing accuracy and cause nearby ink droplets to fuse, and the reduced adherence of the nanoparticles to the PDMS substrate, which makes the electrode less stiff. By adding coupling agents like trimethoxysilane to the PDMS substrate, this flaw may be fixed.

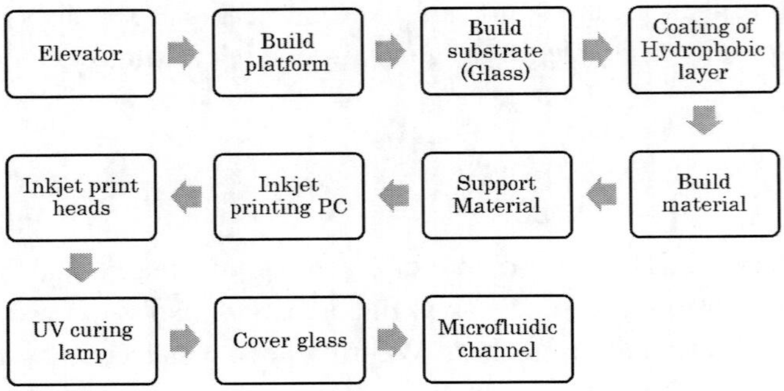

Figure 65. Schematic of Inkjet Printing Procedure (Shen et al. 2012).

6.2. Wax Printing and Wax Dipping

On flexible surfaces, hydrophilic designs may be produced using a desktop wax printer employing coloured solid wax cubes as ink. The schematic of wax printing is shown in Figure 66. To print images on a substrate, the wax printer heats the wax cube, melting it. The solid wax is then allowed to pass through the porous cellulose substrates and create a hydrophobic barrier by being cooked in an oven or on a hotplate. The resultant hydrophilic channels, which are 500 m wide for smooth flow, are strong but have poor resolution between the hydrophobic boundaries. Platforms made of "wax-on-plastic" have lately been constructed using wax designs on PET substrates. On a polyethylene terephthalate (PET) substrate that is non-cellulosic, non-fibrous, and non-porous, the printability and applicability of wax micropatterns were investigated. For line patterns and channels, the printing resolution was found to be 60 m and 120 m, respectively, with the additional bonus of thermal stability reaching 120°C, or the melting point of wax. The fate of mouse embryonic stem cells (mESCs) in the presence of the microenvironment was then effectively investigated using this technique by successfully creating high density microwells on PET. The stem cells obviously differentiate in the presence of fibronectin and spontaneously renew themselves in the presence of collagen I microenvironments.

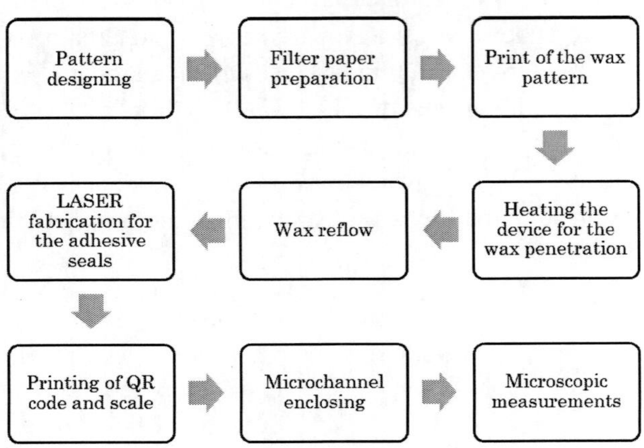

Figure 66. Schematic of Wax printing.

The development of "wax-on-plastic microfluidics," which were used to detect glucose in urine using a flow-based test, was controlled by patterning wax microchannels on a flexible PET substrate. Due to the difference in surface energy between the substrate's (hydrophilic) and borders' (hydrophobic) surfaces, which creates capillary force in these channels, a sub-microliter bioassay may be finished in 40 seconds. The same platform was used to look at the flow dynamic properties of DNA's trinucleotide repeat sequences (TNR), which explain its molecular conformation and structural characteristics. These sequences' capillary flow was shown to be influenced by the kind, dimension, quantity, molecule shape, and metal ion concentration. We can examine the diffusion of two liquids at their contact thanks to the laminar fluid flow in wax-on-plastic microchannels. Serpentine canals may encourage the mixing of liquids without the need for an outside pushing force.

Wax dipping is a technique that is comparable to wax printing. A cellulose substrate is submerged in wax liquid during wax dipping in order to form hydrophobic wax boundaries. This method necessitates the use of a mask in order to transfer a pattern. In one instance, 120–130 °C was used to melt white beeswax pellets into molten wax in a beaker. On a substrate made of cellulose, a piece of an iron mask was inserted in a chopped state. The iron mould was attached to the glass slide and filter paper after being placed between a magnet and the latter. A portion of the whole arrangement was submerged in molten wax for a brief period of time at 120 to 130 °C. The filter paper was taken off the glass slide by removing the iron mask after cooling at ambient temperature. The area that was covered by the mask remained hydrophilic while the exposed area created hydrophobic channels. The greatest resolution obtained using this approach was 639±7 μm. Devices that may be employed in clinical settings for a range of fascinating applications were created using this technology. In one case, their team used paper-based devices to separate blood plasma from whole blood and test plasma proteins in a single step in 2 minutes using a blood sample of 15 to 22 μl.

6.3. Screen Printing

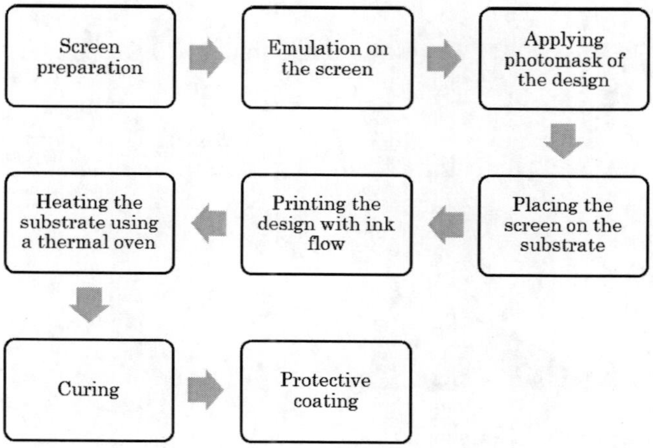

Figure 67. Schematic of Screen printing.

A specially constructed mesh is used in screen printing, a traditional printing technique, to transfer patterns from one surface to another as shown in Figure 67. By selectively passing material through a mesh, certain sections of the mesh become resistive to the substance being transferred. Spread out liquid or slurry material and push it onto a substrate using a squeegee or a flat tool moved back and forth across the screen. Although a variety of devices may be made using screen printing, transferring a design to a substrate necessitates a unique photolithographic screen. Creating separate screens for each kind of design, even though the screens are reusable, is a time-consuming and expensive step in the production process. These gadgets also need a high spatial resolution on mm scale.

6.4. Photolithography

Devices with high resolution features with a 5 μm resolution may be made using photolithography. It is a lengthy, complex process that requires a number of steps and highly qualified personnel. The substrate should be well cleaned before adding photoresist, baking it, exposing it to UV light, etching it, and developing features by removing the photoresist. On a substrate that is coated with photoresist and a patterned mask, patterns are created using a UV light. As it passes through a mask, light selectively cures the exposed photoresist, and the exposed area is then developed for further processing as shown in Figure 68. Paper-based microfluidic devices, paper-based microzone plates, and mobile phone gadgets for real-time diagnostics have all been created as a consequence of the photolithographic techniques used to create paper-based electronics. Furthermore, high resolution (100 nm) conductive patterns for organic field effect transistors and optoelectronic devices are often mapped on plastic substrates. Photolithography has not been extensively used in resource-limited contexts despite its high resolution advantage because it is costly, necessitates expensive equipment with significant environmental impact, and uses extraordinarily clean chemicals. The Canon Global FPA-5550iZ2 (Figure 69) and ASML-EUV lithography NXE-3400B are two of the most effective and extensively used semiconductor lithography tools currently on the market.

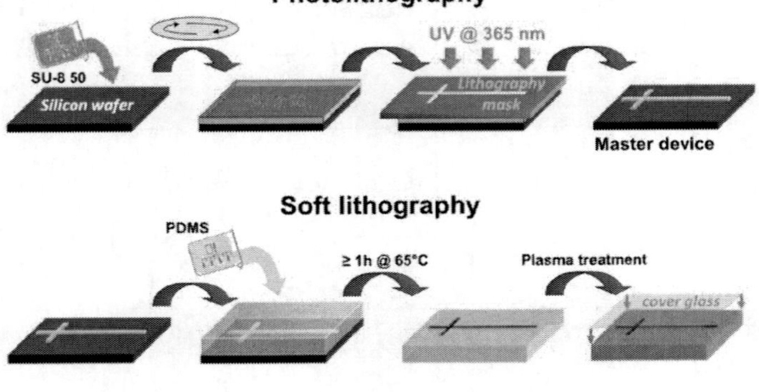

Figure 68. Schematic of Photolithography.

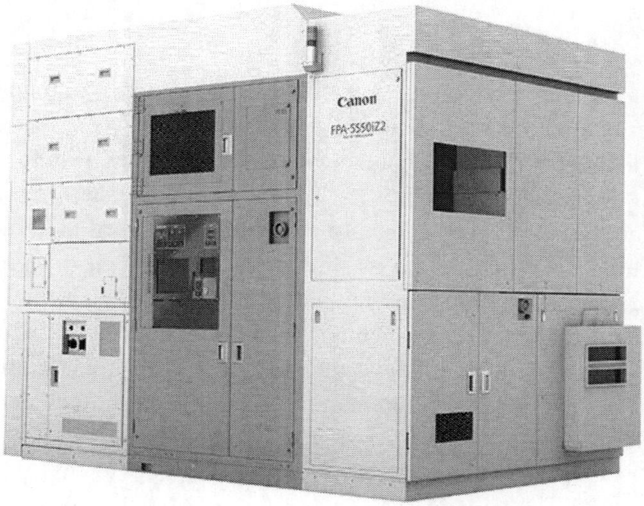

Figure 69. Canon Global FPA-5550iZ2 - Semiconductor Lithography ("FPA-5550iZ2 | Canon Global" n.d.).

6.5. Laser Cutter

On a range of materials, including ceramics, glass, plastic, and papers, a laser cutter may imprint hydrophilic designs using high intensity CO_2 lasers. While wax barriers on cellulose paper have a resolution of 100-300 μm, hydrophilic features on hydrophobic media, such as parchment or wax paper, may be produced with a precision of up to 62 μm utilising paper-based manufacturing. By creating hollow microstructures with 500 μm channel features on paper, the laser cutting method may also be utilised to create microfluidic PADs. Using a laser printer, a CO_2 laser cutter, an office laminator, and common overhead transparencies as a printed substrate, complex multilayer microfluidic devices were created as shown in Figure 70.

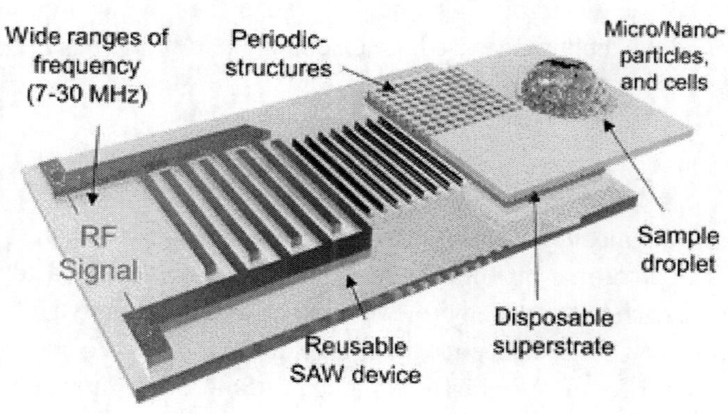

Figure 70. Schematic of Laser Cutter.

6.6. Desktop Cutter

A desktop cutter is a cheap, portable device that uses a sharp, programmable needle to create microchannels on flexible plastic surfaces as small as 200 μm. In a recent example, a transparent, thin PET material was laminated with a cut Parafilm® film to produce flexible two and three-dimensional microfluidic chips. A 3D microfluidic system might perhaps be made using this technique. Since it is often used to cut adhesive-backed patterned vinyl sheets into screens for patterning electrodes, such as through manual screen printing, this tool is also known as a vinyl cutter. For secure handling and transportation, a cellulose filter paper with a central sample loading zone and branching detection zones was sandwiched between a polyester sheet and a laminating bag. It was then heated at 120 °C and compressed using a heat press for 120 seconds. The lowest feature sizes of the vinyl cutter technique might sometimes be several hundred microns, which is a drawback.

6.7. Desktop-Pen Plotter

The pen-plotter, which uses a pen and paper to draw programmable vector images, has been around for almost 40 years. Researchers have looked at the ability of numerous available permanent markers to produce hydrophobic structures on paper surfaces using the pen-plotter method. Utilizing this approach, microfluidic devices were created for colorimetric proof-of-concept glucose detection. The same team created a continuous-ink, multiplexed pen-plotter for the cheap, high-throughput production of paper-based microfluidics using 3D printed multiple pen holders. Despite the feature's size being in the mm range, the continuous plotting method's plotting performance revealed a 400 μm deviation in the x and y axes.

7. Trends of LoC in Various Applications

The development of lab-on-a-chip technology has revolutionised laboratory research by enabling testing to be conducted on micro-devices rather than in a central laboratory. Various application fields and the technology used are discussed below.

7.1. Proteomics

Proteomics is one of the most pressing scientific issues of the post-genome era. The most basic sort of proteomics is proteome profiling, which is the laborious task of identifying all of the proteins expressed in each sample. The proteome provides unique analytical challenges because to its molecular diversity, large concentration range, and proclivity to adsorb to solid surfaces. LOC devices are useful for developing novel techniques to difficult analytical problems such as proteome profiling. LOC devices appear to be progressed in the following four areas in regard to this application: chemical processing, sample preconcentration and cleaning, chemical separations, and mass spectrometry interfaces. The integrated microfluidic LoC chip for proteomics analysis was developed by (Wei et al. 2020) and the schematic is shown in Figure 71.

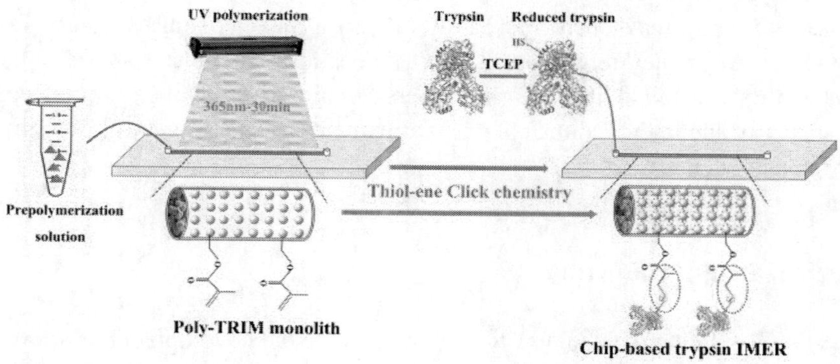

Figure 71. Integrated LoC chip for proteome analysis (Wei et al. 2020).

In the field of proteomics, lab-on-a-chip allows for the examination of proteins while coordinating all processes on a single chip: cell extraction, electrophoresis-based division, processing, and mass spectrometry-based analysis. These combined cycles illustrate the capacity of lab-on-a-chip technology to substantially reduce protein analysis time from hours with naturally evident framework to only a few seconds. They also have a high potential for protein crystallisation, which is an essential field of study that exposes a protein's 3D structure. Researchers may use lab-on-a-chip to regulate all of the hurdles to protein crystallisation concurrently and as rapidly as humanly feasible. Then, much parallelized crystallisation settings may be used to accelerate the discovery of these circumstances for mysterious proteins and analyse their structures using X-beam diffraction.

7.2. Cell Biology

Because of LOC devices, the researcher may handle cell investigations on a microscale. Single-cell handling and manipulation are commonly necessary because the single cell must be positioned or trapped in a precise place when positioned within the microchannel. The microsystem must be able to handle minute volumes of fluid containing minute amounts of analyte. LOC systems provide the capacity to precisely govern the local environment around the cell by having control over the physical or chemical nature of the surface to which the cell attaches, the local pH, and temperature. When the cell must be exposed to a variety of stimuli, many and frequently sophisticated fluid-handling components may be used. These ideas imply that cell location and chemical activation should be coupled. The flow cytometer is one of the most effective early-stage instruments and the most practical LOC device developed to date. Over the last 10 years, two key factors have led to the rapid development of flow cytometer LOC systems: (1) the rising requirement for more high-quality and extensive cellular analysis data, and (2) the enhanced sophistication and accessibility of microfabrication technologies. The LoC system to analyse the cell culture involving cell lysis, cell selection, imaging and biochemical analysis has been developed by (El-Ali, Sorger, and Jensen 2006).

Because microchannels and cells are roughly the same size, cell science quickly became the focus of lab-on-a-chip research. The capacity to handle countless cells fast while also managing them at the single-cell level is demonstrated by lab-on-a-chip. At the microscale level, stream switching may occur incredibly fast, in a matter of milliseconds. For example, high throughput antibodies might be used to detect and isolate a specific cell, such as a fluorescent malignant cell. Lab-on-a-chip offers a number of applications in cell research, including sperm arrangement, control of immature microbe separation, rapid stream cytometry, and, more broadly, cell arranging.

7.3. Medicinal Drug Production

The pharmaceutical industry need new tools to help guide the development of new medications, particularly those that can foresee how potential new therapies would behave in humans based on how well they perform in cells and animals. Some analytical applications of LOC systems in the production and use of biopharmaceuticals appear simple (e.g., analytical systems to monitor and optimise the production of protein drugs such as therapeutic antibodies); others (such as assays based on primary human cells that could predict performance in human clinical trials) are technically more difficult but, in some cases, still feasible. In any case, LOC systems may be used frequently in a very repeatable and adaptable manner. A pharm-on-a-chip module has been developed by (Ai et al. 2019) which monitors the drug toxicity and drug effectiveness.

7.4. Healthcare Sensors and Wearables

LOC biosensors are small devices that utilise biological processes to detect the analytes of interest. A biological recognition component that interacts with the target analyte is closely connected with a physical transducer that turns the biorecognition event into an electrical signal in such systems. LOC biosensors are classified into two types: biocatalytic and bioaffinity devices. Bioaffinity devices function by selectively binding the target analyte to a surface-encapsulated ligand partner (e.g., antibody, oligonucleotide). In hybridization biosensors, for example, single-stranded (ss) DNA probes are anchored to the transducer surface. After binding, the duplex formation may be observed if the suitable hybridization signal is attached. In biocatalytic devices, on the other hand, an immobilised enzyme is used to recognise the target substrate. The LoC device for virus detection integrated with IoT is developed by (Zhu et al. 2020) and the data can be viewed using a smartphone using Wi-Fi connection.

7.5. Genomics

The results of sequencing initiatives like the Human Genome Project have had a significant impact on genomics, which has grown quickly in recent years. Sequencing-related initiatives help to enhance the technology in the area. Analyzing DNA sequences to determine their nucleotide order is the process of sequencing. The Sanger method, which formed the basis for the first sequencing, produces DNA fragments of various lengths with fluorophores matching to various nucleotide ends. The technology was made more compact by swapping out the polyacrylamide gel slab used in conventional DNA sequencing with a tiny capillary loaded with gel. The amount of the necessary DNA sample was reduced due to the capillaries' modest size. Nanopore sequencing, sequencing-by-synthesis (SBS), sequencing-by-hybridization (SBH), and sequencing-by-synthesis are the three next-generation sequencing approaches for very fast DNA sequencing that are presently being developed (SBS). This technology's LOC method makes high-throughput sequencing much less expensive and quicker.

7.6. Diagnosis

Molecular diagnostics from lab-on-a-chip (LoC) greatly aid in the detection and treatment of infectious illnesses. The creation of portable, simple-to-use, affordable, and quick molecular diagnostic technologies is required. Genetic abnormalities, cancer diagnosis, infectious illnesses, pharmacogenomic assays, and noninvasive prenatal testing have all benefited greatly from the use of molecular diagnostics. During the preceding decade, the nucleic acid amplification test (NAAT) industry, has tended to build more easy, rapid, precise, and economical platforms for LoC testing. These systems must be integrated and miniaturised for handling liquids and processing them for speedy diagnosis, low cost, and use by personnel with minimal training. NAAT requirements drove the rapid development of (LOC) technology, which use microfluidic techniques to consolidate various laboratory activities, including as nucleic acid amplification, reagent mixing, and sample processing, on a small chip.

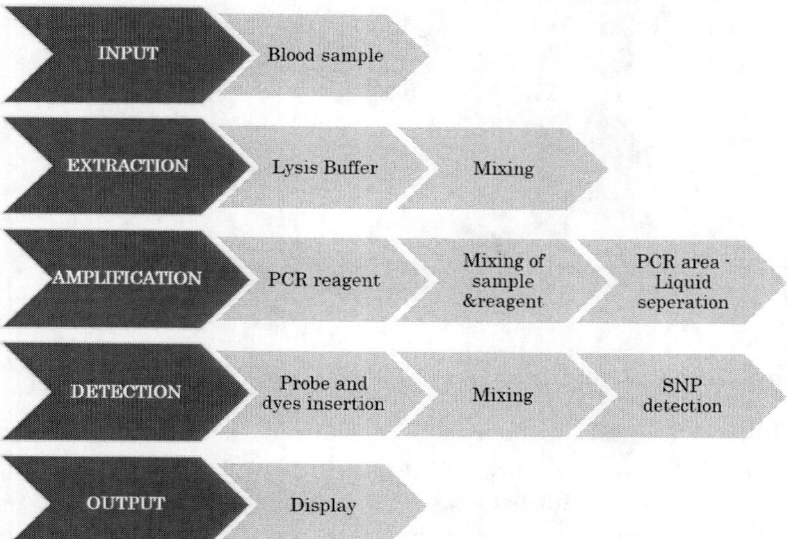

Figure 72. LoC device involving DNA extraction and purification (Huang et al. 2015).

7.6.1. DNA Extraction Using LoC Device

Nucleic acid from live cells has to be isolated for diagnostic testing. The two main phases in the extraction of DNA are cell lysis, which includes rupturing the cell membrane and nucleus membrane, and DNA purification, which involves removing unwanted components such membrane lipids, proteins, and RNA. Devices for lysing LOC cells can use a variety of lysis techniques, including chemical, thermal, ultrasonic, electrical, mechanical, etc. The schematic of LoC device involving DNA extraction and purification is given in Figure 72 (Huang et al. 2015).

7.6.2. DNA Purification Using LoC Devices

The bulk of macroscale DNA purification methods used in extraction columns—methods that depend on DNA adsorption on silica beads in certain buffer conditions—can be applied to microfluidic technology. Therefore, silica beads may be kept out of the washing and elution stages of LOC devices using mechanical barriers or silica-coated magnetic beads that are easy to block with a magnet.

7.6.3. PCR and qPCR Detection Using LoC Devices

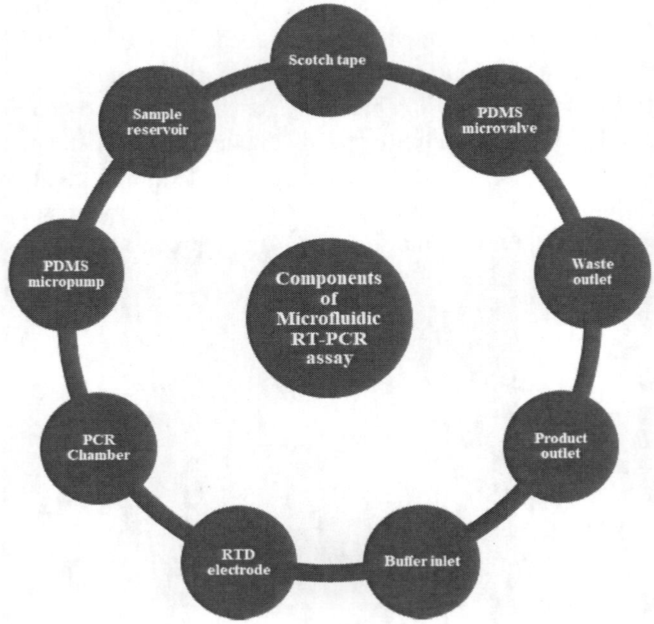

Figure 73. Components involved in Microfluidic RT-PCR assay (Kim et al. 2012).

The PCR test is the most commonly utilised following DNA extraction (Polymerase Chain Reaction). Many PCR uses are similar to sequencing technologies, both directly and indirectly. The amplification of DNA sequences, which aids in the identification of tiny quantities of DNA, is without a doubt one of the uses of direct PCR (e.g., for pathogen detection, like bacteria or virus). The development of multiple LOC PCR apparatus is influenced by the significance of PCR in genetic investigations. Quick heat transfer in rapid and integrated PCR is aided by volume miniaturisation and a high surface-to-volume ratio. A post-analysis is also required for PCR in order to combine PCR and electrophoresis on a single chip for amplicon size detection. At first, agarose gels for longer DNA and polyacrylamide gels for shorter DNA were used for electrophoresis. One of the earliest molecular processes that could be performed on a chip was DNA electrophoresis thanks to the advent of LOCs. This miniaturisation allowed for longer electrophoresis times, the use of fewer chemicals, and the integration of previously known DNA synthesis processes on a chip. The components involved in the microfluidic RT-PCR assay is shown in Figure 73(Kim et al. 2012).

qPCR is a different method made specifically for LOC devices that has the advantages of being quicker (automatic detection during PCR), more sensitive, and more durable. For example, Elvesys developed an ultra-fast qPCR microfluidic system with detection efficiency similar to commercial techniques that are 7 to 15 times slower for the molecular detection of diseases like anthrax and ebola in less than 8 minutes. It uses a fluorescence reader, an ultra-fast temperature controller, and an ultra-fast pressure controller. Digital microfluidics, which deals with emulsion and droplets within LOC devices, is another emerging field. Limits may be improved by using one copy number detection in LOC droplet qPCR, which can be utilised to collect very small amounts of DNA.

7.7. Microarray

DNA microarrays must be used in conjunction with a variety of biosensors to investigate complicated DNA sample expression. The miniaturisation, speed, and accuracy of these LOCs are their primary strengths. This method holds immense promise for applications such as forensic investigation, drug screening, infectious agent identification, differential gene expression measurements, and the detection of genetic disorders. Numerous areas of genetic study are being revolutionised by the use of DNA microarrays.

7.8. Human Implants and Prostheses

People can overcome limitations owing to breakthroughs in nanotechnology therapies, prostheses, implants, and other fields. Due to its ability to make implants and prostheses biocompatible, antibacterial, effective with minimal power consumption, and enhanced in speed and performance, nanotechnology has been employed to improve a broad variety of implants and prostheses. Some of the issues with current materials may be resolved using nanotechnology, such as preventing cell protein from adhering to sensors, which might lead to malfunction, halt functioning, or replacement of the prosthesis. Scientists are using nanotechnology to create nanofibers that aid tissue development and nanocircuits that improve wheelchair technology.

The molecular diagnostics at the nanoscale provide nano-diagnostic data and techniques at the clinical level, which plays a crucial role in the formation of personalized medicine. This personalized medicine and diagnostics lead to the new fields of Point-of-care devices based on LoC systems. This facilitates the remote analysis of various human health factors where there is poor access to resources and facilities prevails. There could be two possible ways in which the LoC system can be developed to address the issues of human health: non-invasive applications (i.e. outside the body) like sensors and invasive applications like drugs, implants or prosthesis as shown in Figure 74.

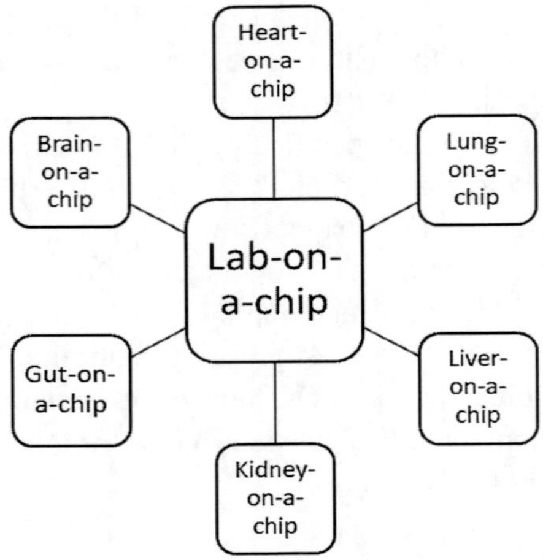

Figure 74. Various organs on a chip used as human prosthesis (Osório, Silva, and Mackay 2021).

7.9. Food Analysis

Growing urbanisation over the last several decades has had a significant impact on the agro-food industry, resulting in the introduction of new technology and products. Chemical agents are being employed in agriculture and industry at an increasing rate due to the rising global food demand. Although these substances have improved process efficiency, the consequences on the environment and human health must be considered as well. Unsafe food items are regularly left uncontrolled and pose major health concerns to consumers, particularly in developing countries. It is crucial to effectively manage the identification and prevention of any food contamination and food deterioration in order to decrease the dangers to the public health while also enhancing production and food quality.

Several cases of pathogen-caused human foodborne illness have been reported. To identify these pathogenic pathogens, cell culture, immunoassays, or polymerase chain reactions are currently employed approaches (PCRs). These approaches cannot be employed for early pathogen surveillance since they are mostly laboratory-based and not at all real-time. They are also extremely difficult to put into action on the ground. However, lab-on-a-chip biosensors have the potential to be incredibly useful in the field because to their capacity to be automated, miniaturised, fast, and exceedingly sensitive. These lab-on-a-chip biosensors can detect diseases in agricultural settings, processing and packaging industries, distribution and delivery networks, and even at the consumer level.

The use of creative sensors to guarantee the safety and quality of food is significantly impacted by developments in a variety of biosensing technologies. In an effort to integrate sophisticated technological advancements into a single platform, several detecting techniques were used. Polymerase chain reaction (PCR) and enzyme-linked immunosorbent assay (ELISA), which are widely employed in the agricultural and food industries and are efficient at identifying a broad variety of diseases, are the most often utilised detection techniques.

8. Limitations of LoC Devices

The bulk of lab-on-a-chip improvements are still not industrially ready. We don't know which manufacturing advances will become ubiquitous in the centre application of super multiplex analysis. In certain circumstances, scaling down raises the signal-to-noise ratio, causing lab-on-a-chip to perform worse than traditional approaches. Without restrictions, frequent handling, and the inescapable openness of LoCs, some individuals may be concerned that the undeveloped public may use them to diagnose potential illnesses in their own place. Furthermore, the lab-on-a-DNA chip has the capability of sequencing the DNA in a more simplified way which may potentially increase the risk of anyone unethically sequencing someone else's DNA even with a drop of any body fluid. Despite the fact that LoC devices are known for its compact as well as effective performance, they require certain hardware to function effectively, such as hardware or stream control frameworks, without which the LoCs are meaningless. The external devices adds to the overall size and expense of the hardware framework that might typically impose constraints on LoC performance like the addition of stream control gear in a device.

Since the LoC research field is a vast area, it's hard to summarise every current device and technology involved in it. The present focus of LoC technology development is mainly on three important areas. The process of industrialising LoC technology in preparation for commercialization. This includes modifying production techniques, developing unique surface treatments, designing flow control systems, and so on. Because of advancements in parallelization and the number of biological activities that can be combined onto a single chip, it is now possible to identify different classes of pathogens that are present in any human fluid by using a single microfluidic cartridge. The deep-seated research on various cutting-edge LoC technologies with considerable potential, such as reading the DNA sequence through nanopores demand more insight to make it practically feasible.

A lot of work is being done to make lab-on-a-chip more user-friendly. Making it feasible to use a smartphone for Elisa tests, cholesterol testing, anaemia diagnosis, or other essential lab-on-a-chip procedures is one example. Furthermore, an intense research is being conducted to improve current technologies for specialised biomedical approaches and diagnostics, such as cell separation, and PCR based DNA sequencing from nanopores. The research focuses on enabling high parallelization through multiplication of PCR chambers, the use of digital microfluidics to perform PCR in micro-droplets, and the use of the most recent developments in molecular biology to perform simultaneous PCR in the same mix in the case of microPCR, one of the most promising technologies for future high throughput diagnostics. Furthermore, decreasing detection thresholds and increasing PCR efficacy while reducing false-positive and false-negative outcomes are both important study fields. Currently, a few labs-on-a-chip are commercially available for specialised applications such as glucose monitoring or specific pathology detection. We may expect labs-on-a-chip to be widely used in hospitals across the world, and eventually in doctor's offices. We can expect lab-on-a-chip technology to deliver at-home real-time health monitoring in the future. Governments and corporations are investing more in labs-on-a-chip since it is becoming clear that these technologies will change our daily lives.

Conclusion

We can now definitely anticipate that lab-on-a-chip will revolutionise diagnostics in the not-too-distant future based on current research and recently emerging technology. Critical applications including glucose monitoring, HIV detection, and heart attack diagnosis have seen the sale of a variety of labs-on-a-chip. To reduce costs, improve ergonomics, and hasten diagnosis, industrial research will aim to combine as many distinct processes as feasible into a single lab-on-a-chip. It is difficult to forecast which materials and technologies will be most beneficial for high throughput diagnostics since technologies have not yet been unified. The solutions will be technological in nature, but they may also include corporate and economic issues in order to function in tandem with current systems like silicon micromachining.

The current limitations of LOC technology include its inability to multiplex, difficulties in identifying and measuring low concentration levels (submicromolar regime), and costly and challenging manufacturing and biofunctionalization techniques. The majority of tests nowadays cannot be performed in complex medium like serum or blood and must instead be thoroughly cleaned and purified before chip analysis. Therefore, basic design issues including signal measurement, sample preparation, and LOC design must be resolved before LOCs can be commercially commercialised. A new pattern shift in LOC design is required to address the intriguing challenges related to nanotechnology. Researchers are striving to create mobile sample preparation technologies that interact with and move through the sample solution as well as bioagent detection/quantification methods. Additionally, the manufacturing complexity and throughput performance of LOC devices will be considerably improved by these advancements. This can be overcome with FRET.

Future developments in molecular biology and microfluidics will be essential for the success of lab-on-a-chip technology. Nanotechnology will become increasingly crucial in linking these two areas as technology advances. Despite the difficulties that come with commercialising a novel technology, examples of these working gadgets are beginning to appear. Lab-on-a-chip technology looks to be gaining traction in the chemical and medical industries in the next years.

References

"3: Chemical Structure of Cellulose | Download Scientific Diagram." n.d. Accessed July 19, 2022. https://www.researchgate.net/figure/Chemical-structure-of-cellulose_fig2_321756571.

Ai, Yongjian, Feng Zhang, Chenlong Wang, Ruoxiao Xie, and Qionglin Liang. 2019. "Recent Progress in Lab-on-a-Chip for Pharmaceutical Analysis and Pharmacological/Toxicological Test." *TrAC Trends in Analytical Chemistry* 117 (August): 215–30. https://doi.org/10.1016/J.TRAC.2019.06.026.

Bang, Seokyoung, Sangcheol Na, Jae Myung Jang, Jinhyun Kim, and Noo Li Jeon. 2016. "Engineering-Aligned 3D Neural Circuit in Microfluidic Device." *Advanced Healthcare Materials* 5 (1): 159–66. https://doi.org/10.1002/ADHM.201500397.

Chau, Mokit, Milad Abolhasani, Héloïse Thérien-Aubin, Yang Li, Yihe Wang, Diego Velasco, Ethan Tumarkin, Arun Ramachandran, and Eugenia Kumacheva. 2014. "Microfluidic Generation of Composite Biopolymer Microgels with Tunable Compositions and Mechanical Properties." *Biomacromolecules* 15 (7): 2419–25. https://doi.org/10.1021/BM5002813/SUPPL_FILE/BM5002813_SI_001.PDF.

Choi, Nak Won, Mario Cabodi, Brittany Held, Jason P. Gleghorn, Lawrence J. Bonassar, and Abraham D. Stroock. 2007. "Microfluidic Scaffolds for Tissue Engineering." *Nature Materials* 6 (11): 908–15. https://doi.org/10.1038/NMAT2022.

El-Ali, Jamil, Peter K. Sorger, and Klavs F. Jensen. 2006. "Cells on Chips." *Nature 2006 442:7101* 442 (7101): 403–11. https://doi.org/10.1038/nature05063.

"File:Polycaprolactone Structure.Png - Wikipedia." n.d. Accessed July 19, 2022. https://en.m.wikipedia.org/wiki/File:Polycaprolactone_structure.png.

"FPA-5550iZ2 | Canon Global." n.d. Accessed August 10, 2022. https://global.canon/en/product/indtech/semicon/fpa5550iz2.html.

Fujii, Teruo. 2002. "PDMS-Based Microfluidic Devices for Biomedical Applications." *Microelectronic Engineering* 61–62 (July): 907–14. https://doi.org/10.1016/S0167-9317(02)00494-X.

Huang, Chao-Wei, Yu-Tsung Lin, Shih-Torng Ding, Ling-Ling Lo, Pei-Hwa Wang, En-Chung Lin, Fang-Wei Liu, and Yen-Wen Lu. 2015. "Efficient SNP Discovery by Combining Microarray and Lab-on-a-Chip Data for Animal Breeding and Selection." *Microarrays 2015, Vol. 4, Pages 570-595* 4 (4): 570–95. https://doi.org/10.3390/MICROARRAYS4040570.

Kim, Yong Tae, Yuchao Chen, Jong Young Choi, Won Jung Kim, Hyun Mi Dae, Jaean Jung, and Tae Seok Seo. 2012. "Integrated Microdevice of Reverse Transcription-Polymerase Chain Reaction with Colorimetric Immunochromatographic Detection for Rapid Gene Expression Analysis of Influenza A H1N1 Virus." *Biosensors & Bioelectronics* 33 (1): 88–94. https://doi.org/10.1016/J.BIOS.2011.12.024.

Liang, Ruixue, Li Wang, Haojie Yu, Amin Khan, Bilal Ul Amin, and Rizwan Ullah Khan. 2019. "Molecular Design, Synthesis and Biomedical Applications of Stimuli-Responsive Shape Memory Hydrogels." *European Polymer Journal* 114 (May): 380–96. https://doi.org/10.1016/J.EURPOLYMJ.2019.03.004.

Neumann, Christiane, and Bastian E. Rapp. 2015. "Fluidic Platforms and Components of Lab-on-a-Chip Devices." *Lab-on-a-Chip Devices and Micro-Total Analysis Systems*, 83–139. https://doi.org/10.1007/978-3-319-08687-3_5.

Osório, Luana A., Elisabete Silva, and Ruth E. Mackay. 2021. "A Review of Biomaterials and Scaffold Fabrication for Organ-on-a-Chip (OOAC) Systems." *Bioengineering 2021, Vol. 8, Page 113* 8 (8): 113. https://doi.org/10.3390/BIOENGINEERING8080113.

Peng, Ran, and Dongqing Li. 2017. "Detection and Sizing of Nanoparticles and DNA on PDMS Nanofluidic Chips Based on Differential Resistive Pulse Sensing." *Nanoscale* 9 (18): 5964–74. https://doi.org/10.1039/C7NR00488E.

"PGA." n.d. Accessed July 19, 2022. https://polymerdatabase.com/Polymer Brands/PGA.html.

"PLGA Polymer: The Ultimate Buying Guide | Polylactide." n.d. Accessed July 19, 2022. https://polylactide.com/plga-polymer-ultimate-buying-guide/.

"Poly Lactic Acid (PLA) Is Gaining Traction in the Market - Polymer Innovation Blog." n.d. Accessed July 19, 2022. https://polymerinnovationblog.com/poly-lactic-acid-pla-is-gaining-traction-in-the-market/.

"Polyanhydrides." n.d. Accessed July 19, 2022. http://polymerdatabase.com/polymer index/polyanhydrides.html.

Shen, Weizhi, Mingzhu Li, Changqing Ye, Lei Jiang, and Yanlin Song. 2012. "Direct-Writing Colloidal Photonic Crystal Microfluidic Chips by Inkjet Printing for Label-Free Protein Detection." *Lab on a Chip* 12 (17): 3089–95. https://doi.org/10.1039/C2LC40311K.

Sood, Ankur, Anuj Kumar, Atul Dev, Vijai Kumar Gupta, and Sung Soo Han. 2022. "Advances in Hydrogel-Based Microfluidic Blood–Brain-Barrier Models in Oncology Research." *Pharmaceutics 2022, Vol. 14, Page 993* 14 (5): 993. https://doi.org/10.3390/PHARMACEUTICS14050993.

"Starch." n.d. Accessed July 19, 2022. https://polymerdatabase.com/polymer classes/Polysaccharide type.html.

Suresh, Vignesh, Ong Qunya, Bera Lakshmi Kanta, Lee Yeong Yuh, and Karen S. L. Chong. 2018. "Non-Invasive Paper-Based Microfluidic Device for Ultra-Low Detection of Urea through Enzyme Catalysis." *Royal Society Open Science* 5 (3). https://doi.org/10.1098/RSOS.171980.

Temiz, Yuksel, Robert D. Lovchik, Govind V. Kaigala, and Emmanuel Delamarche. 2015. "Lab-on-a-Chip Devices: How to Close and Plug the Lab?" *Microelectronic Engineering* 132 (January): 156–75. https://doi.org/10.1016/J.MEE.2014.10.013.

Wei, Zehui, Peiru Fan, Yajie Jiao, Yang Wang, Yanping Huang, and Zhaosheng Liu. 2020. "Integrated Microfluidic Chip for On-Line Proteome Analysis with Combination of Denaturing and Rapid Digestion of Protein." *Analytica Chimica Acta* 1102 (March): 1–10. https://doi.org/10.1016/J.ACA.2020.01.025.

Zhu, Hanliang, Zdenka Fohlerová, Jan Pekárek, Evgenia Basova, and Pavel Neužil. 2020. "Recent Advances in Lab-on-a-Chip Technologies for Viral Diagnosis." *Biosensors and Bioelectronics* 153 (April): 112041. https://doi.org/10.1016/J.BIOS.2020.112041.

Chapter 4

Nanosensors: Material Selection and Synthesis

Learning Outcomes

- The reader will understand the classification and significance of various types of nanosensors, including optical, electrical, magnetic, chemical, biological, mechanical, and thermal sensors.
- The impact of nanomaterials in enhancing sensor attributes such as sensitivity, selectivity, stability, response time, compatibility, cost-effectiveness, scalability, and multi-functionality, can be deeply evaluated by the learner.
- A deep insight of classifying and differentiating the materials used in nanosensors, including metals, semiconductors, polymers, carbon nanotubes, biomolecules, and oxides, based on their properties and suitability for specific sensing applications will be obtained.
- Various synthesis methods employed in nanosensor fabrication, including chemical, physical, biological, hybrid, template-based, and hydrothermal synthesis, and how each method influences sensor performance can be easily comprehended by the learner.
- The learner will gain insights into the practical applications of nanosensors in fields such as energy harvesting, biomedical imaging, environmental monitoring, and medical diagnostics.
- A holistic perspective on the intricate interplay between nanosensor materials selection, synthesis methodologies, and their ultimate functionality in addressing complex real-world challenges will be developed.

Abstract

Due to their outstanding detecting abilities and potential for use in a variety of disciplines, nanosensors have attracted a lot of interest recently. In order to provide a thorough knowledge of nanosensors, this chapter will concentrate on the materials and synthesis techniques needed to attain the necessary sensor performance. Nanosensors and the use of nanomaterials have enormous benefits for academia, business, and the environment. They empower enterprises with effective and dependable monitoring systems, provide researchers new instruments for inquiry, and support environmental sustainability. The importance of nanomaterials in allowing the distinctive features of nanosensors highlights their potential to propel technological innovation in a variety of industries. The chapter outlines several real-time nanosensors and their applications while introducing the categorization and importance of nanosensors. The influence of nanomaterials on characteristics like sensitivity, selectivity, stability, and scalability is what gives nanosensors their effectiveness. For their appropriateness in various sensing modalities, materials including metals, semiconductors, polymers, carbon nanotubes, biomolecules, and oxides are studied. The performance and structure of nanosensors are greatly influenced by the synthesis techniques. In order to understand their contributions to efficient sensor fabrication, a variety of synthesis methods are investigated, including chemical, physical, biological, hybrid, template-based, and hydrothermal. Nanosensors' ability to solve real-world problems is shown by their practical applications across a variety of sectors. The relationship between material choice, synthesis techniques, and sensor functioning is evident throughout the chapter and captures the spirit of contemporary nanosensor research.

Keywords: chemical synthesis, physical synthesis, biological synthesis, hybrid synthesis, template synthesis, hydrothermal synthesis

1. Introduction to Nanosensors

The preceding chapters went into great detail on nanomaterials, covering their history and origins, as well as their types, traits, applications, and manufacturing processes. Although there are still a lot of unanswered questions regarding nanomaterials, the knowledge presented in these chapters is crucial for understanding their potential because they can offer a variety of cutting-edge solutions to current problems, and their use is anticipated to rise in the coming years.

Nanosensors are devices that can detect and respond to changes in the physical or chemical properties of the environment at the nanoscale level. They have potential applications in a wide range of fields, including medicine, environmental monitoring, and defence. However, the development of nanosensors is still in its early stages, and many challenges remain, including the selection and synthesis of appropriate nanomaterials. There are several types of nanosensors, which can be classified based on their sensing mechanism or application. These nanosensors have the potential to revolutionize our understanding of biological processes in vivo, as they provide unprecedented control over intracellular environments that could not be measured before. The illustration of a nanosized sensor is shown in Figure 75 ("What Are Nano-Sized Sensors?" n.d.).

Figure 75. Illustration of nanosized sensors ("What Are Nano-Sized Sensors?" n.d.).

The selection and synthesis of appropriate nanomaterials are critical steps in the development of nanosensors. The properties of the nanomaterials used will determine the sensitivity, selectivity, and response time of the resulting sensor. The ideal nanomaterial should have high surface area, high reactivity, and be easily functionalized to facilitate the detection of the target analyte. By understanding the different types of nanosensors and their applications, researchers can better design and optimize nanosensors for their specific needs. The selection of nanomaterials also depends on the sensing mechanism. Different sensing mechanisms, such as optical, electrical, or magnetic, require specific properties from the nanomaterials used. For example, the choice of nanomaterials used in a sensor that uses surface-enhanced Raman spectroscopy (SERS) will differ from those used in a sensor that uses electrochemical detection.

In this chapter, we will review the selection and synthesis of nanomaterials for nanosensors. We will discuss the properties of different nanomaterials, such as metallic nanoparticles, carbon-based materials, and semiconducting nanoparticles, and their suitability for different sensing mechanisms. We will also highlight recent advances in the synthesis of nanomaterials and their functionalization for use in nanosensors. Overall, this chapter aims to provide a comprehensive overview of the challenges and opportunities in the selection and synthesis of nanomaterials for nanosensors, which will be of interest to researchers and practitioners working in this field.

2. Classification and Significance of Nanosensors

In addition to discussing the selection and synthesis of nanomaterials for nanosensors, it is important to provide an overview of the different types of nanosensors that exist. There are several types of nanosensors, which can be broadly classified based on their sensing mechanism, such as optical, electrical, or magnetic, or based on their application, such as biosensors, gas sensors, or environmental sensors. The components and working of a biosensor is shown in Figure 76 (Sezgintürk 2020).

Optical nanosensors, for example, use the interaction between light and matter to detect changes in the environment, and include surface-enhanced Raman spectroscopy (SERS) sensors and fluorescence-based sensors. Electrical nanosensors, on the other hand, measure changes in electrical properties such as resistance, capacitance, or impedance, and include field-effect transistors (FETs) and nanowire sensors. Magnetic nanosensors, such as magnetic nanoparticles, detect changes in magnetic fields.

Biosensors are a type of nanosensor that are used to detect biological analytes, such as proteins, DNA, or cells. They can be optical, electrical, or magnetic in nature, and can be used for applications such as disease diagnosis, drug discovery, and environmental monitoring. Gas sensors, on the other hand, are used to detect the presence of specific gases, such as carbon monoxide, nitrogen dioxide, or methane, and are used in applications such as air quality monitoring and industrial process control.

Environmental sensors are another type of nanosensor that are used to monitor the physical and chemical properties of the environment, such as temperature, humidity, and pH. They can be used in applications such as smart buildings, agriculture, and food storage.

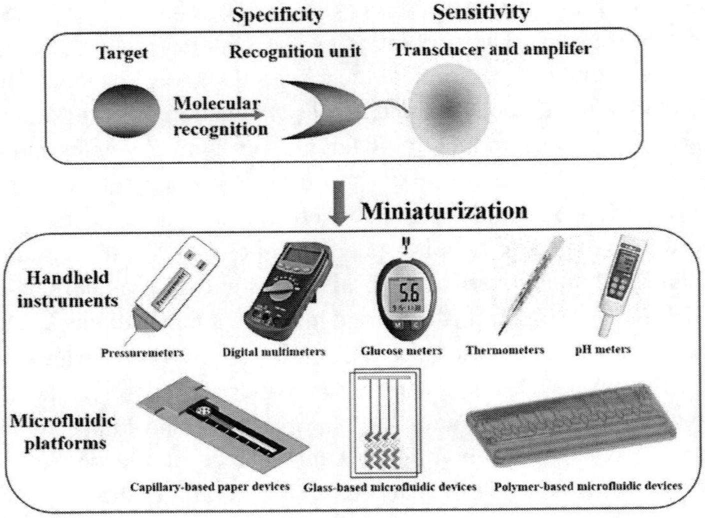

Figure 76. Components and Working of a biosensor (Sezgintürk 2020).

In summary, the field of nanosensors encompasses a wide range of sensor types that differ in their sensing mechanism and application. By understanding the different types of nanosensors and their applications, researchers can better design and optimize nanosensors for their specific needs.

2.1. Optical Nanosensors

Optical nanosensors are a type of nanosensor that utilize the interaction between light and matter to detect changes in the environment. They have the potential for a wide range of applications, such as disease diagnosis, environmental monitoring, and food safety. Optical nanosensors have been made that use the sensitivity of fluorescence to measure things in the intracellular environment in a quantitative way. These measurements are made with the help of small devices that can be put into living cells without causing too much physical damage. These nanosensors are capable of determining a wide variety of intracellular parameters, including concentrations of ions, metabolites, and macromolecules; pH; temperature; and even gene expression among other things. There is a significant amount of untapped potential in optical nanosensors as a non-invasive method of acquiring quantitative information about the intracellular environment. The schematic of an optical fiber nanosensor is given in Figure 77 (Tabbakh et al. 2021).

One common type of optical nanosensor is the surface-enhanced Raman spectroscopy (SERS) sensor. SERS sensors use metallic nanoparticles to enhance the Raman scattering signal of a target molecule, resulting in highly sensitive detection of the molecule. SERS sensors have been used for a wide range of applications, including detection of biomolecules, environmental pollutants, and explosives. Fluorescence-based optical nanosensors are another type of optical nanosensor that utilize the fluorescent properties of certain molecules to detect changes in the environment. For example, quantum dots, which are nanoscale semiconductor particles, can be functionalized to detect the presence of specific biological molecules. Fluorescence-based sensors have the advantage of high sensitivity and specificity, but can also suffer from interference from other fluorescent molecules in the environment.

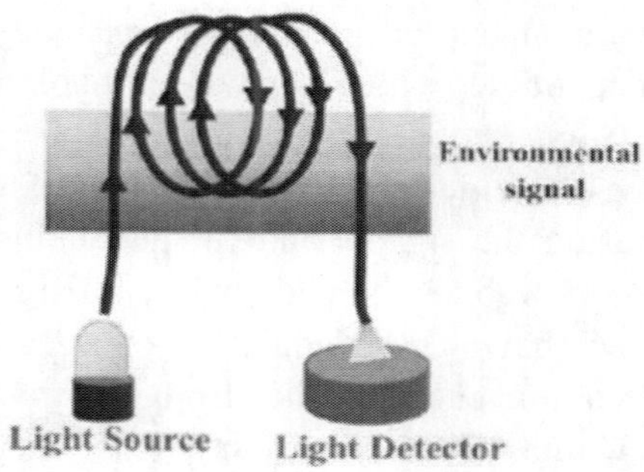

Figure 77. Schematic of Optical Fiber Transducer (Tabbakh et al. 2021).

Optical nanosensors have the potential for high sensitivity and specificity, and can be used for a wide range of applications. The choice of nanomaterials used in an optical nanosensor will depend on the specific application and sensing mechanism used. Ongoing research in the field of optical nanosensors is focused on improving sensitivity, selectivity, and response time, as well as developing new sensing mechanisms and applications.

Optical signal transduction is shown by fluorescence that goes out when gold nanoparticles (AuNPs) are present. When fluorescein isothiocyanates (FITC) are brought close to gold nanoparticles, the fluorescence they give off is severely dimmed, and there is no sign of visible radiation. When this molecule is close to these nanoparticles, its Raman scattering signals get stronger. As a result, it acts like a Raman probe. Surface-enhanced Raman scattering (SERS) is the name given to this phenomenon, which has the potential to be used in the fields of biosensing and imaging. Researchers have shown that they can utilise SERS to identify and analyse biological molecules with a high degree of sensitivity if they bind FITC to gold nanoparticles.

Surface plasmon resonance (SPR), a common methodology based on analytical chemistry, used to observe molecular interactions forces. It deals with the fluctuations of refractive index due to molecules binding to a thin metal surface. Light incident on the surface will excite coherent oscillations of surface electrons that are sensitive to electromagnetic fluctuations at the boundary. These fluctuations may be caused by molecule binding events, so that these may give rise to features in the detected reflectivity spectrum. However, it is challenging to attain high throughput sensitivity, because SPR typically has deprived resolution since the unspecified binding interact with the efficacy of material. An exceptional property of SPR systems unfolds once the light waves intermingle with nanoparticles having smaller size compared to the light wavelength, for example metal nanoparticle. Possible metals that have been utilised are Au, Ag or Cu. The plasmon system that oscillates around nanoparticles is called localized surface plasmon resonance (LSPR). There are different features of metal nanoparticles (dimensions, morphology and shape), sensitivity of LSPR based sensors relies on.

LSPR nanosensors require extremely uniform nanomaterials with a narrow LSPR peak to allow for proper calibration as shown in Figure 78 (Philip and Kumar 2022). The spectral shift of this peak is characteristic for the analyte that caused it. As an illustration, specific silver nanoparticles were used for the detection of amyloid-derived diffusible ligands (ADDL). These are biomarkers for Alzheimer's disease (extracts from human brain) and cancer. The authors used a common principle, a sandwich assay. Such an assay uses a primary protein to capture the analyte and a secondary protein to produce/enhance the signal. Nanosphere lithography was used to synthesize silver nanoparticles with triangular morphology. Mineral substrates provide a platform and functionalization is done by specific protein for ADDL. Once ADDL was captured secondary targeted protein were used to boost LSPR signal. Ultraviolet-visible excitation was used to measure the signals and for qualitative analysis. Optically coupled spectroscope fibres were used for collection and analysis of signals. Silver nanoparticles with specific triangular morphology (having perpendicular cross section of 90 nm and 25 nm height) were selected to increase 35 nm magnetism fields from substrate. It is necessary to notice the distance of captured ADDL within the sandwich assay. This is important since LSPR is strongly depending on the distance. The explained strategy revealed completely different binding coefficients of ADDL protein to secondary protein from brain extracts or tumor. Optical nanosensor detection can be improved by managing the size and form of nanoparticles. LSPR results are helpful for quantitative chemical analysis, distinctive light changes and qualitative analysis for surface-enhanced Raman. The striking advantage of antibodies is that (at least for monoclonal antibodies) every antibody is exactly the same. Thus, unprecedented reproducibility is achieved. The disadvantage is that antibodies are biomolecules and therefore degrade over time and cannot be employed in harsh surroundings. Furthermore, they are relatively expensive and/or time consuming to fabricate.

Nanoparticles used in optical nanosensors act as a signal production source. Optical nanosensors are thus designed in a way to produce a change on nanoparticle surface, in direct proportion to the analyte concentration. Though, such changes often take place on the surface of extremely uniform nanoparticles. This targeted property has not been completely attained because most nanoparticles are stable only after applying stabilizing agents on their surface. Due to particle stability issues, detection in presence of physiological conditions (high salt concentrations) can be an issue for optical detection too. High salt concentrations create harsh conditions for many nanoparticles. Though surface chemistry remains a challenge, nanosensors are ready to bypass these disadvantages. A possible solution could be taking combinatorial approaches (both binding and stabilizing building blocks). Nanosensor signals can be extremely amplified by coupling to nanoparticles having distinctive optical properties. Such sensing approaches are for example useful for early diagnosis of diseases.

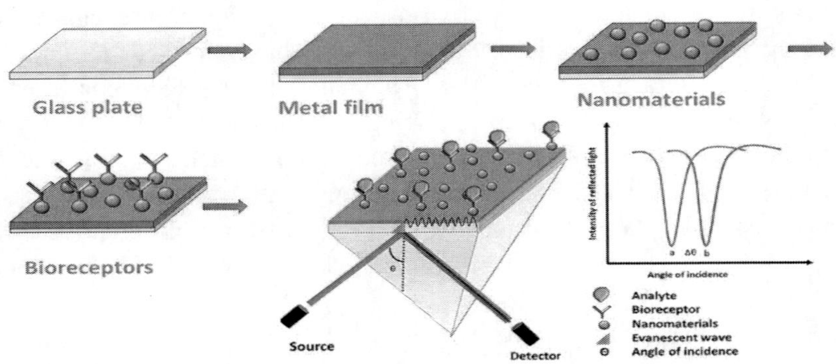

Figure 78. Schematic of Optical LSPR Transducer used in nanosensors (Philip and Kumar 2022).

2.2. Electrical Nanosensors

Electrical recognition is a quickly expanding field with built up, basic and minimal effort in the manufacturing procedures. At present, there are numerous proposed and marketed gadgets in light of the electrochemical approach including those for pathogens, and toxins. Electrochemical detection is so popular due to its extraordinary characteristics including sensitivity, low cost, compatibility with modern miniaturized/lab on chip type strategies, least requirement of power, and requiring no-pre sample processing (no effect from turbidity and color of sample). The standard principle of electrochemical nanosensors is a chemical reaction in which electron release, accept or consume ions. This chemical reaction takes place between a restrained ligand and analyte of interest that measurably affect the transduced signal, such as an electrical current or potential. This electrochemical signal is directly quantified and related to the presence of marker of interest/analyte in the sample solution. Electrochemical detection approaches have different subtypes on the basis of signal types like potentiometry, voltammetry, amperometry, and electrochemical impedance spectroscopic. Sometimes electrochemical measurements are conjugated with immunoassays. The resulting sensors are called immunosensors. These categories depend on the produced and amplified signal types.

Nanosensors based on detection of electrical signals, first and foremost reported nano field-effect transistors (FETs) possess tunable properties and can be responsible for easy and quantitative measurements. Chemical nanosensors based on FET principle utilize rod shaped nanomaterials (nanowires, nanorods, nanoribbons, nanotowers and nanotubes). When targeted analytes cling to the active area, this results in the change of impedance and produces a signal. Nanomaterials with these morphologies provide increased sensitivity and active area for current flow, compared to the activity across the cross section of other nano-scale morphologies on the flat detector surface. The most commonly used nanomaterial is silicon nanowires because they possess high sensitivity and are easy to functionalize or chemically modify on the surface. The schematic of a rGO based FET biosensor is shown in Figure 79 that is capable of detecting the Ebola antigens (Chen et al. 2017).

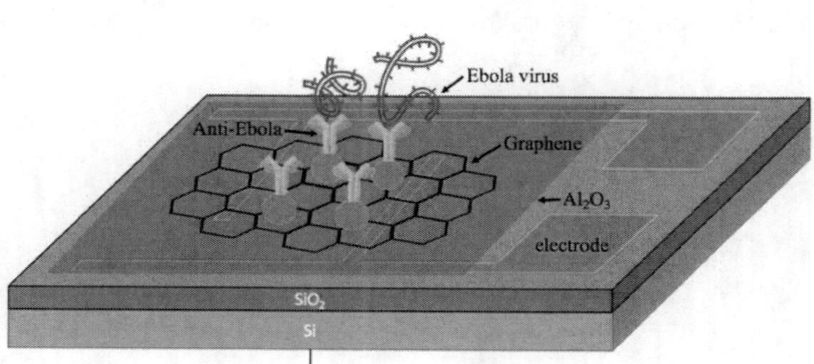

Figure 79. rGO based FET Biosensor to detect Ebola antigens (Chen et al. 2017).

Furthermore, nanosensors can be fabricated and optimized for the detection of proteins on the basis of their conformations. For example conformational changes in calmodulin can be caused by metallic elements. Sensitivity of this kind of sensors relies on the electrical resistance, which originates on the silicon nanowire tips, resulting in a quicker electronic transfer on the silicon nanowire tip as compared to the sidewall. Electrical resistance depends on concentrations ranging from 10 pg L^{-1} to 10 μg L^{-1}. The mechanism of action of such sensors is modification in permittivity and electric resistance within the materials on the surface as a result of macromolecule attachment. Another popular nanosensing material for electrical detection is graphene and its derivatives.

2.3. Magnetic Nanosensors

Magnetic nanosensors are devices that use the magnetic properties of nanoparticles to detect changes in a sample. These sensors are based on the fact that the magnetic properties of nanoparticles change in response to external magnetic fields or when they interact with other particles or molecules.

The basic structure of a magnetic nanosensor consists of magnetic nanoparticles, which are functionalized with ligands or biomolecules that selectively bind to the analyte of interest. When the analyte binds to the functionalized nanoparticles, it induces changes in the magnetic properties of the nanoparticles, which can be detected by measuring changes in the magnetic field or the magnetization of the nanoparticles. There are several types of magnetic nanosensors, including superparamagnetic nanoparticles, magnetic nanowires, and magnetic thin films. Superparamagnetic nanoparticles are the most commonly used type of magnetic nanosensor due to their high sensitivity and biocompatibility. They are typically made from iron oxide (Fe_3O_4) or magnetite ($FeOFe_2O_3$) and have a size range of 5-20 nm as illustrated in Figure 80 (Slimani and Hannachi 2020).

The detection of analytes using magnetic nanosensors can be done using various techniques, including magnetic particle spectroscopy (MPS), magnetoresistive sensors, and magnetic induction sensors. MPS is a highly sensitive technique that measures the harmonic response of magnetic nanoparticles to an applied magnetic field. Magnetoresistive sensors measure the changes in resistance of a magnetic material in response to an external magnetic field. Magnetic induction sensors measure the changes in the magnetic field induced by the magnetic nanoparticles.

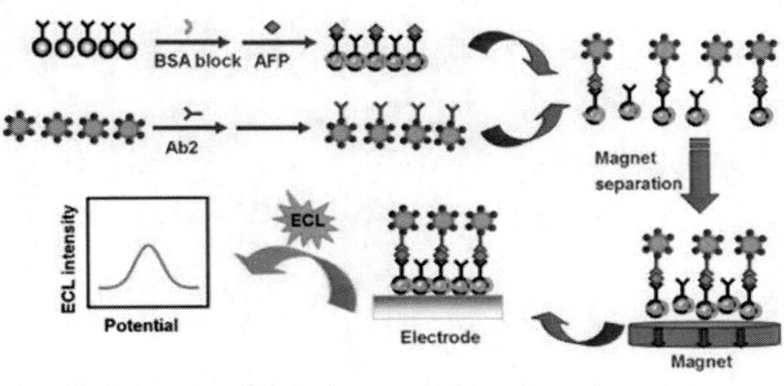

Figure 80. Schematic of ECL immunosensors using magnetic separation (Slimani and Hannachi 2020).

Magnetic nanosensors have several advantages over other types of sensors, including high sensitivity, specificity, and stability. They can be used in a variety of applications, including biomedical diagnostics, environmental monitoring, and food safety testing. Magnetic nanosensors have the potential to revolutionize the field of sensing and have already demonstrated significant promise in various fields.

2.4. Chemical Nanosensors

Nanosensors are analytical tools that react selectively and reversibly to a chemical agent's concentration or activity in a sample. Chemical nanosensors are often used to determine the quality or quantity of the problematic analytes. The chemical sensor's surface detects the analyte as a consequence of a straightforward physical or chemical interaction. This contact is transformed into a constant electrical signal that a transducer, a machine, can measure. The quantity of analyte in the sample is reflected in the output signal. These signal transformation techniques often take into account thermal, electrochemical, optical, and auditory factors. Chemical or physical qualities are transformed into valuable information by the sensor. This definition states that a chemical sensor is a device that produces an electrical signal in response to the chemical make-up or quantity of an analyte molecule. The perfect chemical sensor should be compact, simple, affordable, and trustworthy.

2.5. Biological Nanosensors

Biological nanosensors are devices that are designed to detect and measure specific biomolecules, such as proteins, DNA, and RNA, at the nanoscale level. These sensors rely on the unique properties of biological molecules to interact with other molecules and produce a measurable signal, which can be used to detect the presence or concentration of the target molecule. Biological nanosensors typically consist of two main components: a recognition element and a transducer. The recognition element is a molecule or biomolecule that selectively binds to the target molecule, while the transducer converts the binding event into a measurable signal.

The recognition element can be a variety of biomolecules, such as antibodies, aptamers, enzymes, or receptors. For example, an antibody-based sensor may use an antibody that specifically binds to a particular protein to detect its presence. Similarly, an aptamer-based sensor may use a short DNA or RNA sequence that folds into a specific conformation to bind to a target molecule. The transducer converts the binding event into a measurable signal, which can be electrical, optical, or mechanical in nature. For example, an electrical transducer may use a field-effect transistor (FET) to measure changes in the electrical conductivity of the sensor when the recognition element binds to the target molecule. An optical transducer may use a fluorescent molecule that emits light when it binds to the target molecule.

Biological nanosensors have several advantages over traditional sensors. They can be highly specific, as the recognition element can be tailored to bind to a specific molecule with high affinity and selectivity. They can also be very sensitive, detecting molecules at low concentrations. Additionally, biological nanosensors can be designed to work in complex environments, such as in biological fluids or living cells. Biological nanosensors have a wide range of potential applications in various fields, such as medical diagnostics, environmental monitoring, and food safety. They can be used to detect biomarkers of diseases, monitor the concentration of pollutants in the environment, or detect pathogens in food.

2.6. Mechanical Nanosensors

Mechanical nanosensors are devices that detect and measure mechanical forces and deformations at the nanoscale level. These sensors typically consist of a mechanical structure that can deform in response to an applied force, and a transducer that converts the mechanical deformation into a measurable signal. There are several types of mechanical nanosensors, such as cantilever sensors, nanowire sensors, and resonant sensors. Each type of sensor has its own unique design and mechanism of operation, but they all rely on the ability to measure changes in the mechanical properties of the sensor.

Cantilever sensors are thin, flexible beams that are anchored at one end and free to move at the other end as shown in Figure 81 (Ansari and Cho 2008). When a force is applied to the free end of the cantilever, it bends and produces a measurable deflection. The magnitude of the deflection is proportional to the applied force, and can be measured using various transduction methods, such as optical interferometry or piezoresistive sensing.

Nanowire sensors are similar to cantilever sensors, but they are typically much smaller in size and have a cylindrical or rectangular cross-section. The deflection of the nanowire in response to an applied force can be measured using various methods, such as electrical conductivity changes or optical interferometry.

Resonant sensors are designed to measure changes in the resonant frequency of a mechanical structure, such as a cantilever or a nanowire. When a force is applied to the structure, it changes the resonant frequency, which can be measured using various techniques, such as optical or electrical readouts. Mechanical nanosensors have several advantages over traditional sensors. They can be highly sensitive, detecting forces and deformations at the nanoscale level. They can also be highly selective, as the mechanical properties of the sensor can be tailored to respond to specific types of forces or deformations. Additionally, mechanical nanosensors can be used in a wide range of applications, such as biological sensing, materials science, and nanotechnology.

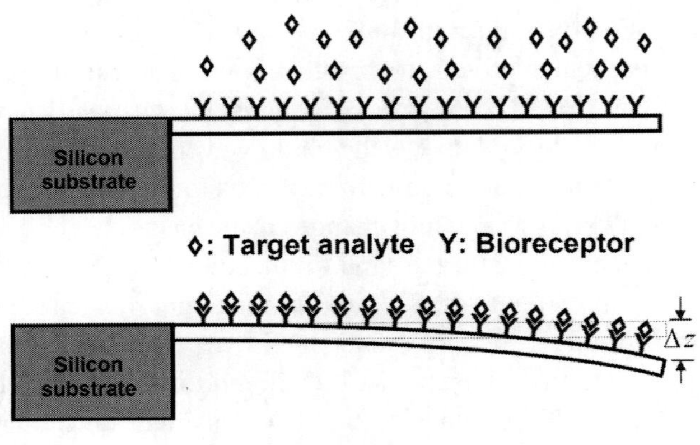

Figure 81. Schematic of Cantilever array biosensor (Ansari and Cho 2008).

2.7. Thermal Nanosensors

Thermal nanosensors are devices that are designed to detect and measure changes in temperature at the nanoscale level. These sensors typically consist of a thermally sensitive material that undergoes a change in its physical or chemical properties in response to a change in temperature, and a transducer that converts the change in the material into a measurable signal.

The thermally sensitive material can be a variety of materials, such as metals, semiconductors, or polymers, that exhibit a change in their electrical, optical, or mechanical properties in response to a change in temperature. For example, a metal may undergo a change in its electrical resistance when it is heated, while a polymer may expand or contract in response to a change in temperature. The transducer converts the change in the thermally sensitive material into a measurable signal, which can be electrical, optical, or mechanical in nature. For example, an electrical transducer may use a Wheatstone bridge circuit to measure changes in the electrical resistance of the sensor, while an optical transducer may use a fluorescent molecule that emits light when it is heated.

Thermal nanosensors have several advantages over traditional sensors. They can be highly sensitive, detecting changes in temperature at the nanoscale level. They can also be highly selective, as the properties of the thermally sensitive material can be tailored to respond to specific temperature ranges or changes. Additionally, thermal nanosensors can be used in a wide range of applications, such as temperature sensing in electronic devices, thermal imaging in biomedical imaging, and thermal analysis in materials science. Some of the examples of thermal nanosensors and their applications are given below.

2.7.1. Thermoelectric Nanogenerators for Energy Harvesting

Thermoelectric nanogenerators (TENGs) are devices that convert heat energy into electrical energy using thermoelectric materials as shown in Figure 82 (Kim et al. 2020). These materials exhibit a phenomenon known as the Seebeck effect, where a temperature difference across the material generates a voltage. TENGs can be used to harvest energy from waste heat sources, such as the heat generated by electronic devices or the human body. For example, researchers at the Georgia Institute of Technology developed a flexible TENG that can be worn on the skin to generate power from body heat, which can be used to power wearable devices.

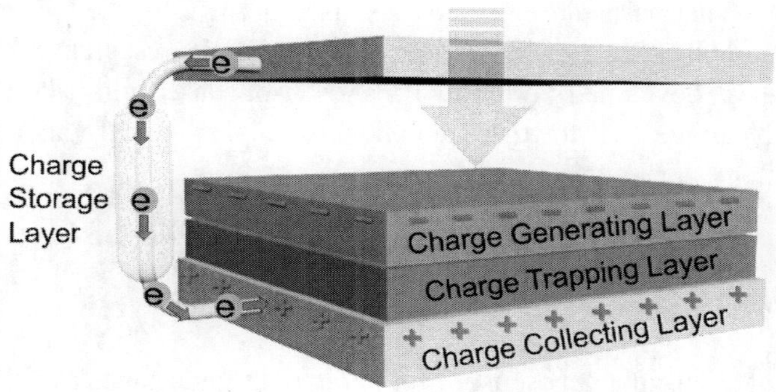

Figure 82. Schematic of a triboelectric TENG used in power generation (Kim et al. 2020).

2.7.2. Plasmonic Nanothermometry for Biomedical Imaging

Plasmonic nanothermometry is a technique that uses plasmonic nanoparticles to measure temperature changes at the nanoscale level. Plasmonic nanoparticles exhibit a phenomenon known as localized surface plasmon resonance, where the absorption and scattering of light is strongly dependent on the temperature of the particle. By measuring the changes in the optical properties of the nanoparticles, the temperature can be calculated. Plasmonic nanothermometry has potential applications in biomedical imaging, such as for monitoring temperature changes in cancer cells during hyperthermia treatment. For example, researchers at the University of Cambridge developed a plasmonic nanothermometer that can measure temperature changes with high spatial and temporal resolution in live cells. Figure 83 shows the schematic of Hydrogen nanothermometry to deduce Pd nanoparticles temperature (Tiburski, Nugroho, and Langhammer 2022).

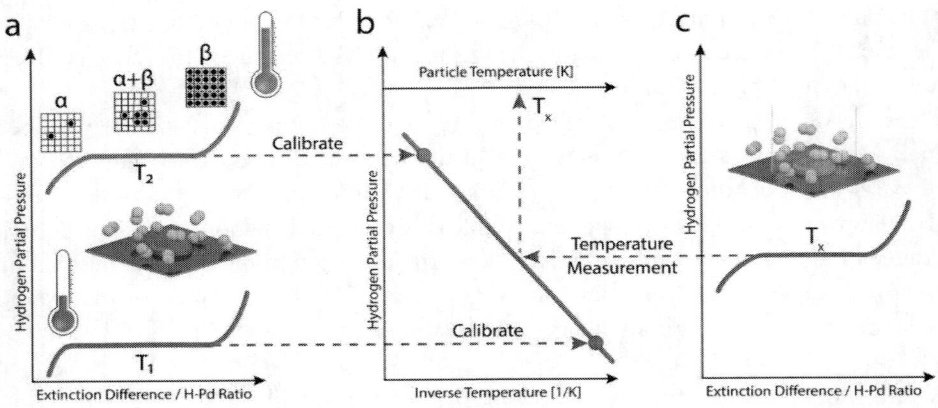

Figure 83. Hydrogen nanothermometry to deduce Pd nanoparticles temperature a) Pressure composition isotherms of Pd and H b) Phase transition plot c) Temperature determination of Pd nanoparticle (Tiburski, Nugroho, and Langhammer 2022)

2.7.3. Thermal Conductivity Measurements of Nanomaterials

Thermal conductivity is an important property of materials that affects their performance in various applications, such as in electronic devices and thermal management systems. However, measuring the thermal conductivity of nanomaterials is challenging due to their small size and complex structures. Thermal nanosensors can be used to measure the thermal conductivity of nanomaterials with high accuracy and sensitivity. For example, researchers at the University of California, Berkeley developed a thermal nanosensor based on a suspended graphene nanoribbon, which can measure the thermal conductivity of materials with a thickness as small as 10 nm.

3. Impact of Nanomaterials Deployed in Nanosensors

The selection of materials in a nanosensor is of paramount importance as it directly impacts the sensor's performance, sensitivity, selectivity, stability, and response time. Here are some key factors that illustrate the significance of different materials used in nanosensors:

3.1. Sensitivity

The material choice significantly affects the sensitivity of a nanosensor. Certain materials, such as nanomaterials (e.g., nanoparticles, nanowires), possess high surface area-to-volume ratios, leading to increased interaction with analytes and enhanced sensitivity. Materials with unique electrical, optical, or magnetic properties can also facilitate highly sensitive detection and measurement of target signals.

3.2. Selectivity

Selectivity refers to a sensor's ability to differentiate between specific target analytes and interfering substances. Materials with specific binding capabilities or chemical reactivity can be utilized to enhance the selectivity of nanosensors. For example, functionalizing the sensor surface with biomolecules like antibodies or aptamers allows for specific recognition and binding of target molecules, ensuring accurate and selective detection.

3.3. Stability

Stability is a critical aspect for the long-term performance and reliability of nanosensors. The chosen materials should exhibit robustness against environmental factors (such as temperature, humidity, pH) and provide resistance against chemical or physical degradation. Stability ensures that the sensor maintains its performance over extended periods without significant changes in its properties.

3.4. Response Time

The response time of a nanosensor refers to the time taken for the sensor to detect and provide a measurable signal in response to a stimulus. Materials with fast response times, such as semiconductors or highly conductive materials, can enable rapid and real-time detection of analytes. Faster response times are especially crucial for applications requiring quick monitoring and analysis.

3.5. Compatibility

The compatibility of materials used in nanosensors is crucial when considering the application environment. For example, in biomedical applications, the materials should be biocompatible to ensure safe interactions with biological systems. Compatibility with fabrication techniques and integration processes is also important for manufacturing nanosensors with consistent performance.

3.6. Cost and Scalability

The cost and scalability of materials are significant factors for commercial viability. Utilizing materials that are readily available, cost-effective, and compatible with large-scale production methods can facilitate the widespread adoption of nanosensors.

3.7. Multi-functionality

Certain materials can offer multifunctionality, allowing the integration of multiple sensing mechanisms into a single sensor platform. For instance, using a material that exhibits both electrical and optical properties enables the sensor to detect analytes through different sensing modalities, enhancing versatility and expanding the range of detectable signals.

The choice of materials in nanosensors is driven by the specific sensing requirements, performance targets, and application contexts. By carefully selecting and engineering materials, researchers can optimize nanosensor performance and enable a wide range of applications across diverse fields, including healthcare, environmental monitoring, food safety, and security.

4. Classification of Materials Used for Nanosensors

Nanosensors have revolutionized various fields, ranging from healthcare to environmental monitoring and beyond. These sensors operate at the nanoscale, allowing for highly sensitive and selective detection of analytes. A critical aspect of nanosensor design is the careful selection of materials, as they significantly influence the sensor's performance. The classification of materials used for nanosensors is a crucial consideration that impacts their functionality, sensitivity, and overall capabilities.

4.1. Metals

Metallic nanoparticles have gained significant attention in the field of nanosensors due to their unique properties, including their surface plasmon resonance (SPR), high electrical conductivity, and catalytic activity. These properties make metallic nanoparticles highly suitable for various sensing applications, such as chemical and biological sensing, environmental monitoring, and medical diagnostics.

One of the key properties of metallic nanoparticles, particularly gold and silver nanoparticles, is their ability to exhibit surface plasmon resonance (SPR). SPR is a phenomenon where the conduction electrons in the metal nanoparticles oscillate collectively in response to incident light, leading to the absorption and scattering of light at specific wavelengths. This phenomenon can be harnessed for sensing applications, as changes in the local environment near the nanoparticles can affect the SPR, enabling the detection of analytes.

Gold nanoparticles (AuNPs) have been extensively used in nanosensor development due to their unique optical properties, biocompatibility, and ease of synthesis. AuNPs can be synthesized in a controlled manner, allowing precise tuning of their size, shape, and surface properties, which directly impact their sensing capabilities. The localized surface plasmon resonance (LSPR) of AuNPs can be tuned by altering their size, shape, and surrounding medium, making them highly sensitive to changes in their local environment.

In a study published in the journal "Nature Nanotechnology," Kim et al. (2009) demonstrated the use of gold nanoparticles for the detection of single-nucleotide polymorphisms (SNPs), which are genetic variations associated with diseases. The researchers functionalized the surface of gold nanoparticles with DNA probes specific to the target SNPs. The binding of complementary DNA sequences induced aggregation of the nanoparticles, leading to a visible color change, which served as a detection signal. Huang et al. (2013) reported in the journal "Analytical Chemistry" the development of a highly sensitive and selective nanosensor based on gold nanoparticles for the detection of mercury ions (Hg2+). The researchers functionalized the gold nanoparticles with thymine-rich DNA sequences, which exhibited strong affinity towards Hg2+. The presence of Hg2+ induced the aggregation of gold nanoparticles, resulting in a color change that could be detected by the naked eye or spectroscopic techniques.

Silver nanoparticles (AgNPs) have also shown promise in nanosensor applications due to their strong SPR and excellent antimicrobial properties. AgNPs can be synthesized in various sizes and shapes, allowing for tunable sensing properties. In a study published in the journal "ACS Nano," Liu et al. (2011) developed a sensitive and selective nanosensor using silver nanoparticles for the detection of hydrogen peroxide (H_{2O2}), a reactive oxygen species implicated in various diseases. The researchers modified the surface of silver nanoparticles with horseradish peroxidase (HRP) enzymes. In the presence of H_2O_2, HRP catalyzed the reduction of silver ions, leading to the aggregation of silver nanoparticles and a color change, enabling H_2O_2 detection.

Khan et al. (2017) reported in the journal "Nanoscale" the development of a silver nanoparticle-based nanosensor for the detection of heavy metal ions. The researchers functionalized the silver nanoparticles with mercaptoacetic acid, which selectively bound to heavy metal ions. The presence of heavy metal ions induced aggregation of the silver nanoparticles, resulting in a change in the optical properties that could be detected using spectroscopic techniques.

These examples highlight the potential of metallic nanoparticles, such as gold and silver nanoparticles, in nanosensor applications. Their unique properties, including SPR, high electrical conductivity, and catalytic activity, enable sensitive and selective detection of analytes. Ongoing research continues to explore novel approaches and materials to enhance the performance and broaden the applications of metallic nanoparticles in nanosensors.

4.2. Semiconductors

Semiconductor nanoparticles have emerged as crucial building blocks in the development of nanosensors due to their unique electronic and optical properties. These nanoparticles, often referred to as quantum dots (QDs), nanowires, or nanocrystals, offer advantages such as size-dependent tunability, high photostability, and strong fluorescence, making them highly attractive for various sensing applications. Semiconductor nanoparticles find applications in fields like environmental monitoring, biomedical diagnostics, and optoelectronics.

Quantum dots are semiconductor nanoparticles with unique electronic properties that arise from quantum confinement effects. These nanoparticles exhibit size-dependent emission spectra, where the emission wavelength can be tuned by controlling their size or composition. This tunability allows for the precise design of nanosensors with desired optical properties and sensitivity.

In a study published in the journal "Nature Biotechnology," Michalet et al. (2005) demonstrated the use of quantum dots as highly sensitive and photostable fluorescent probes for biological imaging. The researchers functionalized quantum dots with antibodies specific to cellular targets and successfully visualized individual biomolecules in live cells with high sensitivity and spatial resolution. Kim et al. (2017) reported in the journal "Sensors" the development of a quantum dot-based nanosensor for the detection of glucose. The researchers synthesized glucose-sensitive quantum dots by conjugating glucose oxidase enzymes onto the surface of the quantum dots. In the presence of glucose, the enzymatic reaction produced hydrogen peroxide, which quenched the fluorescence of the quantum dots, enabling glucose detection with high sensitivity.

Nanowires are elongated semiconductor structures with diameters in the nanometer range and lengths that can extend up to micrometers. They exhibit unique properties such as high aspect ratios, large surface areas, and excellent charge transport characteristics. These properties make nanowires promising for nanosensors, enabling sensitive and selective detection of analytes. Patolsky et al. (2004) published a study in the journal "Science" describing the use of silicon nanowires for ultrasensitive detection of biological molecules. The researchers functionalized the nanowires with receptor molecules specific to target analytes. The binding of the target molecules to the nanowires induced a change in electrical conductance, enabling the detection of proteins and DNA with high sensitivity. Cui et al. (2010) reported in the journal "Nature Nanotechnology" the development of a nanowire-based nanosensor for the detection of single virus particles. The researchers functionalized silicon nanowires with specific antibodies and demonstrated the detection of individual influenza virus particles with high sensitivity and specificity.

Nanocrystals, also known as quantum nanocrystals or colloidal nanocrystals, are small semiconductor particles with dimensions ranging from a few to tens of nanometers. These nanocrystals exhibit size-dependent optical and electrical properties, making them suitable for various sensing applications. Liu et al. (2018) published a study in the journal "Analytical Chemistry" describing the development of a nanocrystal-based nanosensor for the detection of heavy metal ions. The researchers functionalized cadmium-based nanocrystals with specific ligands that selectively bound to heavy metal ions. The presence of heavy metal ions induced a change in the fluorescence emission of the nanocrystals, enabling their sensitive detection. Zhang et al. (2012) reported in the journal "ACS Nano" the development of a nanocrystal-based nanosensor for the detection of explosives. The researchers functionalized lead sulfide nanocrystals with specific receptors that selectively recognized explosive molecules. The binding of the explosives induced a change in the electrical conductance of the nanocrystals, allowing their sensitive and selective detection.

These examples highlight the significant role of semiconducting nanoparticles, including quantum dots, nanowires, and nanocrystals, in nanosensor development. The unique electronic and optical properties of these nanoparticles enable sensitive, selective, and tunable detection of various analytes, opening up exciting possibilities for advancements in sensing technologies. Ongoing research continues to explore new materials, synthesis techniques, and functionalization strategies to enhance the performance and expand the applications of semiconducting nanoparticles in nanosensors.

4.3. Polymers

Polymer nanoparticles have gained significant attention in the field of nanosensors due to their versatility, tunability, and biocompatibility. These nanoparticles offer advantages such as easy synthesis, tailorable properties, and functionalization capabilities, making them ideal candidates for various sensing applications. Polymer nanoparticles find applications in fields like biomedical diagnostics, environmental monitoring, and drug delivery systems.

Polymer nanoparticles can be functionalized with specific receptors or probes to enable the detection of target analytes. The functionalization can be achieved through the attachment of ligands, antibodies, enzymes, or DNA/RNA aptamers onto the nanoparticle surface. These functionalized polymer nanoparticles can then selectively bind to the target analytes, leading to detectable changes in the nanoparticle's optical, electrical, or mechanical properties. Chen et al. (2013) published a study in the journal "ACS Nano" describing the development of a polymer nanoparticle-based nanosensor for the detection of cancer cells. The researchers functionalized polymeric nanoparticles with antibodies specific to cancer biomarkers and demonstrated their selective binding to cancer cells. The binding event induced changes in the fluorescence properties of the nanoparticles, enabling sensitive and specific detection of cancer cells. Huynh et al. (2015) reported in the journal "Chemical Communications" the development of a polymer nanoparticle-based nanosensor for the detection of heavy metal ions. The researchers functionalized polymer nanoparticles with chelating ligands that selectively bound to specific metal ions. The binding event induced changes in the electrical conductance of the nanoparticles, allowing for the sensitive and selective detection of heavy metal ions.

Polymer nanoparticles have also been used in environmental monitoring applications, enabling the detection of pollutants, toxins, and contaminants in water, air, and soil. Xu et al. (2017) published a study in the journal "Analytical Chemistry" describing the development of a polymer nanoparticle-based nanosensor for the detection of mercury ions in water. The researchers functionalized polymer nanoparticles with thiol ligands, which selectively bound to mercury ions. The binding event induced changes in the fluorescence properties of the nanoparticles, enabling sensitive and selective detection of mercury ions in water samples. Wang et al. (2016) reported in the journal "ACS Applied Materials & Interfaces" the development of a polymer nanoparticle-based nanosensor for the detection of volatile organic compounds (VOCs) in air. The researchers functionalized polymer nanoparticles with receptors that selectively interacted with specific VOCs. The interaction induced changes in the electrical conductance of the nanoparticles, allowing for the sensitive detection of VOCs in air samples.

Polymer nanoparticles have shown promise in biomedical diagnostics, enabling sensitive and specific detection of disease biomarkers. Choi et al. (2018) published a study in the journal "Biosensors and Bioelectronics" describing the development of a polymer nanoparticle-based nanosensor for the detection of cardiac troponin I, a biomarker for myocardial infarction. The researchers functionalized polymer nanoparticles with antibodies specific to cardiac troponin I and demonstrated their selective binding to the biomarker. The binding event induced changes in the electrical conductance of the nanoparticles, enabling the sensitive and specific detection of cardiac troponin I in serum samples. Zhao et al. (2017) reported in the journal

"Nanomedicine" the development of a polymer nanoparticle-based nanosensor for the detection of cancer-associated microRNAs. The researchers functionalized polymer nanoparticles with DNA probes complementary to the target microRNAs. The binding event induced changes in the fluorescence properties of the nanoparticles, enabling the sensitive and specific detection of cancer-associated microRNAs in biological samples.

These examples demonstrate the significant role of polymer nanoparticles in nanosensor development. The tunable properties, ease of functionalization, and biocompatibility of polymer nanoparticles make them versatile platforms for the sensitive and specific detection of various analytes in biomedical, environmental, and other sensing applications. Ongoing research continues to explore new polymer materials, functionalization strategies, and detection mechanisms to advance the performance and broaden the applications of polymer nanoparticle-based nanosensors.

4.4. Carbon Nanotubes

Carbon nanotubes (CNTs) have garnered significant attention in the field of nanosensors due to their exceptional electrical, mechanical, and chemical properties. CNTs are cylindrical structures made of rolled-up graphene sheets, and they exhibit unique one-dimensional nanostructures. These properties make CNTs attractive for various sensing applications, including gas sensing, biosensing, and environmental monitoring.

Carbon nanotubes can be utilized in nanosensors as sensing elements or as platforms for functionalization with specific receptors, enzymes, or probes. Their high surface-to-volume ratio, excellent electrical conductivity, and sensitivity to changes in the surrounding environment make them ideal for the detection of various analytes. Kong et al. (2000) published a study in the journal "Science" describing the use of single-walled carbon nanotubes (SWCNTs) as ultrasensitive gas sensors. The researchers demonstrated that SWCNTs can detect trace amounts of gases, such as ammonia and nitrogen dioxide, through changes in their electrical conductance. The adsorption of gas molecules onto the surface of the nanotubes alters their conductivity, allowing for sensitive gas detection. Liang et al. (2013) reported in the journal "Nature Communications" the development of a carbon nanotube-based nanosensor for the detection of small molecules. The researchers functionalized carbon nanotubes with DNA aptamers specific to adenosine, a small molecule involved in various biological processes. The binding of adenosine to the aptamers induced changes in the electrical properties of the nanotubes, enabling the sensitive and selective detection of adenosine.

Carbon nanotubes have been extensively explored in biosensing applications due to their biocompatibility, high sensitivity, and compatibility with biological systems. They can be functionalized with biomolecules or enzymes to enable the detection of specific targets, such as proteins, DNA, or cells. Cui et al. (2019) published a study in the journal "Advanced Materials" describing the development of a carbon nanotube-based nanosensor for the detection of cancer biomarkers. The researchers functionalized carbon nanotubes with antibodies specific to cancer antigens and demonstrated their selective binding to cancer cells. The binding event induced changes in the electrical conductance of the nanotubes, enabling the sensitive detection of cancer cells. Zheng et al. (2016) reported in the journal "Nano Letters" the development of a carbon nanotube-based nanosensor for DNA detection. The researchers functionalized carbon nanotubes with DNA probes complementary to the target DNA sequences. The hybridization of the target DNA with the nanotube-bound probes induced changes in the electrical properties of the nanotubes, allowing for the sensitive and specific detection of DNA.

Carbon nanotubes have been employed in environmental monitoring applications for the detection of various pollutants, heavy metals, and toxins. Banerjee et al. (2019) published a study in the journal "ACS Applied Materials & Interfaces" describing the development of a carbon nanotube-based nanosensor for the detection of heavy metals in water. The researchers functionalized carbon nanotubes with specific ligands that selectively bound to heavy metal ions. The binding event induced changes in the electrical conductance of the nanotubes, enabling the sensitive detection of heavy metals in water samples. Yang et al. (2014) reported in the journal "Environmental Science & Technology" the development of a carbon nanotube-based nanosensor for the detection of volatile organic compounds (VOCs) in air. The researchers functionalized carbon nanotubes with polymers that selectively interacted with specific VOCs. The adsorption of VOCs onto the nanotubes induced changes in their electrical properties, allowing for the sensitive detection of VOCs in air samples.

These examples demonstrate the significant role of carbon nanotubes in nanosensor development. Their exceptional properties, such as high electrical conductivity, sensitivity to changes in the environment, and compatibility with biological systems, enable the sensitive and selective detection of various analytes in diverse applications. Ongoing research continues to explore new functionalization strategies, hybrid nanomaterial systems, and detection mechanisms to advance the performance and broaden the applications of carbon nanotube-based nanosensors.

4.5. Biomolecules

Biomolecules play a vital role in nanosensor development by providing specificity and recognition capabilities for target analytes. These molecules, such as proteins, enzymes, antibodies, DNA/RNA, and aptamers, possess unique binding properties that can be leveraged for the sensitive and selective detection of various targets. Biomolecule-based nanosensors have applications in areas like biomedical diagnostics, environmental monitoring, and food safety.

Proteins are versatile biomolecules that can be used in nanosensors for specific target recognition. Antibodies, in particular, are widely employed due to their high affinity and specificity towards antigens. Sharma et al. (2020) published a study in the journal "ACS Sensors" describing the development of an antibody-based nanosensor for the detection of cancer biomarkers. The researchers immobilized antibodies onto the surface of nanoparticles and demonstrated the selective binding of cancer biomarkers. This binding event induced changes in the optical or electrical properties of the nanoparticles, enabling the sensitive detection of cancer biomarkers. Yguerabide et al. (2000) reported in the journal "Analytical Chemistry" the use of antibody-coated quantum dots as fluorescent nanosensors for the detection of pathogens. The researchers functionalized quantum dots with antibodies specific to the target pathogens, allowing for their selective capture and fluorescence-based detection.

Enzymes are highly efficient catalysts that can be utilized in nanosensors for the detection of specific substrates or analytes. Enzymatic reactions can generate measurable signals, such as changes in fluorescence, color, or electrical properties, enabling sensitive detection. Roda et al. (2016) published a study in the journal "Analytical Chemistry" describing the development of an enzyme-based nanosensor for glucose detection. The researchers incorporated glucose oxidase enzymes into nanomaterials, such as nanoparticles or nanofibers, and demonstrated their ability to selectively react with glucose. The enzymatic reaction produced measurable changes in the optical or electrical properties of the nanomaterials, enabling the sensitive detection of glucose. Chen et al. (2018) reported in the journal "ACS Sensors" the development of an enzyme-functionalized nanomaterial-based nanosensor for the detection of organophosphates, which are toxic compounds. The researchers immobilized organophosphatase enzymes onto the surface of nanoparticles and demonstrated their selective catalytic activity towards the target analytes. The enzymatic reaction induced changes in the electrical conductance of the nanoparticles, allowing for the sensitive detection of organophosphates.

DNA/RNA and aptamers are nucleic acid-based biomolecules that can be used as recognition elements in nanosensors. They can be engineered to specifically bind to target molecules, enabling highly selective detection. Liu et al. (2017) published a study in the journal "Nano Letters" describing the development of a DNA aptamer-based nanosensor for the detection of small molecules. The researchers functionalized nanoparticles with DNA aptamers that selectively bound to the target analytes. The binding event induced changes in the optical properties of the nanoparticles, enabling the sensitive and selective detection of the small molecules. Stoltenburg et al. (2017) reported in the journal "Sensors" the development of an aptamer-based nanosensor for the detection of toxins. The researchers designed aptamers with high affinity towards the target toxins and incorporated them into various nanomaterials, such as graphene oxide or gold nanoparticles. The binding of toxins to the aptamers induced changes in the electrical or optical properties of the nanomaterials, enabling the sensitive detection of toxins.

Biomolecule-based nanosensors offer high sensitivity, specificity, and biocompatibility, making them valuable tools for various sensing applications. The integration of biomolecules with nanomaterials allows for the development of highly sensitive and selective nanosensors that can be tailored for different targets and applications. Ongoing research continues to explore novel biomolecules, innovative sensing platforms, and integration strategies to enhance the performance and expand the applications of biomolecule-based nanosensors.

4.6. Oxides

Oxide materials have a significant role in nanosensor development due to their diverse properties, such as high surface area, tunable conductivity, and surface reactivity. Oxide nanomaterials, including metal oxides and metal oxide composites, offer advantages for sensing applications, including gas sensing, biosensing, and environmental monitoring.

Oxide nanomaterials are widely utilized in nanosensors as active sensing elements or as platforms for functionalization with specific receptors or probes. Their unique properties enable sensitive and selective detection of various analytes by monitoring changes in electrical, optical, or electrochemical signals. Zhang et al. (2017) published a study in the journal "Nano Letters" describing the development of metal oxide-based nanosensors for gas detection. The researchers synthesized tin dioxide (SnO_2) nanoparticles and demonstrated their high sensitivity towards volatile organic compounds (VOCs). The adsorption of VOC molecules onto the surface of SnO_2 nanoparticles induced changes in their electrical conductivity, enabling the sensitive detection of VOCs. Liu et al. (2015) reported in the journal "Chemical Reviews" the use of metal oxide nanoparticles for biosensing applications. The researchers developed metal oxide nanomaterials, such as titanium dioxide (TiO_2) and zinc oxide (ZnO), functionalized with biomolecules for the detection of specific targets. The binding of target biomolecules to the nanomaterials induced changes in their optical or electrical properties, allowing for sensitive and selective biosensing.

Oxide nanomaterials have been extensively employed in environmental monitoring applications to detect pollutants, heavy metals, and toxic gases in air, water, and soil. Lv et al. (2020) published a study in the journal "ACS Applied Materials & Interfaces" describing the development of metal oxide nanosensors for heavy metal detection. The researchers synthesized iron oxide (Fe_3O_4) nanoparticles and functionalized them with chelating ligands for selective binding to heavy metal ions. The binding of metal ions induced changes in the electrical properties of the nanoparticles, enabling the sensitive detection of heavy metals in water samples. Zhang et al. (2018) reported in the journal "Environmental Science & Technology" the use of metal oxide nanomaterials for gas sensing in environmental monitoring. The researchers developed composite nanomaterials, such as indium oxide (In2O3) and graphene oxide, for the detection of toxic gases. The adsorption of gas molecules onto the nanomaterials induced changes in their electrical conductivity, allowing for the sensitive detection of toxic gases in air samples.

Oxide nanomaterials have shown promise in biomedical applications, including disease diagnostics, drug delivery, and bioimaging. Zhang et al. (2019) published a study in the journal "ACS Nano" describing the development of metal oxide nanosensors for cancer diagnostics. The researchers synthesized manganese oxide (MnO_2) nanoparticles and functionalized them with tumor-targeting ligands. The binding of target cancer cells to the nanomaterials induced changes in their magnetic resonance imaging (MRI) contrast, enabling the sensitive detection of cancer cells in vivo. Yang et al. (2016) reported in the journal "Biomaterials" the use of metal oxide nanomaterials for drug delivery systems. The researchers developed mesoporous silica nanoparticles coated with iron oxide (Fe3O4) for controlled drug release. The magnetic properties of the iron oxide allowed for targeted drug delivery and monitoring using magnetic resonance imaging.

Oxide nanomaterials offer a wide range of applications in nanosensor development, from gas sensing and environmental monitoring to biomedical diagnostics and therapeutics. Their unique properties, including high surface area, tunable conductivity, and reactivity, enable sensitive and selective detection of analytes. Ongoing research aims to further enhance the performance of oxide-based nanosensors by exploring new oxide materials, improving functionalization strategies, and developing advanced sensing mechanisms.

5. Synthesis Methods for Nanosensors

The field of nanosensors has emerged as a transformative technology with wide-ranging applications in healthcare, environmental monitoring, and beyond. The development and fabrication of these miniature devices heavily rely on synthesis methods that enable the controlled assembly and integration of nanomaterials with customized properties. The synthesis methods employed in the fabrication of nanosensors, focusing on their underlying principles, advantages, and limitations are been explored. By comprehending these techniques, researchers and engineers can unlock the full potential of nanosensors, pushing the boundaries of sensing technology and driving innovation.

5.1. Chemical Synthesis

Chemical synthesis plays a crucial role in the preparation of nanomaterials used in nanosensors, enabling the controlled fabrication of materials with tailored properties and dimensions. This section provides comprehensive background data and context regarding the chemical synthesis methods commonly employed in the synthesis of nanomaterials for nanosensors, supported by scientific reports.

One of the widely used chemical synthesis techniques is the sol-gel method, which involves the hydrolysis and condensation of precursor molecules to form a colloidal suspension, followed by a gelation process. This method allows for the synthesis of various nanomaterials, such as metal oxides (e.g., silica, titanium dioxide), metal nanoparticles (e.g., gold, silver), and semiconductor nanoparticles (e.g., quantum dots). For example, Li et al. (2019) reported the synthesis of zinc oxide nanowires using a sol-gel approach, demonstrating their potential as sensitive gas sensors.

Another commonly employed method is chemical vapor deposition (CVD), which involves the decomposition of precursor molecules in the gas phase to deposit thin films or nanomaterials onto a substrate. CVD enables the synthesis of nanomaterials with controlled composition, structure, and morphology. For instance, Liu et al. (2020) utilized CVD to grow graphene on a metal substrate, resulting in a highly sensitive and selective nanosensor for detecting toxic gases as shown in Figure 84 (Saeed et al. 2020).

In addition, hydrothermal synthesis is extensively employed for the preparation of nanomaterials, particularly metal oxides and metal sulfides as shown in Figure 85 (Girija et al. 2019). This method involves the reaction of precursor solutions under high-temperature and high-pressure conditions. Liu et al. (2018) demonstrated the hydrothermal synthesis of manganese oxide nanowires, which exhibited excellent electrochemical sensing properties for detecting neurotransmitters.

Furthermore, the precipitation method is widely used for the synthesis of nanoparticles by rapidly mixing precursor solutions to induce nucleation and growth. This approach allows for the control of particle size and morphology. For example, Zhang et al. (2021) employed the precipitation method to synthesize silver nanoparticles, which were subsequently incorporated into a nanosensor for the rapid detection of foodborne pathogens.

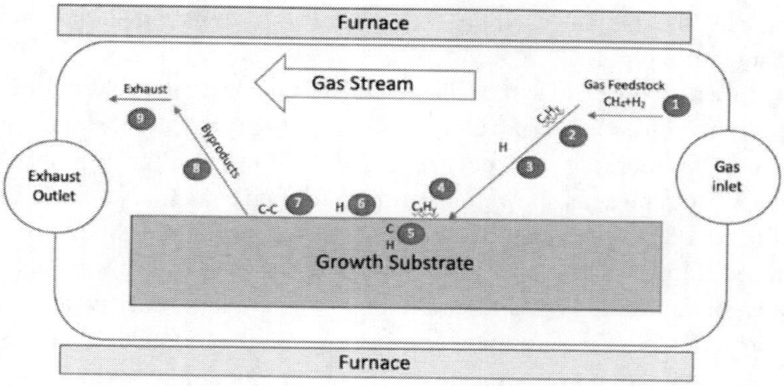

Figure 84. Schematic of Graphene synthesis using CVD process (Saeed et al. 2020).

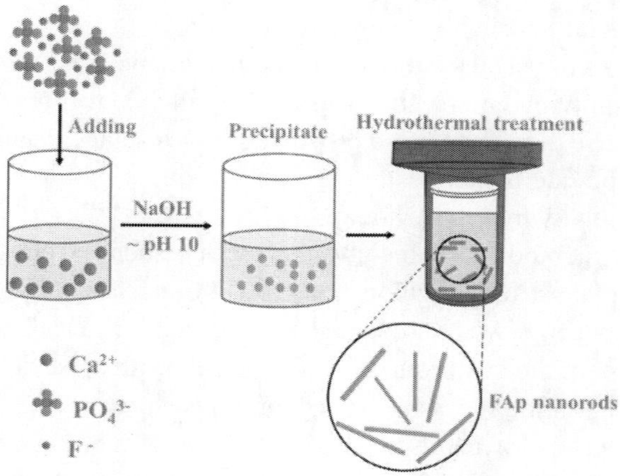

Figure 85. Schematic of hydrothermal synthesis of FAp nanoparticles (Girija et al. 2019).

Template-assisted synthesis is another notable technique, where a template with specific dimensions and morphology guides the growth of nanomaterials. This method enables the fabrication of nanostructures with precise geometries and high aspect ratios. For instance, Zhang et al. (2017) utilized a template-assisted synthesis approach to prepare one-dimensional nanowire arrays of tin oxide, demonstrating their potential as efficient gas sensors.

These are just a few examples of the diverse chemical synthesis methods employed in the preparation of nanomaterials for nanosensors. The choice of synthesis method depends on the desired nanomaterial properties, scalability, and target applications. By utilizing these techniques, researchers have achieved significant advancements in the development of nanosensors with enhanced sensitivity, selectivity, and response characteristics.

5.2. Physical Synthesis

Physical synthesis methods are commonly employed in the preparation of nanomaterials for nanosensors, offering precise control over the size, shape, and composition of the synthesized materials. This section provides comprehensive background data and context regarding the physical synthesis techniques used in the fabrication of nanomaterials for nanosensors, supported by scientific reports.

One of the primary physical synthesis methods is physical vapor deposition (PVD), which involves the deposition of material atoms or molecules onto a substrate through physical processes such as evaporation or sputtering. PVD allows for the synthesis of thin films, coatings, and nanostructures as shown in Figure 86 (Safavi et al. 2021). For example, Huang et al. (2018) utilized PVD techniques to fabricate a nanostructured silver film as a sensitive surface-enhanced Raman scattering (SERS) substrate for chemical sensing applications.

Another widely used method is laser ablation, where a high-energy laser beam is focused on a target material, leading to its vaporization and subsequent condensation to form nanoparticles. Laser ablation synthesis allows for the production of various nanoparticles, such as metal nanoparticles and semiconductor quantum dots. Gondal et al. (2016) reported the synthesis of silver nanoparticles using laser ablation, which were employed as a sensing material for detecting volatile organic compounds (VOCs).

Electrodeposition is a physical synthesis technique that involves the deposition of materials onto a substrate by applying an electric current. It allows for the controlled growth of nanomaterials with precise dimensions and properties. For instance, Wang et al. (2019) utilized electrodeposition to fabricate nickel oxide nanowires, which exhibited excellent gas sensing performance for detecting hydrogen sulfide gas.

Physical synthesis methods also include techniques such as physical template-assisted synthesis, where a template structure guides the growth of nanomaterials, and physical milling or grinding, which can be employed to reduce the size of bulk materials to the nanoscale. These methods offer versatility in the synthesis of nanomaterials with specific structures and sizes.

These examples highlight the wide range of physical synthesis methods used in the preparation of nanomaterials for nanosensors. The choice of method depends on factors such as desired material properties, scalability, and target applications. Through the utilization of physical synthesis techniques, researchers have achieved significant advancements in the development of nanosensors with enhanced sensing capabilities and performance.

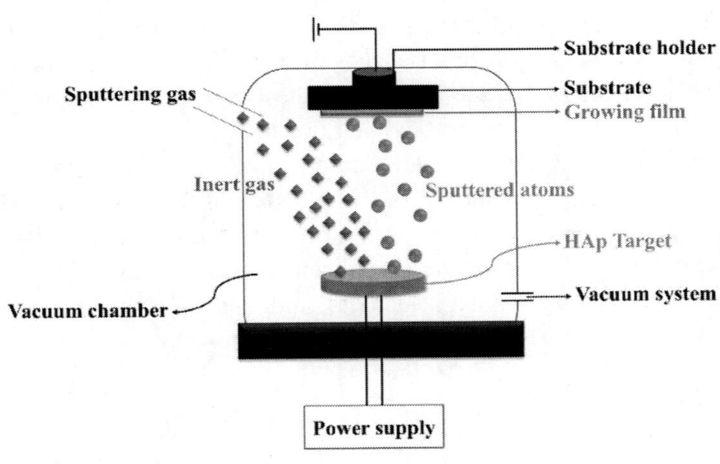

Figure 86. Illustration PVD process (Safavi et al. 2021).

5.3. Biological Synthesis

Biological synthesis methods have gained significant attention in the preparation of nanomaterials for nanosensors, offering environmentally friendly and sustainable approaches. This section provides comprehensive background data and context regarding the biological synthesis techniques employed in the fabrication of nanomaterials for nanosensors, supported by scientific reports.

One of the notable biological synthesis methods is microbial-mediated synthesis, where microorganisms such as bacteria, fungi, or algae are employed to produce nanomaterials. These microorganisms act as natural bioreactors, facilitating the reduction or oxidation of metal ions to form nanoparticles. For example, Riddin et al. (2016) reported the microbial synthesis of silver nanoparticles using bacterial strains, demonstrating their potential as sensing materials for detecting heavy metals.

Plant-mediated synthesis is another prominent approach, where plant extracts or plant-derived materials are utilized to reduce metal ions and form nanoparticles. Plant extracts contain various biomolecules such as flavonoids, terpenoids, and phenolic compounds that act as reducing and stabilizing agents. Yadav et al. (2021) demonstrated the synthesis of gold nanoparticles using Aloe vera leaf extract, which exhibited excellent sensing properties for the detection of environmental pollutants. Figure 87 illustrates the biological synthesis of Ag nanoparticles from plant extracts (Khan et al. 2022).

Enzymatic synthesis is a specific biological approach that involves the use of enzymes to catalyze the synthesis and assembly of nanomaterials. Enzymes can control the shape, size, and composition of the synthesized nanomaterials with high precision. For instance, Luo et al. (2017) utilized horseradish peroxidase enzyme to synthesize polyaniline nanofibers, which were integrated into a nanosensor for the detection of hydrogen peroxide.

Biological synthesis methods offer several advantages, including mild reaction conditions, scalability, and eco-friendly characteristics. These methods allow for the synthesis of diverse nanomaterials, including metal nanoparticles, metal oxides, and carbon-based nanomaterials, with controlled properties and functionalities.

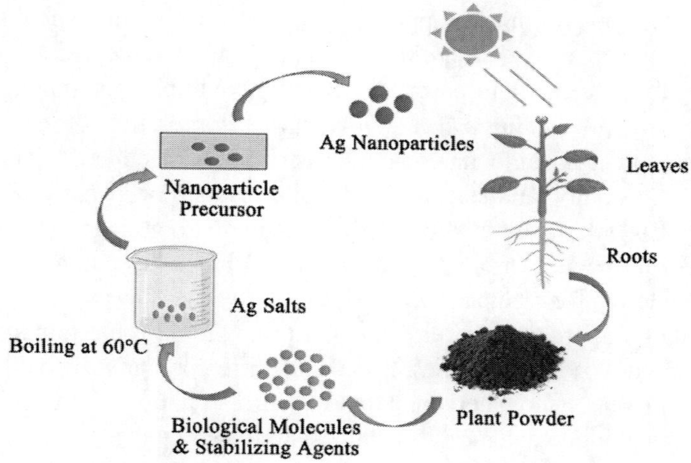

Figure 87. Biological synthesis of Ag nanoparticles from plant extracts (Khan et al. 2022).

Furthermore, the use of biological templates, such as viruses and proteins, provides a template-assisted synthesis approach for the fabrication of nanomaterials with specific structures and properties. Viral templates, in particular, have been utilized to produce nanowires, nanorods, and nanotubes. For example, Lee et al. (2019) employed a tobacco mosaic virus template to synthesize gold nanowires, which were integrated into a nanosensor for the detection of biological molecules.

These examples illustrate the diverse applications of biological synthesis methods in the preparation of nanomaterials for nanosensors. By harnessing the potential of biological systems, researchers have achieved significant advancements in the development of sustainable and bio-inspired nanosensors with unique properties and enhanced sensing capabilities.

5.4. Hybrid Synthesis

Hybrid synthesis methods, combining multiple approaches, have emerged as versatile strategies in the preparation of nanomaterials for nanosensors, offering enhanced control over material properties and functionalities. This section provides comprehensive background data and context regarding hybrid synthesis techniques employed in the fabrication of nanomaterials for nanosensors, supported by scientific reports.

One prominent hybrid synthesis approach is the combination of chemical and physical methods. This involves the utilization of chemical synthesis techniques to prepare nanomaterial precursors, followed by physical methods for their subsequent manipulation or assembly. For instance, Li et al. (2018) reported a hybrid synthesis approach where they initially prepared gold nanoparticles via chemical reduction and then employed physical methods such as centrifugation and self-assembly to fabricate a highly sensitive plasmonic nanosensor for detecting biomolecules.

Another hybrid synthesis method involves the integration of biological components with chemically or physically synthesized nanomaterials. This approach combines the advantages of biological systems, such as their ability to control the size and shape of nanomaterials, with the versatility and scalability of chemical or physical synthesis methods. For example, Liu et al. (2020) utilized a hybrid synthesis approach by combining the biological synthesis of silver nanoparticles using a plant extract with subsequent physical deposition onto a substrate, resulting in a highly efficient nanosensor for detecting environmental pollutants.

Additionally, hybrid synthesis can involve the incorporation of functional molecules, polymers, or biomolecules onto nanomaterials surfaces. This integration can enhance the sensing properties, stability, and selectivity of nanosensors. For instance, Zhang et al. (2019) employed a hybrid synthesis method by functionalizing carbon nanotubes with biomolecules such as enzymes, antibodies, or DNA strands, leading to the development of highly specific and sensitive biosensors for medical diagnostics.

Hybrid synthesis approaches offer a flexible and tailored route to engineer nanosensors with desired properties, combining the strengths of different synthesis methods. They enable precise control over nanomaterial composition, structure, surface properties, and functionalization, leading to enhanced sensing capabilities.

These examples illustrate the diverse applications of hybrid synthesis methods in the preparation of nanomaterials for nanosensors. By combining different synthesis approaches, researchers have achieved significant advancements in the development of highly sensitive, selective, and functional nanosensors for various applications, ranging from healthcare to environmental monitoring.

5.5. Templates Synthesis

Template synthesis is a versatile approach used in the preparation of nanomaterials for nanosensors, allowing for precise control over the size, shape, and structure of the synthesized materials as shown in Figure 88 (Wade and Wegrowe 2005). One of the primary template synthesis methods is the use of porous templates, where a template with well-defined pores or cavities guides the growth of nanomaterials. The template can be a solid material, such as anodized aluminum oxide (AAO) or track-etched membranes, or a soft template, such as micelles or emulsion droplets. For example, Wang et al. (2017) utilized AAO templates to synthesize one-dimensional nanostructures of metal oxides, demonstrating their potential as high-performance gas sensors.

In addition to porous templates, biological templates such as viruses, proteins, or DNA can be employed in template synthesis. These biological templates provide a highly controlled and programmable platform for the synthesis of nanomaterials with specific shapes and sizes. For instance, Lee et al. (2020) utilized tobacco mosaic virus (TMV) as a template to synthesize metallic nanowires, which were integrated into a nanosensor for the detection of heavy metal ions.

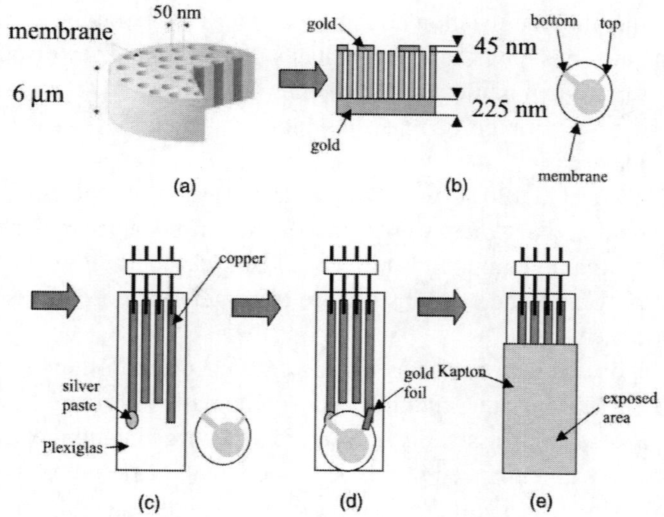

Figure 88. Illustration of Template synthesis of nanomaterials (a) membrane, (b) sputtered gold contacts, (c) sample holder, (d) membrane on holder with contacts, and (e) Kapton isolation (Wade and Wegrowe 2005).

Furthermore, sacrificial templates can be utilized, where a template material is selectively removed after the synthesis process, leaving behind the desired nanomaterials. This approach allows for the fabrication of hollow or porous nanomaterials. For example, Zhang et al. (2018)

employed sacrificial template synthesis to prepare hollow mesoporous silica nanoparticles, which exhibited excellent drug delivery and sensing capabilities.

Template synthesis methods offer the advantage of precise control over nanomaterial properties, including size, shape, composition, and surface structure. They allow for the fabrication of nanomaterials with tailored functionalities and enhanced sensing performance. These examples highlight the diverse applications of template synthesis methods in the preparation of nanomaterials for nanosensors. By utilizing various types of templates, researchers have achieved significant advancements in the development of nanosensors with unique architectures and improved sensing capabilities for applications ranging from environmental monitoring to biomedical diagnostics.

5.6. Hydrothermal Synthesis

Hydrothermal synthesis is a widely used method for the preparation of nanomaterials in the field of nanosensors. This approach involves the use of a high-temperature and high-pressure aqueous environment to promote the nucleation, growth, and crystallization of nanomaterials. This section provides comprehensive background data and context regarding hydrothermal synthesis techniques employed in the fabrication of nanomaterials for nanosensors, supported by scientific reports.

Hydrothermal synthesis offers several advantages for the preparation of nanomaterials. Firstly, it provides control over the size, morphology, and crystallinity of the synthesized nanomaterials by adjusting the reaction conditions such as temperature, pressure, and reaction time. Secondly, the aqueous environment facilitates the homogeneous dispersion of precursor materials, leading to uniform nanoparticle growth. Thirdly, the hydrothermal method allows for the synthesis of a wide range of materials including metal oxides, semiconductors, and carbon-based nanomaterials.

For example, Wang et al. (2021) employed a hydrothermal synthesis approach to fabricate zinc oxide (ZnO) nanorods, which were subsequently used as a sensing material for the detection of volatile organic compounds (VOCs). The hydrothermal method provided precise control over the aspect ratio and crystal structure of the ZnO nanorods, resulting in enhanced gas sensing performance.

Furthermore, hydrothermal synthesis can be combined with other techniques to achieve additional functionalities. One such example is the combination of hydrothermal synthesis with a self-assembly process. Self-assembly allows for the organization of nanomaterials into hierarchical structures with enhanced properties. Chen et al. (2019) reported the hydrothermal synthesis of iron oxide nanoparticles, followed by a self-assembly process to fabricate hierarchical flower-like structures. These hierarchical structures exhibited improved electrochemical sensing performance for the detection of heavy metal ions.

Hydrothermal synthesis has been widely applied in the preparation of nanomaterials for various sensing applications, including gas sensors, biosensors, and environmental sensors. It offers a scalable and cost-effective method for the synthesis of nanomaterials with tailored properties and high sensitivity.

These examples highlight the versatile applications of hydrothermal synthesis in the preparation of nanomaterials for nanosensors. Through the utilization of hydrothermal techniques, researchers have achieved significant advancements in the development of highly sensitive and selective nanosensors for a range of applications.

Conclusion

This chapter provides a comprehensive overview of nanosensors, their significance in various fields, the materials used to fabricate nanosensors, and the synthesis methods employed for their preparation. Nanosensors, with their nanoscale dimensions and unique properties, have revolutionized sensing technology, enabling highly sensitive and selective detection of target analytes.

The chapter begins by discussing the significance of nanosensors in diverse applications, including environmental monitoring, healthcare diagnostics, and industrial processes. Nanosensors offer advantages such as high sensitivity, miniaturization, rapid response, and compatibility with different sensing platforms, making them highly desirable for advanced sensing applications.

Next, the chapter delves into the materials utilized in nanosensor fabrication. Various nanomaterials such as metal nanoparticles, metal oxides, carbon-based nanomaterials, and quantum dots are explored for their unique properties and suitability for specific sensing applications. The selection of materials depends on factors such as target analyte, desired sensitivity, selectivity, and stability of the nanosensor.

The chapter further provides an in-depth analysis of the synthesis methods used to prepare nanomaterials for nanosensors. It covers three major categories: chemical synthesis, physical synthesis, and biological synthesis. Chemical synthesis methods include sol-gel, precipitation, and vapor deposition techniques, offering precise control over nanomaterial properties. Physical synthesis methods encompass techniques such as physical vapor deposition, laser ablation, and template-assisted synthesis, enabling the fabrication of nanomaterials with controlled size, shape, and structure. Biological synthesis methods exploit natural systems such as microorganisms, plants, enzymes, and biological templates to produce nanomaterials, offering environmentally friendly and sustainable approaches.

Each synthesis method is discussed in detail, highlighting their advantages, limitations, and specific examples of nanosensors fabricated using these techniques. The chapter emphasizes the importance of tailoring the synthesis method based on the desired nanosensor properties and sensing requirements.

In conclusion, this chapter provides a comprehensive understanding of nanosensors, their significance, the materials used for their fabrication, and the various synthesis methods employed. It serves as a valuable resource for researchers, engineers, and scientists working in the field of nanosensors and sensing technology, facilitating the development of advanced sensing platforms for a wide range of applications.

References

Ansari, Mohd Zahid, and Chongdu Cho. 2008. "Design and Analysis of a High Sensitive Microcantilever Biosensor for Biomedical Applications." *BioMedical Engineering and Informatics: New Development and the Future - Proceedings of the 1st International Conference on BioMedical Engineering and Informatics, BMEI 2008* 2: 593–97. https://doi.org/10.1109/BMEI.2008.308.

Chen, Yantao, Ren Ren, Haihui Pu, Xiaoru Guo, Jingbo Chang, Guihua Zhou, Shun Mao, Michael Kron, and Junhong Chen. 2017. "Field-Effect Transistor Biosensor for Rapid Detection of Ebola Antigen." *Scientific Reports 2017 7:1* 7 (1): 1–8. https://doi.org/10.1038/s41598-017-11387-7.

Girija, E. K., S. Karthi, D. Karthickraja, G. A. Kumar, D. K. Sardar, and C. Santhosh. 2019. "Synthesis of NIR Emitting Rare Earth Doped Fluorapatite Nanoparticles for Bioimaging Applications." *Current Physical Chemistry* 9 (2): 80–93. https://doi.org/10.2174/1877946809666190708131511.

Khan, Naeem, Shahid Ali, Sadia Latif, Asif Mehmood, Naeem Khan, Shahid Ali, Sadia Latif, and Asif Mehmood. 2022. "Biological Synthesis of Nanoparticles and Their Applications in Sustainable Agriculture Production." *Natural Science* 14 (6): 226–34. https://doi.org/10.4236/NS.2022.146022.

Kim, Dong Wook, Ju Hyun Lee, Jin Kon Kim, and Unyong Jeong. 2020. "Material Aspects of Triboelectric Energy Generation and Sensors." *NPG Asia Materials 2020 12:1* 12 (1): 1–17. https://doi.org/10.1038/s41427-019-0176-0.

Philip, Abin, and A. Ruban Kumar. 2022. "The Performance Enhancement of Surface Plasmon Resonance Optical Sensors Using Nanomaterials: A Review." *Coordination Chemistry Reviews* 458 (May): 214424. https://doi.org/10.1016/J.CCR.2022.214424.

Saeed, Maryam, Yousef Alshammari, Shereen A. Majeed, and Eissa Al-Nasrallah. 2020. "Chemical Vapour Deposition of Graphene—Synthesis, Characterisation, and Applications: A Review." *Molecules 2020, Vol. 25, Page 3856* 25 (17): 3856. https://doi.org/10.3390/MOLECULES25173856.

Safavi, Mir Saman, Frank C. Walsh, Maria A. Surmeneva, Roman A. Surmenev, and Jafar Khalil-Allafi. 2021. "Electrodeposited Hydroxyapatite-Based Biocoatings: Recent Progress and Future Challenges." *Coatings 2021, Vol. 11, Page 110* 11 (1): 110. https://doi.org/10.3390/COATINGS11010110.

Sezgintürk, Mustafa Kemal. 2020. "Introduction to Commercial Biosensors." *Commercial Biosensors and Their Applications: Clinical, Food, and Beyond*, January, 1–28. https://doi.org/10.1016/B978-0-12-818592-6.00001-3.

Slimani, Yassine, and Essia Hannachi. 2020. "Magnetic Nanosensors and Their Potential Applications." *Nanosensors for Smart Cities*, January, 143–55. https://doi.org/10.1016/B978-0-12-819870-4.00009-8.

Tabbakh, Thamer, Noha Alotaibi, Zahrah A. Almusaylim, Sundos Alabdulkarim, N.Z. Jhanjhi, Nawaf Bin Darwish, Thamer Tabbakh, et al. 2021. "Optoelectronics and Optical Bio-Sensors." *Optoelectronics*, February. https://doi.org/10.5772/INTECHOPEN.96183.

Tiburski, Christopher, Ferry Anggoro Ardy Nugroho, and Christoph Langhammer. 2022. "Optical Hydrogen Nanothermometry of Plasmonic Nanoparticles under Illumination." *ACS Nano* 16 (4): 6233–43. https://doi.org/10.1021/ACSNANO.2C00035/ASSET/IMAGES/MEDIUM/NN2C00035_M013.GIF.

Wade, T. L., and J. E. Wegrowe. 2005. "Template Synthesis of Nanomaterials." *EPJ Applied Physics* 29 (1): 3–22. https://doi.org/10.1051/EPJAP:2005001.

"What Are Nano-Sized Sensors?" n.d. Accessed May 29, 2023. https://www.azosensors.com/article.aspx?ArticleID=1331.

Chapter 5

Nanosensor Fabrication Methods

Learning Outcomes

- The reader will have a deep insight into various fabrication techniques used for fabricating a nanosensor.
- The significance of each fabrication technique used for manufacturing different nanosensors based on their unique application can be clearly learned.
- The reader will have detailed knowledge about the working principle, instrumentation and the selection of appropriate fabrication techniques to manufacture the nanosensor with desired parameter or property or behavior.
- The reader will be able to choose the particular fabrication technique appropriately for fabricating various parts of a nanosensor owing to their unique requisites, thus adding more value to their research.

Abstract

As they enable the measurement of physical and chemical processes at the nanoscale, nanosensors are a key instrument in the area of nanotechnology. A variety of methods are employed to acquire the exact control over the size, shape, and composition of the materials utilised in the manufacture of nanosensors. Top-down lithography, which involves patterning a material with light or electrons, is one of the most popular methods for making nanosensors. This procedure may be carried out at the nanoscale, enabling the construction of very intricate structures. Etching is another top-down method that involves removing material to produce a design. Self-assembly and chemical synthesis are examples of bottom-up methods that are vital in the creation of nanosensors. These methods make it possible to build things from the ground up, beginning with atoms and molecules. When making nanoscale electrical devices, this method is very helpful since it can produce exceedingly homogeneous structures. Along with these ways, there are other hybrid strategies that blend top-down and bottom-up strategies to produce the desired results. For instance, self-assembled nanostructures may have their characteristics altered via lithography or etching. Technological advancements in manufacturing have made it possible to create devices with ever-more complicated and sophisticated capabilities, which has fueled the development of nanosensors. Nowadays, nanosensors are employed for a variety of purposes, such as sensing in physics, chemistry, and biology. The field of nanosensors is anticipated to expand and change as manufacturing processes continue to increase, resulting in the creation of new and better nanosensors with distinctive characteristics and capacities. In this chapter, many nanosensor production methods have been thoroughly covered.

Keywords: nanofabrication, lithography, mask, resist, strain engineering, self-assembly, template

1. Introduction to Nanofabrication

There has been a huge increase in interest in creating submicron (1 μm to 100 nm) and nanostructures in recent years (100–1 nm range). This interest is a result of both basic and practical considerations. At a deeper, more basic level, nanostructures provide an intriguing tool for investigating the mechanical, thermal, electrical, magnetic, and optical characteristics of materials at the nm scale. These include significant quantum-mechanical phenomena that result from the confinement of charged carriers in structures like quantum wells, wires, and dots, such as conductance quantization, bandgap alteration, coulomb blockade, etc. On the practical side, nanostructures may significantly enhance the functionality of sensors and electronic/optical systems. Researchers in the device field have mostly focused on creating nm-sized transistors since it is expected to be technically challenging to extend Moore's law beyond 32 nm resolution. Additionally, optical sources and detectors of diameters of nm show enhanced properties not possible in bigger systems (e.g., lower threshold current, improved dynamic behavior, and improved emission line width in quantum dot lasers). New opportunities are opened up by these advancements for next-generation computing and communication equipment. Beyond standard optical lithography, decreasing dimensions in the sensor region may result in significant gains in sensitivity and selectivity.

Nanofabrication is the process used to make nanostructures with very complicated structural designs and excellent functioning, such as nanosensors. Nanofabrication is not mainly a production activity; rather, it is a research activity with an emphasis on developing new materials and processes. The only industry adopting nanoscale manufacturing techniques on a large scale right now is the semiconductor industry, whose device layouts have shrunk to the size of a single nanometer. A top-down fabrication is one such strategy. Like carving from a block of stone, top-down production is a process. The basic bulk material is cut into the necessary shape. The bottom up manufacturing process is the second frequently utilised technique. Building a house from the ground up is similar to the bottom up fabrication, except that nanostructures are built by introducing atoms or molecules one at a time rather than bricks one at a time. The most common top-down manufacturing method is nanolithography. In this process, the material required for the nanostructure is protected by a mask, and the exposed material is removed by being etched. There are several methods. In this process, the material required for the nanostructure is protected by a mask, and the exposed material is removed by being etched. There are many different methods for etching. Depending on the level of resolution required for the nanostructure, the foundation material may be etched mechanically using UV light, x-rays, or electron beams or chemically using acids. Etching may also be done in a huge range of different methods mechanically and chemically. A typical phase in top-down production procedures is etching. The top-down strategy is a spin-off of conventional lithography and micromachining methods. The bottom-up strategy, on the other hand, involves more chemical engineering and material science and is based on fundamentally dissimilar ideas.

Use of self-assembled nanostructures in the case of biomolecules is one of the logical approaches to use the top-down approach to nanosensors. On a broader scale of self-assembly, this process is even relevant to planets and galaxies. Similar to how self-assembly occurs in nature, molecular assembly is one of the synthetic processes used in the bottom-up method. In a controlled, synthetic laboratory setting with chemical assistance, the molecular construction of nanostructures is often equivalent to the self-assembly. These are able to manufacture the necessary nanoscale structure for usage in a variety of applications by reproducing the nanostructures and carefully assembling themselves.

This chapter covers a variety of nanofabrication methods, including scanning probe techniques, strain engineering, self-assembling nanostructures, epitaxial development of nanomaterials, and template production.

2. Nanofabrication Techniques

In order to promote the development of next-generation bio-sensing and bio-actuation technologies, it is becoming more and more crucial to have the ability to design and describe conjugate nano-biological systems that interface solid nanostructures and biological polymers. The fabrication of relevant solid-state components, such as micro- or nano-electrodes, nano-engineered coatings, nanowires, or nanoparticles; immobilisation of biomolecules on the surfaces to create desired bioconjugates; and monitoring nano-biological interfaces are all multidisciplinary studies that fall under this umbrella. The best manufacturing, bio-functionalization, and characterization approaches are often closely connected to one another. It is obvious that the requirements of the solid state components of the system would determine the choice of nanofabrication techniques, which in turn would be greatly influenced by the detection method, which is in turn influenced by the type of biopolymers involved and the goal of monitoring the interface. While some chemical synthesis methods and techniques, such as the thermal, sputter, or electrochemical deposition of roughened metal films, are frequently straightforward and inexpensive, the majority of them face difficulties due to a lack of control over the location of the structures and a lack of reproducibility from sample to sample.

2.1. Electron Beam Lithography Nanofabrication

Electron beam lithography is a specialised approach that is used in the modern electronics industry and integrated circuit research that is based on nanofabrication. This is because these two fields need the production of very minute patterns. This is made possible by the very small spot size of the electron beam, while the resolution of optical lithography is restricted by the wavelength of the light that is used for the exposure process. Diffraction can no longer be used to determine the lithographic resolution since the wavelength of the electron beam has become so very small. the method that was developed in order to produce the very first scanning electron microscopes. An electron beam is used to scan a surface that has been covered with a resist coating that is sensitive to electrons. The desired pattern is produced at the location along the path of the electron beam where energy is deposited.

Electron beam lithography is an intriguing alternative technique that may be used to create nanostructures as shown in Figure 89 (Peters et al. 2015). Utilizing an electron beam, it exposes an electron-sensitive resist such as poly (methyl methacrylate) (PMMA) dissolved in trichlorobenzene (positive) or poly chloromethylstyrene (negative). The e-beam cannon is often a component of a scanning electron microscope; however, transmission electron microscopes (TEM) may also be used (SEM). The possible resolutions are restricted to be more than 10 nm due to electron scattering in the resist, despite the fact that it is easy to manufacture electron wavelengths on the order of 1. Beam control and pattern generation may both be achieved by using a computer interface. Because of this, serial electron beam lithography has a throughput that is not as high as it might be. Although this does not provide much of a challenge when it comes to the construction of devices that are used to investigate fundamental microphysics, it does place significant restrictions on large-scale nanofabrication. Combining electron beam lithography with other processes including as lift-off, etching, and electrodeposition enables the fabrication of a wide variety of nanostructures.

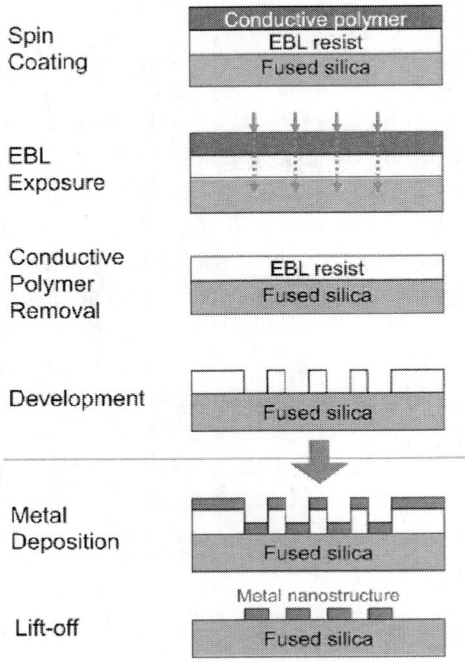

Figure 89. Schematic of electron beam lithography (Peters et al. 2015).

2.2. Nanoimprint Lithography

Because optical lithography is so expensive and has a limited resolution, Chou and his students came up with nanoimprint lithography in 1995 as a low-cost and high-throughput alternative to photolithography and e-beam lithography (EBL) for researchers who require high resolution patterning. This was done as a result of the limitations that optical lithography imposes on its resolution. Nanoimprint lithography, commonly known as NIL, is a form of replication that has proved to offer a resolution that is unmatched by many other technologies, all while allowing for the development of micro and nanostructures in parallel and in a very short amount of time. On the one hand, this makes it possible to enter domains where many similar structures are required for statistical analysis or when enormous regions are covered by nanostructures. On the other hand, this also makes it possible to enter domains where there are enormous regions covered by nanostructures. This was often difficult to do because of the subpar output of the lithographic research equipment. Molding, on the other hand, has been able to achieve resolution levels that are far higher than high-end photolithography, which is used in the industrial production of central processing units (CPUs) and memory chips. Because of this, NIL has the potential to become a viable approach for nanolithography in the near future. In addition to these advantages, moulding also provides a number of additional advantages. For instance, a range of specialised functional materials and polymers may be patterned by mechanically displacing material in order to generate a three-dimensional resist pattern as shown in Figure 90 (Li et al. 2021). This can be done without causing the materials or polymers to lose their chemical capabilities. Additionally, this ability may be used to produce intricate structures if one constructs devices with embedded channels.

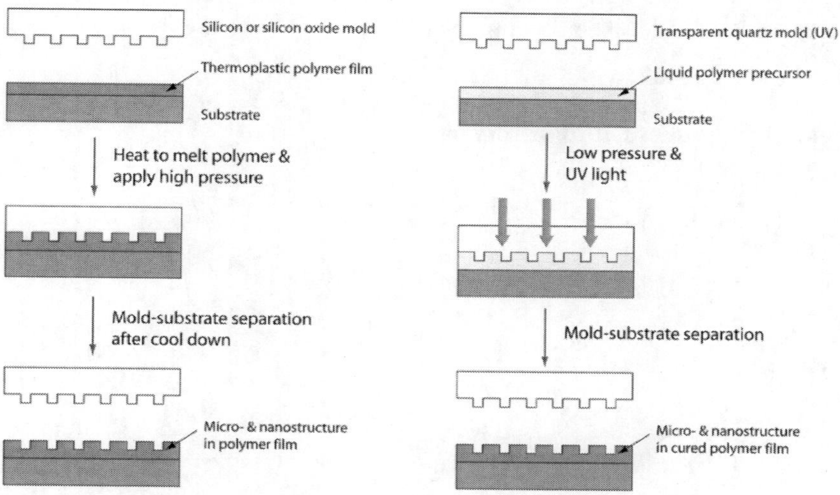

Figure 90. Schematic of nanoimprint lithography (Li et al. 2021).

Imprinting techniques are necessary for moulding operations because they make use of the differences that may be drawn between the mechanical properties of a structured stamp and those of a moulding substance. In order to form the material, a hard stamp is pushed into the viscous moulding fluid. In order to achieve an acceptable process time and yield, this is often done under pressure with moulding materials that have a viscosity that is on the lower end of the scale. The majority of the materials that we work with in hot embossing techniques are thermoplastics, which are materials whose mechanical properties can be frequently and reversibly altered from a solid into a fluid state by changing the temperature. This allows us to create a variety of different designs and patterns. The rheological processes that have been discussed here for thermoplastic materials may be regarded of as equivalent to those that have been described for thermoset or UV-curable materials as long as the thermomechanical features can be tweaked without having an influence on the chemical ones.

For historical reasons, the term "NIL" was first used to refer to the hot embossing lithography (HEL) method. Additionally, the term "NIL" was used as a synonym for "thermal NIL." In spite of this, the term "nanoimprint lithography" (NIL) now has a broader definition that includes many more developed variations of the technique, in addition to the two fundamental categories of nanoimprint lithography, such as the roll imprint process, laser-assisted direct imprint, reverse imprint lithography, substrate conformal imprint lithography, ultrasonic NIL, and so on (UV-NIL and hot embossing lithography). The most significant advantage of NIL is its capacity to pattern three-dimensional and large-area structures at scales ranging from microns to nanometers, as well as its potential to do so at a high throughput and low cost in comparison to other lithography processes and the next generation of lithography with nanoscale resolution, such as extreme ultraviolet lithography and electron beam lithography (EUVL). This opens the door to a wide variety of applications, including those in the domains of data storage, nano-optoelectronics, optical components, NEMS and NOEMS, and many more. In addition, NIL has recently shown extraordinary potential and commercial prospects in a range of application domains. These areas include optical technology, biological technology, flexible display technology, LED technology, and OLED technology. After more than ten years of effort, a variety of industrial suppliers are now delivering a range of standard tools and materials. This has been accomplished.

2.3. Extreme Ultraviolet Lithography

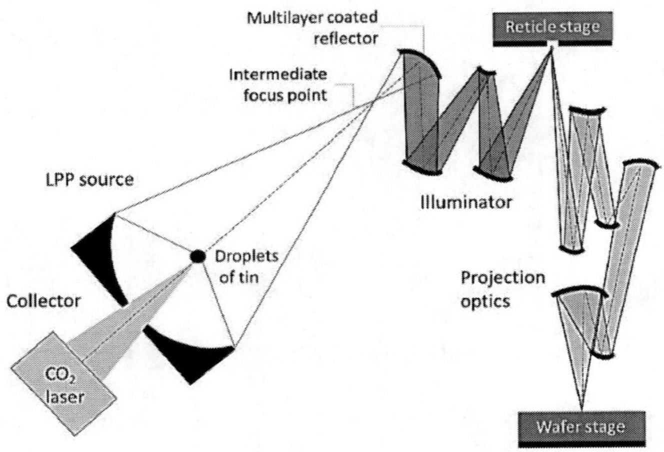

Figure 91. Schematic of Extreme Ultraviolet Lithography (Manouras and Argitis 2020).

The wavelength of 13.5 nm is the one that is used in extreme ultraviolet (EUV) lithography, which is a kind of soft X-ray method as shown in Figure 91 (Manouras and Argitis 2020). The latest generation of EUV scanners may achieve resolutions as low as 22 nanometers at the half-pitch. Plasma is produced using a high-power laser in a particular configuration in order to serve as the EUV light source. This plasma functions as the light source for the EUV. As a direct result of this, light with an extremely restricted spectrum that can be produced in a vacuum is produced. The room is fitted with a number of multilayer mirrors, each of which makes use of interlayer interference in order to reflect light. The usage of EUV was supposed to begin at the 45/40 nm manufacturing node, but there have been multiple delays since the throughput of the power supply is not adequate. It was given a lot of attention since avoiding duplicate patterning at 16/14 nm was essential in order to use 193 nm immersion, which was the reason why it was so important. The very first EUV systems designed to work with 7-nanometer processes have begun shipping. It is conceivable to do single patterning at 7 nm using EUV, but doing so requires a significant amount of time when using the resists that are now available, and it may lead to undesired random or stochastic flaws in the patterns, which may reduce yield. Even with EUV, it will be necessary to do duplicate patterning on the layers that are essential for 5 nm. Because of double patterning, feature pitches may be lowered before processing. This has the potential to cut down on the amount of errors, despite the fact that the procedure takes more time and requires more people to complete.

2.3.1. Working of EUV

EUV sources are constructed using a laser-produced-plasma (LPP) technology, and one of the numerous components that go into the construction of these sources is a carbon dioxide (CO_2) laser. The laser that drives the source may be found in the sub-fab, which is located under the fabrication floor. The laser consists of a power amplifier as well as a seed laser, which produces the pre-pulse and the main pulse. Lasers with a power output of 20 kW are typically used in today's EUV sources. The true EUV source may be found on the production floor of the fab. A vacuum chamber, a droplet generator, and a collector make up the source, which is attached to the EUV scanner. Together, these components form the EUV scanner. The process takes place in a vacuum because so much of the surrounding environment absorbs the extreme ultraviolet light.

The container that creates the droplets is quite small. In the course of its functioning, the droplet generator receives an addition of tin, which is then subjected to heating. After leaving the droplet generator, a procession of very minute droplets of tin passes through a filter before entering the vacuum chamber of the source. The droplets are falling at a rate of 50,000 times per second while having a diameter of 25 microns. There is a camera located aboard this vessel. Within the chamber, a droplet can be seen moving over part of it. After that, the camera sends a command to the laser in the sub-seed fab, telling it to send a laser pulse into the main vacuum chamber. It is known as the pre-pulse in the industry. The next step is going to be the most difficult. After being hit by the pre-pulse laser, the previously spherical tin droplet transforms into the shape of a pancake. Following this, the laser unit will fire once more, which will symbolise the initial pulse. When the primary pulse hits the droplet of tin in the form of a pancake, it causes it to vaporise. When this occurs, the tin vapour undergoes a transformation into plasma, which subsequently emits extreme ultraviolet light with a wavelength of 13.5 nm. The purpose is to hit a droplet with as much accuracy as possible. This has an effect on the conversion efficiency, which may be thought of as the proportion of laser power that is transformed into EUV light (CE). Once the EUV light has been produced, the photons will subsequently collide with the collector, which is a multi-layer mirror. Following its passage via a secondary focus device and subsequent reflection off of the collector, the light finally enters the scanner.

The light from the scanner is reflected off of a complex pattern that consists of 10 different surfaces or multi-layer mirrors. The light is initially directed via an illuminator that may be adjusted. This produces a shape like a pupil, which allows just the appropriate quantity of light to pass through and illuminate the EUV mask. The reflecting mask is then put through a process in which it is illuminated by EUV light. The light is reflected back off of the six multilayer mirrors that are a part of the projection optics. The wafer is illuminated by light that arrives at an angle of 6% when it eventually hits it. Each multi-layer mirror reflects around 70% of the light that strikes it. According to the various estimates, the transmission rate of the EUV scanner itself is barely 4%. Next, an illumination is applied to the photoresist that is on the wafer.

2.3.2. Masks Used in EUV Lithography Technique

An EUV mask consists of an LTEM substrate that is 6 inches square and 0.25 inches thick, and it has been etched with a circuit design. It is covered in an absorbent substance and coated with a material that is reflective of EUV light. If the patterned mask, which is sometimes referred to as the reticle, does not provide a high fidelity replication of the circuit design in accordance with the specifications, any computer chip that has the pattern imprinted on it will not work correctly. Consequently, there can be no room for mistake in the production process, or it must offer a means through which problems may be located and corrected. EUV masks are fundamentally different from optical masks in a variety of ways, and these differences may be broken down into many categories. Reflection is present in the EUV mask, just as it is in all of the other EUV optical components; in contrast, transmission is present in optical masks. Second, in contrast to optical masks, which receive light with its centre centred on the normal, the light incidence on the mask has its centre centred on a chief angle that is not often incident (usually 6 degrees off the normal). As a consequence of this, the characteristics of the scene that are aligned perpendicular to the light plane are lighted or "shadowed" in an inconsistent way. The third characteristic of optical masks is protective coverings that are transmissive and have the appearance of pellicle. They are suspended around 6 millimetres in height above the etched circuit architecture. EUV masks do not have this pellicle because it is difficult to create a sufficiently transmissive membrane (i.e., one that is 100 nm thick) that has the structural integrity to span the mask's 6-inch square surface. This is one of the reasons why EUV masks do not have this pellicle. The primary purpose of the pellicle is to stop particles from colliding with the surface of the mask, keeping them in an out-of-focus plane and preventing them from imaging on the wafer. This is accomplished by preventing collisions between the particles and the surface of the mask. Because there is no pellicle present, it is imperative that the surface of the mask be kept free of any and all particles at all times.

2.3.2.1. Mask Fabrication

The fabrication of EUV masks is a labor-intensive process that requires many steps as mentioned in Figure 92 (Kim and Ahn 2018). The technique will begin by using an LTEM substrate as its starting point. On this substrate, the peak-to-valley surface flatness must be no more than 50 nanometers (nm). An application of a backside metallic coating, which is typically formed of chromium nitride, is produced in order to make it possible to employ electrostatic chucks (ESC) in the exposure apparatus (CrN). The primary advantage of employing ESCs is that they remove the need that a clamp be brought into close contact with the patterned side of the mask. As a consequence of the inevitable physical contact with the surface, particles will always be produced; these particles may or may not constitute defects.

After the backside deposition is complete, the mask is cleaned and examined for any imperfections in the substrate polishing that occurred on the front side. After being cleaned and inspected, the substrate is given a multilayer coating before being used. A substrate that has many coatings is what's known as an EUV mask blank. Because maintaining a high level of purity is essential, the apparatus that is used to coat mask blank multilayer coatings is distinct from the apparatus that is used to cover standard EUV optics. Ion beam deposition is the technique of choice at this time (IBD). An IBD apparatus alternately focuses a stream of argon ions onto targets constructed of silicon and molybdenum in order to spit atoms and clusters of atoms onto an LTEM substrate. This is done in order to get the desired physical effect. The method has to be carefully managed in order to achieve the desired level of surface roughness and avoid the creation of particles that may get encapsulated inside the multilayer stack. After the ML stack has been constructed, a capping layer is produced by depleting ruthenium using the same IBD method. After that, the newly fabricated mask blank is checked for any imperfections before having its reflectivity measured.

The production of an EUV mask blank is finished off with the application of an absorber as the last step. It is necessary for this material to take in EUV light rather than reflect it. This seems to be of little importance given that the vast majority of materials strongly absorb EUV light. As was said at the beginning of this section, the light incidence on the EUV mask is significantly different from the usual. The ensuing shadowing may result in the transmission of patterns in an asymmetrical manner. One strategy for addressing this issue is to minimise the impact of mask shadowing by fabricating an absorber that is as thin as possible. One other tactic is to adjust the size of the features on the mask so that they take into consideration the shadowing. The research that is being done on absorber materials and compensation mechanisms is now quite busy and very secretive. After the ML stacking process is finished, antireflective coatings (ARC) are occasionally added in order to improve the contrast for particular inspection equipment. The blank is then sent to a mask shop, which employs photoresist, electron beam writing, and etching equipment to design the absorber in a manner that is analogous to the production of optical masks. An aerial imaging tool that duplicates the mask's performance in an exposure tool is used to verify the patterned mask for any problems that could be present in it. Using this procedure, any faults that are detected are corrected locally using focused ion beam etching and deposition, provided that this is practicable. The mask is either given the go-ahead for use in production after the post-repair assessment of the aerial imaging tool, or it is flagged for more repair work or to be discarded.

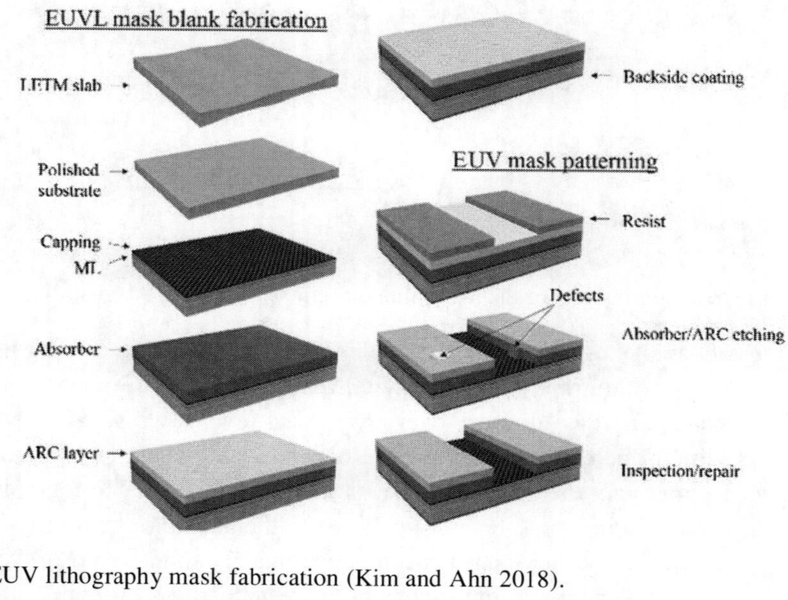

Figure 92. EUV lithography mask fabrication (Kim and Ahn 2018).

2.3.3. Resists Used in EUV Lithography Technique

Lithography is the process of applying circuit designs from a mask to semiconductor devices that are at different stages of manufacture. A thin coating of the material to be patterned has been put onto a circular silicon-based wafer. A chemical emulsion known as photoresist that is photosensitive is applied to the wafer. The resist coating on the wafer is exposed to light in the form of a circuit pattern, which causes the lit resist to change chemically. The unexposed material may also be removed using negative tone resist methods. The etched wafer with the patterned resist coating removes the underlying material where the resist was not present. After a final chemical stripping procedure to remove the resist, the lithographic transfer process is finished. Specifically, the coat, expose, and develop processes in the centre of the figure make up the photolithography process as shown in Figure 93 (Yilbas, Al-Sharafi, and Ali 2019).

Resist manufacturers discovered that 248 nm resist systems had acceptable lithographic performance when exposed to 13.5 nm light after the start of commercial EUV activities. These resist systems were chemically enhanced; a base polymer is catalysed via a series of deprotection processes by a photo-acid generator, rendering exposed areas of the film developer-soluble. Chemically amplified resist was used to resolve 50 nm lines on the SEMATECH LBNL MET in 2004 (effectively matching results from non-chemically amplified resist obtained in 1990), but the initial insertion target for EUV at around 45 nm features was missed due to the general lack of readiness of EUV technology, necessitating the development of higher resolution materials. There is more information on the chemistry of photoresist in general in this book. The criteria for EUV resists and the difficulties in developing commercial, high-resolution photoresists for EUV are the main topics of this section.

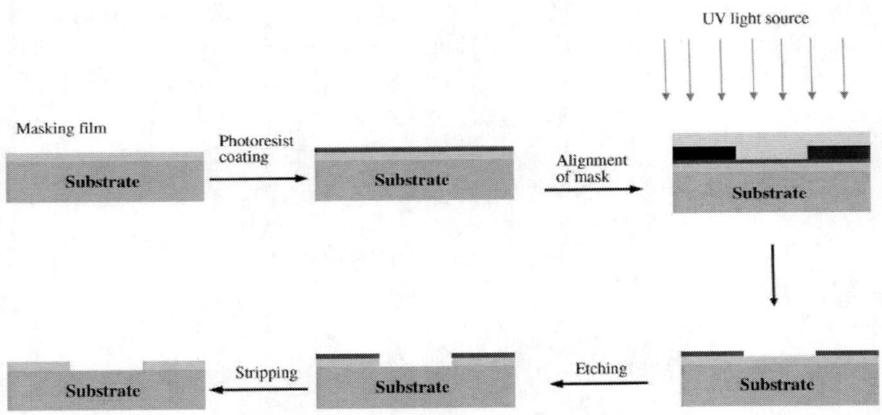

Figure 93. Schematic representation of the patterning using resists (Yilbas, Al-Sharafi, and Ali 2019).

The obvious advantage of EUV is a 15 wavelength reduction when compared to the most advanced 193 nm laser lithography technology currently available. Given that Moore's law assumes that transistor densities "double" every two years (every two years is referred to as a "node" or "generation"), such improvements in resolution would allow for the extendibility of technology across numerous generations. The decision to replace optical lithography with EUV lithography is ultimately being made economically since the technology promises to be usable for numerous generations, during which time the economical "single exposure" method may be maintained. The accompanying improvement in functionality brought on by ongoing transistor scaling may be used by manufacturers of semiconductor devices to market better, more potent goods for comparable or higher profits. Without the clockwork progress of technology, this business model fails. The industry is prepared to make significant expenditures rather than deal with such an unpleasant situation.

2.4. X-Ray Lithography

X-ray lithography is a technique that may be used to create nanoscale structures, such as those that are used in nanosensors. A high-energy X-ray beam is used in this process, which involves exposing a photoresist material on a substrate in a selective manner. The surface of the substrate is given a patterned appearance once the exposed parts of the photoresist have been chemically developed. The patterned surface may then serve as a mask for an etching process that is performed on the substrate in order to construct the desired nanoscale structures. These forms could comprise minuscule transistors, electrodes, and other functional components when they are implemented in nanosensors. X-ray lithography is a powerful method that may be used for the production of detailed and precise structures on the nanoscale scale.

X-ray lithography is a technique that is used in the production of nanosensors. Nanosensors are devices that can detect and measure events that occur at the nanoscale level in the fields of physics, chemistry, or biology. During the procedure, X-rays are used as a source of radiation to imprint patterns on a material that is used as a substrate. These patterns are then used in the production of a nanosensor device as shown in Figure 94 ("X-Ray Lithography - Integrated Circuit Fabrication" n.d.). As the first stage in the process, a mask consisting of a thin coating of an X-ray-obstructing material is first generated. This coating is termed a mask. The required pattern for the nanosensor, which may consist of components that are just a few nanometers in size, is patterned into the mask before the nanosensor is fabricated. Next, a layer of photoresist, which is a substance that is sensitive to X-rays and is coated in a thin layer on top of the substrate material, which is often a semiconductor such as silicon.

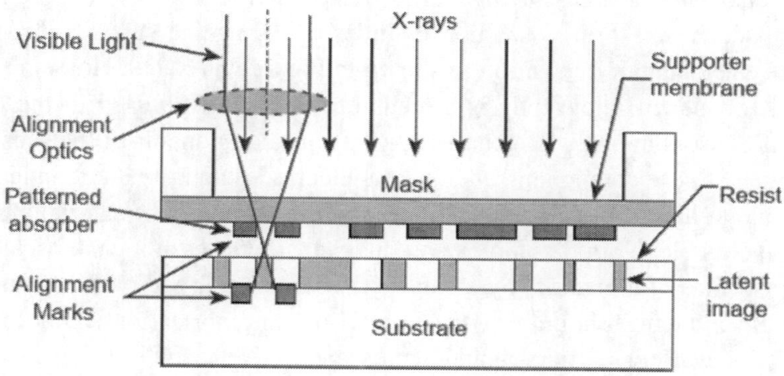

Figure 94. Schematic of X-Ray Lithography ("X-Ray Lithography - Integrated Circuit Fabrication" n.d.).

The X-rays are then directed through the mask and onto the photoresist while the substrate, mask, and photoresist are all enclosed in a vacuum chamber. The photoresist will be exposed to X-rays via the transparent portions of the mask, which will modify the photoresist's chemical composition. The photoresist is exposed in this way to produce an inverted picture of the mask pattern. Following exposure, the substrate is cleaned with a developer solution, which dissolves the photoresist in the regions that were not exposed to X-rays. As a result, the substrate has a layer of photoresist that is patterned and is utilised as a mask for the etching process that follows.

When the photoresist has been removed, the substrate material is etched away from those locations. Chemical or physical techniques, such as dry or wet etching, may be used to accomplish this. A substrate layer with a pattern matching the mask pattern is the end product.

The final result is the patterned substrate material after the leftover photoresist has been removed. The nanosensor device, which may comprise components like nanowires, nanoparticles, or other structures particularly made to detect and quantify the required physical, chemical, or biological events, can then be made using this patterned substrate. Because it enables exact control over the size and positioning of the features on the substrate, X-ray lithography is a potent method for developing nanosensors. Highly sensitive and precise nanosensors may be made because to the tiny feature sizes that X-ray lithography is able to attain. Large-scale arrays of nanosensors may also be made using X-ray lithography, which is advantageous for uses like biosensing or environmental monitoring.

In conclusion, X-ray lithography is a critical process in the creation of nanosensors, which are instruments that detect and quantify events at the nanoscale scale in physics, chemistry, or biology. The process makes use of X-rays as a radiation source to imprint patterns on a substrate material, which is then utilised to make a nanosensor device. The method is exact and may be used to build massive arrays of nanosensors, which are beneficial for a variety of uses.

2.5. Ion Beam Lithography

Ion beam lithography is a kind of nanofabrication technique that is used in the process of fabricating precise patterns and structures at the nanoscale. A beam of charged particles, most often ions, is used in this process, which includes changing the surface of a material via a process known as etching or deposition as shown in Figure 95 (Nahal, Hosseini, and Mahjour-Shafiei 2021). Ion beam lithography is a technique that may be used in the context of the creation of nanosensors in order to construct the various components of the sensor in a precise and exact manner. These components include the electrodes and the interconnects. Using ions makes it possible to have a more precise control over the process of material alteration, which in turn makes it feasible to build complex and difficult structures at the nanoscale. The process of altering the surface of a material by means of a beam of charged particles, most often ions, is referred to as ion beam lithography. It is a kind of nanofabrication technology. Using this technique, a wide variety of micro- and nanoscale devices, such as nanosensors, may be manufactured. When compared to other techniques of nanofabrication, ion beam lithography has the significant benefit of being able to produce patterns and structures at the nanoscale that are very accurate and precise.

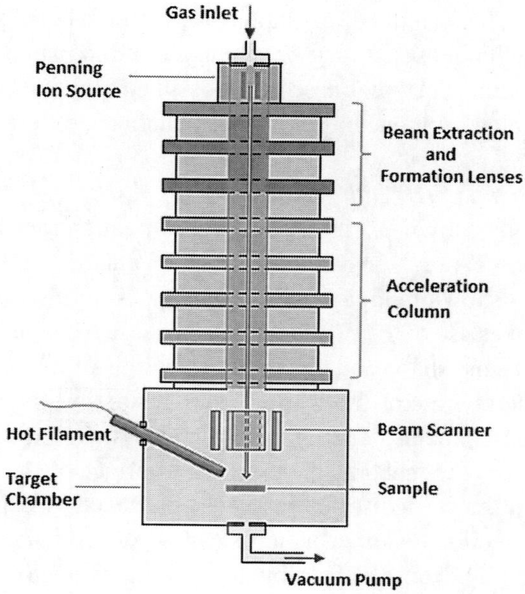

Figure 95. Schematic of Ion Beam Lithography (Nahal, Hosseini, and Mahjour-Shafiei 2021).

An essential component of ion beam lithography is the concept of directing an ion beam towards a target material, which may take the form of either a substrate or a deposited layer. Changing the ion energy may allow for more or less control over the degree to which the material is changed throughout the process. Depending on the amount of energy utilised and the kind of ions that are used in the process, ion beam lithography may either etch or deposit the material. In the context of the manufacture of nanosensors, ion beam lithography may be used to fabricate various components of the sensor. These components may include the electrodes, interconnects, and other essential structures. Because of the high level of precision and accuracy it has, ion beam lithography makes it possible to create elaborate and sophisticated structures with dimensions as small as sub-micron or even sub-nanometer. This high degree of control over the process of material modification is particularly important for the creation of nanosensors, which often need precise and perfect control over the form and electrical properties of their components.

2.5.1. Types of Ion Beam Lithography Techniques

The creation of nanosensors may be accomplished using a variety of ion beam lithography methods. Focused ion beam lithography (FIB), which includes utilising a focused beam of ions to etch or deposit material on the surface of a substrate, is one of the most used methods. The focused ion beam is an excellent option for fabricating nanosensors because it may be utilised to produce precise patterns or structures with great precision and resolution.

Ion beam deposition (IBD), which involves depositing material onto the surface of a substrate using an ion beam, is another popular ion beam lithography method. For the fabrication of thin films or other structures that call for high-quality, homogenous deposition, IBD is especially helpful. It is possible to produce very exact and accurate nanoscale structures by controlling the thickness, content, and electrical characteristics of the deposited material using the ion beam.

Ion beam mixing, ion beam induced crystallisation, and ion beam sputtering are a few more ion beam lithography techniques that may be employed in the manufacture of nanosensors in addition to the ones mentioned above. These methods all employ ion beams to alter a material's surface in certain ways, and each one has particular benefits and drawbacks.

2.5.1.1. Focused Ion Beam (FIB) Lithography

Ion beam lithography is known as focused ion beam lithography because it entails the application of a focused beam of ions to the surface of a substrate in order to either etch or deposit material (FIB) as shown in Figure 96 (Pimpin and Srituravanich 2012). The fabrication of nanosensors often makes use of the FIB technique because it enables a high level of precision and accurate control over the shape and electrical properties of the sensor components. FIB is frequently utilised in the process of fabricating nanosensors to produce essential components such as electrodes, interconnects, and other structures that require precise and accurate patterning. This can be accomplished by patterning the surface of the nanosensor using atomic force microscopy. Either the needed structure may be produced by depositing material onto the substrate using the focused ion beam, or patterns can be etched into the substrate or deposited layer using the focused ion beam. It is possible to vary the degree of the etching or deposition process by regulating the energy of the ion beam. This permits the production of patterns and structures at the nanoscale that are accurate and correct.

FIB's excellent precision and accuracy make it especially effective in the production of nanosensors. Highly sensitive and selective sensors may be made using the focused ion beam to build structures with sub-micron and even sub-nanometer dimensions. By varying the kind and energy of the ions utilised in the process, the focused ion beam may also be used to alter the electrical characteristics of the sensor components, such as the conductivity and resistance.

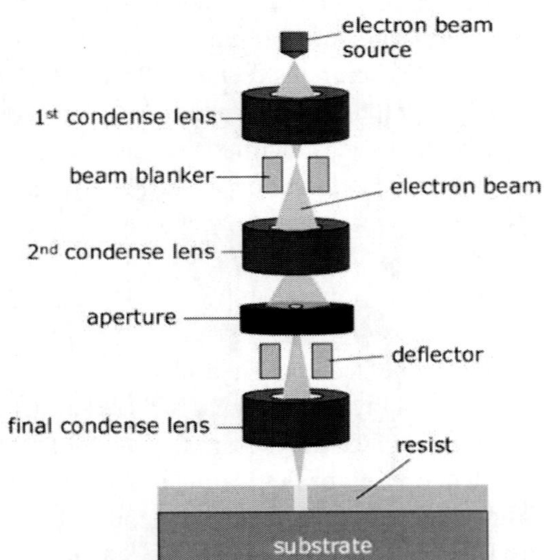

Figure 96. Schematic of Focused Ion Beam Lithography (Pimpin and Srituravanich 2012).

The adaptability of FIB in the creation of nanosensors is another benefit. The construction of basic metal interconnects to the formation of very detailed and complicated structures are just a few of the many uses for the focused ion beam. For the manufacture of nanosensors, which often calls for a mix of precise and accurate patterning, deposition, and etching, this adaptability makes FIB an invaluable tool. Despite its benefits, FIB has several drawbacks that should be carefully examined before using it to the creation of nanosensors. For instance, FIB may be more costly than other nanofabrication processes due to the equipment and process expenses involved. The process may take a long time, and using the equipment and carrying out the essential steps may need specific knowledge and experience.

2.5.1.2. Ion Beam Deposition

Ion beam deposition, also known as IBD, is a method of nanofabrication that involves bombarding a substrate with ions in order to create thin coatings on the substrate. This method is also known by its acronym. The precise control that IBD affords over the thickness and composition of the deposited films enables the development of sensor components with the requisite electrical, optical, and mechanical properties. Because of this, IBD is often used in the manufacturing of nanosensors.

IBD is often used in the building of nanosensors to manufacture thin film layers that operate as active or passive components of the sensor. These layers may work as either an active or a passive portion of the sensor. When making electrodes or interconnects, for example, conductive metals like gold or platinum may be deposited onto the substrate using IBD. This allows the metals to be connected to one another. IBD may also be used to deposit insulating layers or to adjust the electrical properties of the sensor components by depositing dielectric materials such as oxide or nitride. Both of these uses involve regulating the electrical characteristics of the components.

The precision and accuracy of the process of making nanosensors may be attributed, at least in part, to the ability of IBD to control the amount of energy, the kind of ions used, and the flow of those ions throughout the method. The control that we have allows us to deposit films that are homogeneous and homogenous, which is necessary for the nanosensors to function well and is made feasible by the control. IBD may also be used in concert with other nanofabrication techniques, such as lithography or etching, in order to generate structures at the nanoscale that are more intricately detailed and complex.

When it comes to the production of nanosensors, IBD is a process that, despite its many advantages, has a few limitations that need to be carefully considered and taken into consideration before proceeding. For instance, the procedure may be time-consuming and need specialised knowledge as well as competence in order to run the equipment and carry out the essential activities. As a result of the price of the IBD equipment as well as the costs associated with the process, it is possible that it will be more expensive than other techniques of nanofabrication.

To summarise, ion beam deposition is an essential process that is often used in the manufacturing of nanosensors. Because of its adaptability and capacity to produce homogenous and uniform films, as well as its accuracy and precision in manipulating the composition and thickness of the deposited films, it is an important instrument for the creation of extremely sensitive and selective nanosensors that can be utilised in a variety of different contexts.

2.5.1.3. Ion Beam Mixing

A thin film material is mixed at the atomic level by the employment of ions in the ion beam mixing technique used to create nanosensors. This procedure may be used to change the chemical makeup of the film and forge new chemical bonds, which will enhance the film's performance and qualities. The thin layer is often bombarded with high-energy ions during the ion beam mixing process, such as metal or noble gas ions. The atomic species in the film are mixed as a result of the ions' collisions with the film's atoms. New chemical bonds are created as a result, and a new compound material with enhanced characteristics is produced.

Ion beam mixing has the advantage of allowing for uniform mixing of the thin film, which is difficult to achieve using standard production procedures. This is one of the benefits that ion beam mixing offers. As a direct consequence of this, the end product has more consistent performance. Ion beam mixing is often used in the production of nanosensors because it permits accurate and reproducible manipulation of the material features of the sensor film as shown in Figure 97 (Jain and Agarwal 2011). This is one of the reasons why this technique is so popular. The use of the ion beam mixing approach allows for the incorporation of extra components, such as catalytic chemicals, into the film, which may be done for the purpose of improving the performance of the sensor. Nanograins, nanocomposites, and nanolayered structures are some examples of the types of nanostructured materials that may be produced by the use of ion beam mixing. These materials have superior properties compared to the original film. Ion beam mixing, in general, is an essential stage in the process of manufacturing nanosensors because it permits the synthesis of high-performance materials with increased properties. This phase is critical because it enables the production of nanosensors.

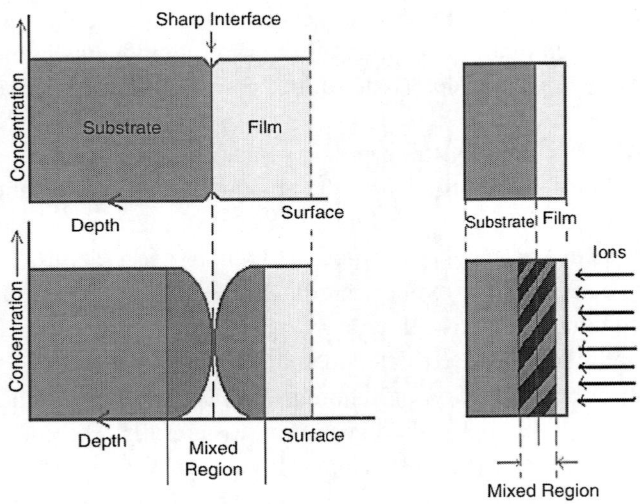

Figure 97. Schematic of Ion Beam Mixing (Jain and Agarwal 2011).

2.5.1.4. Ion Beam Induced Crystallization

Ion beams are used in the Ion Beam Induced Crystallization (IBIC) method, which is a technique for inducing crystallisation in films that are either thin and amorphous or nanocrystalline. This technique is often used in the production of nanosensors due to the fact that it may be utilised to manufacture high-quality crystalline thin films that have increased electrical and optical features. Ion bombardment at a high energy level causes a localised heating action in IBIC, which involves a thin coating as shown in Figure 98 (Lang et al. 2012). This action is caused by the formation of a layer. When the material that is either amorphous or nanocrystalline is heated, it causes the atoms to rearrange themselves into a crystalline structure. Through careful manipulation of the ion beams, it may be possible to improve the electrical and optical properties of the homogenous crystalline structure.

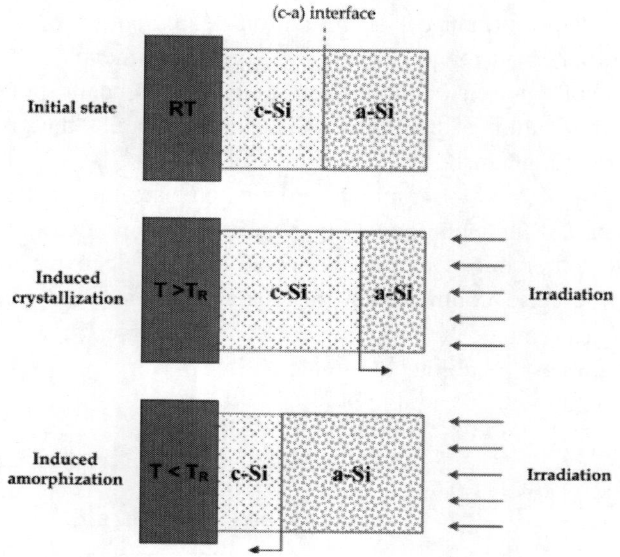

Figure 98. Schematic representation of the ion-induced crystallization and amorphization (Lang et al. 2012).

The performance of the nanosensor may be increased, resulting in better sensitivity, stability, and repeatability thanks to the improved characteristics of the crystalline material. Additionally, the material's crystalline structure may be exploited to forge certain forms, such nanograins or nanolayered structures that have better qualities than the original film. IBIC may also be used to make nanostructured materials with better qualities than the original film, such as nanograins, nanocomposites, and nanolayered structures. This is accomplished by adjusting the dosage rate, energy, and angle of the ion beams, which may be utilised to produce nanoscale objects with certain features.

Ion Beam Induced Crystallization, in general, is a crucial procedure in the production of nanosensors because it enables the development of high-performance materials with enhanced characteristics. This results in the creation of nanosensors that are more sensitive, reliable, and repeatable for a variety of applications.

2.5.1.5. Ion Beam Sputtering

Ion beam sputtering is a method for creating nanosensors that uses energetic ions to remove material from the sensor layer. High-energy ions are used in the process to bombard the target material, which causes material to be expelled off the surface and deposited onto a substrate as shown in Figure 99 (Patrick 2018). This may be used to produce nanostructured materials like nanograins, nanocomposites, or nanolayered structures, or it can be used to deposit thin films onto the substrate. Ion Beam Sputtering is used to produce accurate, homogeneous thin films with controllable attributes, such as conductivity, sensitivity, and stability, in the context of nanosensor production. By incorporating additional substances or components into the film, the method may also be utilised to change the film's physical characteristics.

The ability to make high-quality films with exceptional control over the composition, thickness, and shape of the film is one of the main benefits of ion beam sputtering. As a result, the end product has more constant performance, which is vital for applications like chemical sensing where accurate performance is essential. Ion Beam Sputtering is a flexible method that may be utilised with a broad range of materials, including metals, ceramics, and polymers, making a wide range of nanosensor applications possible. Ion beam sputtering, in general, is a crucial step in the fabrication of nanosensors because it enables the production of accurate, homogeneous films with enhanced characteristics.

The capacity of ion beam lithography to produce precise and exact patterns and structures at the nanoscale is one of the technique's key benefits in the production of nanosensors. This is crucial for the development of extremely sensitive and selective nanosensors since these devices often need precise control over the shape and electrical characteristics of their constituent parts. Furthermore, the employment of ions enables more exact control over the material alteration process, enabling the construction of very complex structures with sub-nanometer dimensions.

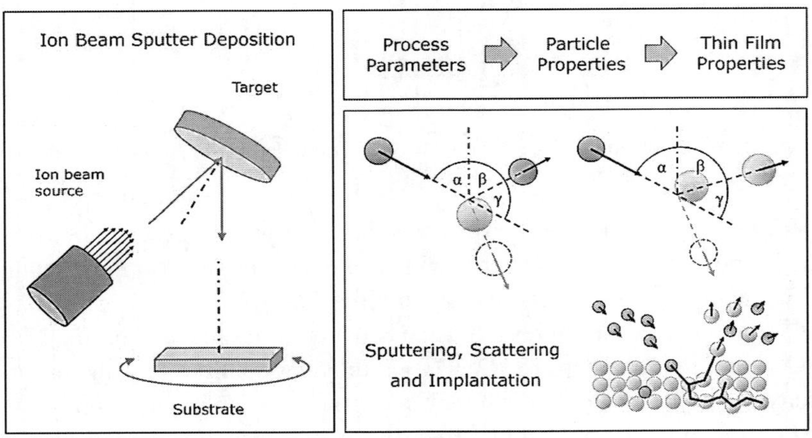

Figure 99. Schematic of ion beam sputter deposition (Patrick 2018).

Ion beam lithography also has the benefit of being versatile, since it may be used to a variety of tasks, ranging from the production of simple metal interconnects to the development of very detailed and complicated structures. Ion beam lithography is a useful method for the manufacture of nanosensors because of its flexibility. Nanosensors often need a mix of accurate patterning, deposition, and etching. Ion beam lithography has a number of restrictions as well, which should be carefully taken into account when using it to the creation of nanosensors. Ion beam lithography, for instance, might be more costly than other methods for nanofabrication due to the equipment expenses as well as the related process costs. The process may take a long time, and using the equipment and carrying out the essential steps may need specific knowledge and experience. Furthermore, ion beam lithography may not be appropriate for making certain kinds of nanosensors, such those that need to have huge regions of material deposited on them or those that need to be produced in big quantities.

Ion beam lithography is still a crucial and popular method for creating nanosensors despite these drawbacks. The procedure is versatile and capable of producing intricate and complex structures at the nanoscale, which makes it a great tool for the creation of extremely sensitive and selective nanosensors for a variety of applications. Ion beam lithography, which offers precise and perfect control over the geometry and electrical characteristics of sensor components, is a critical technology in the manufacturing of nanosensors. It is a useful technique for the creation of extremely sensitive and selective nanosensors for a variety of applications due to its adaptability and capacity to produce intricate and complicated structures at the nanoscale. Ion beam lithography is still a vital and popular method for creating nanosensors, despite the expenses and constraints it comes with.

2.6. Direct Epitaxial Growth

Both molecular-beam epitaxy (MBE) and metalorganic chemical vapour deposition (MOCVD), which are both atomic-precision deposition processes, have been proved to be helpful tools for the construction of a wide variety of quantum confinement structures and devices (quantum well lasers, photodetector, resonant tunnelling diodes, etc.). In spite of the fact that quantum wells and superlattices are the structures that lend themselves to these methods the most easily, quantum wires and dots have also been made by adding additional phases such as selective growth and etching. Due to the fact that epitaxial growth is an established and well-established area, it will not be covered in this chapter. This is because the chapter will not discuss the construction of quantum well and superlattice structures. Instead, we are going to put our attention on developing quantum wire and dot nanostructures by using basic epitaxial techniques.

In order to fabricate quantum wires and dots, epitaxial layers have been employed in a variety of different ways and in a number of different methods. The most straightforward approach is to use e-beam lithography followed by etching of an epitaxially produced layer (e.g., InGaAs on GaAs substrate) However, because of the potential for damage and/or contamination during the lithography process, this method is not especially well suited for the production of active devices (e.g., quantum dot lasers). In order to fabricate quantum wires and dots without the need for the quantum confinement structure to be lithographed and etched first, a variety of methods that involve the regrowth of epitaxial layers over nonplanar surfaces, such as step-edge, cleaved-edge, and patterned substrate, have been utilised. These methods have been successful in producing the desired results. The formation of these non-planar surface templates may be accomplished by a variety of processes, such as the cleaving of materials along crystallographic planes or the etching of masks.

Further epitaxial growth on top of these structures results in the production of a collection of planes with varying growth rates. These growth rates are dependent on the geometry as well as the surface diffusion and adsorption processes. These effects have the potential to significantly increase or decrease the growth rate on certain planes, which can result in the lateral patterning and confinement of epitaxial layers that have been deposited, as well as the emergence of quantum wires in V-grooves and dots in inverted pyramids. This can be caused by a combination of factors, such as the orientation of the epitaxial layers and the planes on which they are occurring. The top and bottom surfaces grow at a faster rate than the side walls, which may be seen to be expanding more slowly. The thicker InGaAs layer at the bottom of the V-groove forms a quantum wire as a consequence. This quantum wire is surrounded on all sides by a layer that is thinner and has a wider bandgap than the layer it is created from. To construct the window and groove templates for the quantum wires and dots, e-beam lithography is still necessary, despite the fact that it is quite straightforward to manufacture quantum wells using any of these approaches.

2.7. Strain Engineering

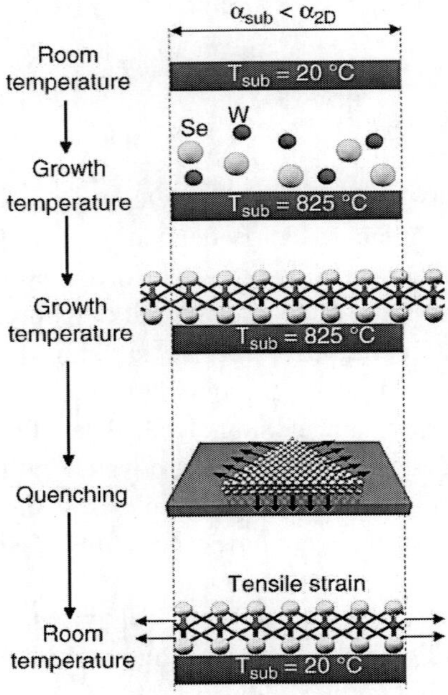

Figure 100. Schematic of thermal expansion induced strain engineered 2D semiconductor material (Peng et al. 2020).

In more recent times, the formation of quantum wires and dots has been accomplished by the application of strain-induced self-assembly. Self-assembly is the term used to describe the process by which a stressed two-dimensional (2-D) system may reduce its energy by transforming into a morphology that is three-dimensional. The InxGa1xAs/GaAs system, which has a substantial lattice mismatch of 7.2%, is the material combination that is used for this method the most often. Ge dots on Si substrate have also recently received a lot of attention due to their potential applications. This method requires a substrate and an epitaxially produced layer with varying lattice parameters. Using this method, one may construct an array of quantum dots or wires. When the lattice constants of the substrate and the epitaxial layer are considerably different, only the first few monolayers that are deposited crystallise into the form of an epitaxial strained layer with the same lattice constants. This occurs extremely early in the deposition process. When a particular thickness threshold is exceeded, the organised structure begins to disassemble, and the spontaneous production of randomly dispersed islets of regular shape and comparable size (typically 30 nm in diameter) begins to take place. These islets have the same size and shape as one another. Figure 100 shows the schematic of thermal expansion induced strain engineered 2D semiconductor material (Peng et al. 2020).

The most popular term given to this kind of development pattern is the Stranski-Krastanow mode. Quantum dot size, separation, and height are all affected by the deposition parameters (i.e., the total amount of material deposited, the growth rate, and the temperature), as well as the material combinations. It is evident that this is a very practical technique to manufacture perfect crystalline nanostructures over a large area without the need for lithography or etching, and it is clear that this method has a high degree of applicability. The unpredictability of the quantum dot distribution is one of the most significant drawbacks of this approach. It is important to highlight that in addition to making quantum dots, this technique may also be used to make quantum wires via strain-relaxation bunching at the step edges.

2.8. Scanning-Probe Techniques

The discovery of scanning probe microscopy in the 1980s sparked a revolution in the field of atomic-scale imaging and spectroscopy. STMs and AFMs, which stand for scanning tunnelling microscopes and atomic force microscopes, respectively, have found widespread use in a variety of scientific disciplines, in particular physics, chemistry, material science, and biology. Since the beginning, researchers have taken into account the possibility that such probes may be used for atomic-scale manipulation, lithography, and nanomachining. This field of research has made substantial progress over the last 10 years. In this section, after providing a brief introduction to scanning probe microscopes, we will discuss several major machining and nanolithography processes that have been used to make nm-sized objects.

The movement of an atomically sharp tip that is in close proximity to or in touch with a surface may be subnanometer-precisely regulated by scanning probe microscopy (SPM) devices. This allows for the examination of surfaces with a higher level of detail. In order to achieve such a high level of accuracy, piezoelectric positioners are often used. High-resolution photographs may be obtained by raster scanning the tips over a surface while monitoring how they interact with the surface at the same time. In scanning tunnelling microscopy systems, a bias voltage is applied to the sample, and the tip is positioned in such close proximity to the surface that it causes a tunnelling current to develop across the gap. Since the tunnel current is very sensitive to the distance between the tip and the surface, it is possible to map the surface topography with atomic-scale precision by scanning the tip in the x-y plane while recording the tunnel current. This is possible because the tunnel current is very sensitive to the distance between the tip and the surface. In a mode of operation that is more usual, a feedback loop links the amplified current signal to the z-axis piezoelectric positioner. This connection helps to keep the current constant and, as a result, the distance constant during the scanning process. A depiction of the topography of the surface may be made by recording the vertical placement of the tip at each x-y point in this configuration. This will allow one to map out the surface.

Because the STM system requires the generation of a tunnelling current, it can only be used with conductive materials and surfaces. Atomic force microscopy is a technique that was developed as an alternate method for visualising conducting or nonconducting surfaces. In atomic force microscopy (AFM), the tip of the instrument is brought into contact with the surface while it is still linked to a flexible cantilever. One is able to detect the force that is operating between the cantilever's tip and the surface by sensing the cantilever's deflection. A topographic depiction of the surface may be obtained by plotting the deflection as a function of the x-y position of the point of interest.

In a method of operation that is more usual, a feedback loop is used to maintain a constant deflection while topographic data is acquired from the vertical displacement of the cantilever. This allows the data to be collected in a more efficient way. Some scanning probe systems combine the AFM and STM modes into a single instrument. This is accomplished by positioning the tip of the instrument on a cantilever that also contains an electrical connection. This configuration makes it possible to control or observe tunnelling currents in addition to surface forces. Systems that use AFM are typically operated in air, but systems that use STM may either be operated in ultrahigh vacuum (known as UHV STM) or in air. When a scanning probe system is utilised in air, water that has been adsorbed onto the surface of the sample builds up beneath the tip of the system, generating a meniscus that is sandwiched between the surface and the tip. Several of the techniques for scanning probes that are described farther down make considerable use of the water meniscus in various ways.

2.8.1. Scanning Probe Induced Oxidation Technique

Scanner probes that operate in air and are biassed at a voltage that is sufficiently high have the potential to cause localised oxidation of a variety of materials on a nanoscale scale. A tip bias of 2 to 10 V is commonly used while writing at speeds ranging from 0.1 to 100 metres per second in environments with humidity levels ranging from 20 to 40 percent. The water meniscus that develops at the point of contact is thought to function as an electrolyte, which enables the skewed tip to anodically oxidise a small portion of the surface. This is supported by the hypothesis. The most common example of this theory being put into practise is the oxidation of silicon that has first been hydrogen-passivated. The dipping of silicon surfaces in HF solution produces a hydrogen atom passivated surface most of the time. Etching may be done either wet or dry, and it is feasible to use silicon surfaces that have oxide patterns etched on them as a mask for the process. Patterns with a line width of 10 nm have been successfully transferred from silicon substrates to other materials using this method. Local anodization of metals such as titanium and aluminium has also been accomplished via the use of this technique. The anodization of silicon that has been deposited is a fascinating variation of this approach. On top of many different substances, amorphous silicon may be deposited when the temperature is kept low enough. The silicon layer that was deposited may be created and used in a variety of ways; for example, it can be used as the gate of a 0.1-micron CMOS transistor, or it can be used as a mask to pattern an underlying film. Both of these options are possible. The lack of reproducibility that results from anodizing tip wear is this method's primary shortcoming. However, it has been shown that by utilising AFM in the noncontact mode, this problem may be solved. The illustration of the local modification of the nanomaterials by thermal scanning probe technique is given in Figure 101 (Howell et al. 2020).

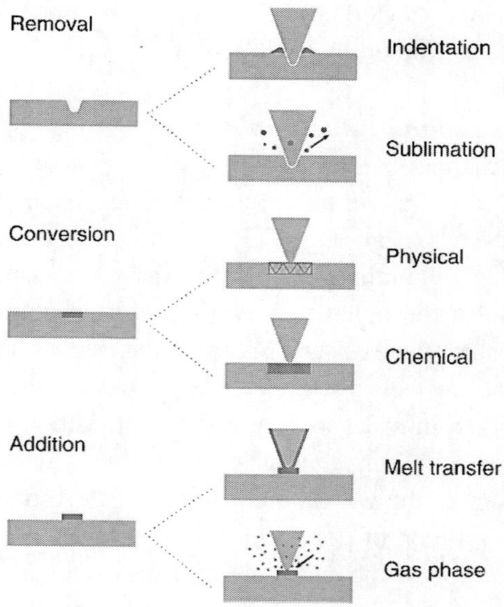

Figure 101. Schematic of local modifications by thermal scanning probe technique used in nanoengineering (Howell et al. 2020).

2.8.2. Dip-Pen Nanolithography Technique

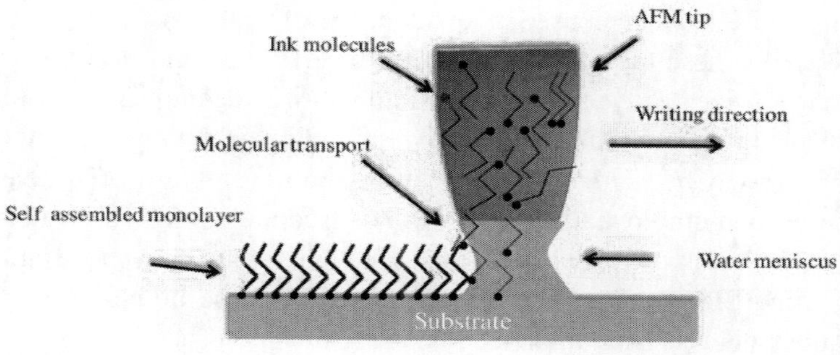

Figure 102. Schematic of 'Dip-Pen Nanolithography technique (Goyal et al. 2014).

In dip-pen nanolithography (DPN), a surface is inked with an interesting chemical and the tip of an AFM operating in air is brought into contact with it. Similar to a fountain pen, the ink molecules flow from the tip onto the surface. As shown in Figure 102, the water meniscus that naturally develops between the tip and the surface allows for the diffusion and transportation of the molecules (Goyal et al. 2014). By dipping the tip into a solution with a low concentration of the molecules, followed by a drying stage, inking may be accomplished (e.g., blow-drying with compressed difluoroethane). With this approach, line widths as small as 12 nm and spatial resolution as little as 5 nm have been obtained. Conducting polymers, gold, dendrimers, DNA, organic dyes, antibodies, and alkanethiols are some of the species that can be patterned using DPN. Alkanethiols have also been used as an organic monolayer mask for the following etching of the exposed silicon substrate after a gold layer had been removed. A heated AFM cantilever may also be used to manage the deposition of a solid organic ink. Sheehan et al. recently described a method in which heated AFM probes were used to write 100 nm lines of octadecylphosphonic acid (melting point 100 C).

2.8.3. Probe Resist Exposure and Lithography

To generate electrons in a manner similar to that of which e-beam lithography reveals a resist, biassed SPM tips may be used. This lithography method has been implemented in a number of different systems, such as STMs that operate with a continuous current, AFMs that do not need direct contact, and AFMs that maintain a constant tip-resist force and current. When using AFM cantilevers in systems, one of the benefits is that imaging and alignment operations may be carried out without exposing the resist. This is a significant advantage. Scanning probe lithography and resists that are well documented for e-beam lithography, such as PMMA or SAL601, have been combined in order to give dependable sub-100 nm lithography. This was accomplished by combining the two techniques. The following are the stages involved in this procedure. After being cleaned, the wafers are dipped in HF solution in order to eliminate the native oxide (in the case of silicon or poly). After that, a layer of resist with a thickness of 35 to 100 nm is spin-coated on top of the surface. Exposure is achieved by moving the tip of the scanning probe microscope (SPM) over the surface while applying a bias voltage that is high enough to induce the tip to emit electrons (a few tens of volts). Following the exposing step, the resist is developed using the standard developing solutions. It was possible to manufacture features with widths of less than 50 nm by using this technology.

2.8.4. Other Common Scanning Probe Techniques

Scanning probe systems have been used in a wide range of nanofabrication processes that have been shown. Some of these are proof-of-concept experiments, and it is not yet known how useful they will be as repeatable and feasible manufacturing techniques. A STM/AFM tip may be used to mechanically process a substrate by functioning as a plough or engraving tool, for instance. Although it may be used to shape resist for a future etch, lift-off, or electrodeposition process, it is more often employed to directly build structures in the substrate. By raising the tip temperature above a polymeric substrate material's glass transition temperature, mechanical nanomachining using SPM probes may be enabled. SPM-based high-density data storage on polycarbonate substrates has been implemented using this method.

Applying voltage pulses over 3 V makes it simple to create electric fields strong enough to cause atom emission from the tip. Material has been moved from the tip to the surface and vice versa using this phenomena. In this way, metal mounds (10–20 nm) such as Au, Ag, and Pt have been deposited or removed from a surface. The same method has been used to remove individual atoms off a semiconductor surface and re-deposit them in other locations. It has also been possible to manipulate nanoparticles, molecules, and single atoms on top of a surface by simply pushing or sliding them with the SPM tip. The STM chemical vapour deposition method may also deposit metals locally. In this method, the STM chamber is filled with a precursor organometallic gas. The precursor gas is split into a thin layer of metal by a voltage pulse that is delivered between the tip and the surface. Using SPM technologies, local electrochemical etching and electrodeposition are also feasible. First, a droplet of an appropriate solution is applied to the substrate. A voltage is then provided once the STM tip has been inserted into the droplet. The tip is covered with wax so that only the very end is exposed to the fluid in order to decrease Faradaic currents. Using this method, features with a size of under 100 nm have been produced.

It is impracticable for mass production to use a single tip to serially create the necessary alteration in a surface since this results in highly sluggish manufacturing procedures. However, a lot of the scanning probe methods that have been established so far may also be carried out by a group of tips, which would speed up throughput and make them more competitive with other parallel nanofabrication methods. Scanner arrays in one and two dimensions have been used to show this method for imaging, lithography, and data storage. We could see these methods being used as standard manufacturing procedures in the industry as a result of the development of bigger arrays with improvements in individual control of force, vertical position, and current.

2.9. Self-Assembly Nanofabrication Technique

Colloidal nanoparticles are brought together by the process of self-assembly, which results in the formation of the desired structure. This aggregation may take place naturally (entropically) as a result of the constraints imposed by thermodynamic minima (energy minimization), or it may take place chemically as a result of the complimentary binding of organic molecules and supramolecules. Both of these possibilities are possible as a result of the restrictions imposed by thermodynamic minima (molecular self assembly). The process of molecular self-assembly is one of the most important strategies that are used in biology for the formation of complicated functional structures. Because the target structures for these methods need to be thermodynamically stable, the structures that these methods often produce are largely defect-free and have the ability to repair themselves. Self assembly is a powerful bottom-up assembly and manufacturing approach that is not only confined to molecules or the nanodomain and can be carried out on practically any size scale (multiscale ordering). The Langmuir–Blodgett approach to the monolayer self-assembled deposition is illustrated in Figure 103 (Galeotti, Pisco, and Cusano 2018).

Another enticing element of this approach is its capacity to integrate the self-assembly qualities of organic molecules with the electrical, magnetic, and photonic capabilities of inorganic components. Template manufacture is another bottom-up approach that may be used to build nanostructures. This process involves the deposition of material onto nanotemplates utilising a variety of techniques, such as electroplating and CVD. Due to the ease and adaptability of electrochemistry for plating and surface finishing of a wide variety of materials, the electrochemical manufacturing of various metallic nanostructures based on various templates has recently become quite popular. This popularity can be attributed to the fact that electrochemistry is used to produce diverse metallic nanostructures. For example, extensive arrays of nanostructures have been created in nanoporous templates, such as porous alumina, by the process of electrochemical deposition. The metal nanowires that are created as a result of this template-based deposition process typically have a diameter of just 25 nm and a length of only a few micrometres. Producing the nanotemplates necessary for the construction of nanostructures often involves the use of self-assembly techniques. In the next sections, we will discuss a variety of key manufacturing techniques, including self-assembly, template manufacturing, and others.

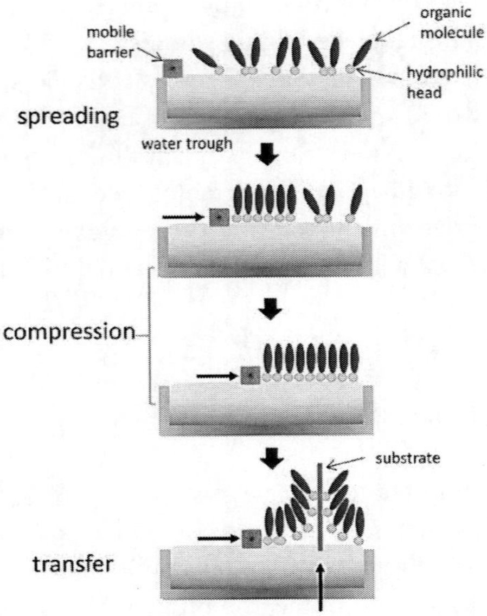

Figure 103. Schematic of the Langmuir–Blodgett approach to the monolayer self-assembled deposition (Galeotti, Pisco, and Cusano 2018).

2.9.1. Types of Self-Assembly Nanofabrication

The primary factor that propels the process of self-assembly is the spontaneous aggregation of colloidal nanoparticles, which may take place either physically or chemically. The potential for spontaneous self-assembly may be realised by taking use of the tendency of monodispersed colloidal spheres of nano- or submicron-sized to organise themselves into a face-centered cubic lattice. The goal of the system to achieve a state that is thermodynamically stable is the force that is driving this process (minimum free energy). Complex three-dimensional structures may be formed by a variety of different forces, including gravitational, convective, and electrohydrodynamic ones, in addition to the naturally occurring process of thermal self-assembly. The process of chemical self-assembly begins with the attachment of a single-molecule organic layer to the colloidal particles, which is referred to as a self-assembled monolayer (SAM). The next step involves the self-assembly of these components into complex structures through the process of molecular recognition and binding.

2.9.2. Physical Self-Assembly

This entropy-driven technique relies on colloidal particles spontaneously assembling into a reasonably stable structure through noncovalent interactions. For instance, colloidal polystyrene spheres can form a three-dimensional structure on a substrate that is held vertically in the colloidal solution. The spheres gather into a hexagonal close-packed (hcp) form when the solvent evaporates. The polymer sphere size controls the density and size of the interstitial pores. After creating the hcp arrays, the polymer spheres may be etched into smaller sizes, changing the template pore separations. Large patterned regions may be created using this method in a rapid, easy, and affordable manner. The spontaneous construction of silicon photonic-bandgap crystals on chips, which can reflect light coming from any direction within a certain wavelength range, is a typical example. This technique involves assembling a thin layer of silica colloidal spheres on a silicon substrate. A silicon wafer is placed vertically in a vial containing an ethanolic solution of silica spheres to accomplish this. The movement of silica spheres is aided by a temperature differential across the vial. Once such a template has been created, silicon may be utilised in LPCVD to fill the interstitial gaps, allowing silicon's high refractive index to provide the required bandgap. The physical self-assembly of nanogel assisted by cross linking is shown in Figure 104 that is capable of producing a nanocarrier system (Kendre and Satav 2019).

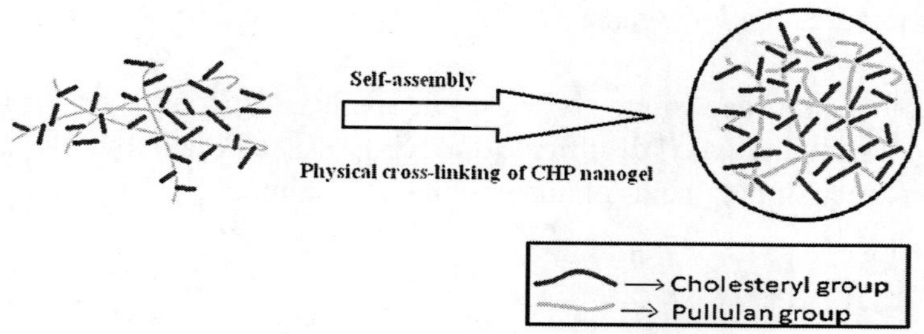

Figure 104. Representation of physical self-assembly of nanogel by cross linking (Kendre and Satav 2019).

Colloidal particles may also be applied on patterned substrates (template-assisted self-assembly, TASA). The foundation of this technique is the idea that when an aqueous colloidal dispersion is allowed to dewey from a solid surface that has already been patterned, the colloidal particles are trapped by the recessed regions and assemble into aggregates with shapes and sizes determined by the geometric confinement provided by the template. Conventional contact mode photolithography may be used to create the patterned arrays of templates because it allows the user control over the size and form of the templates, which enables the construction of complicated structures from colloidal particles. The fluidic cell's cross-section as employed in TASA. To contain the colloidal particles' water dispersion, the fluidic cell features two parallel glass substrates. A 2-D array of templates is used to shape the surface of the bottom substrate. The capillary force acting on the liquid propels the colloidal spheres over the surface of the bottom substrate until they are really caught by the templates when the aqueous dispersion is allowed to dewet gradually throughout the cell. The greatest number of colloidal particles permitted by geometric confinement will fill the template if the colloidal dispersion's concentration is high enough. A variety of challenging-to-generate polygonal and polyhedral aggregates may be created using this technique.

2.9.2.1. Chemical Self-Assembly

The organic and supramolecular SAMs are essential to the colloidal particle self-assembly process. Solid substrates that have been chemically adsorbed with robust organic molecules may be referred to as SAMs. They often have a hydrophobic (nonpolar) tail that is rather long and extends outward, as well as a hydrophilic (polar) head that can attach to a variety of solid surfaces. SAMs are formed either when a substrate is immersed in a solution of the molecule in water or an organic solvent (referred to as the liquid phase), or when the substrate is exposed to an environment that contains the molecule (gas phase). The final film is a dense collection of molecules that are organised in such a way as to expose the end group. The SAM's longevity is directly proportional to the effectiveness of its surface anchoring, which in turn is determined by the nature of the substrate. SAMs have garnered a lot of interest because to the fact that the end group may be functionalized to produce finely ordered molecular arrays for a number of applications. These applications range from simple ultrathin insulators and lubricants to complex biological sensors. Through the use of organic or supramolecular SAMs as binding and recognition sites, chemical self-assembly may generate intricate three-dimensional structures from colloidal nanoparticles. A schematic representation of the chemical self-assembly of sequence-dependent peptide using perylene bismide is given in Figure 105 (Cui et al. 2018).

The following are the two organic monolayers that are used the vast majority of the time:

1. Compounds of organosilicon on glass and oxidised silicon
2. Gold doped with dialkyl disulfides, dialkyl thiols, and dialkyl thiols
3. The presence of fatty acids on metal oxides like alumina
4. DNA

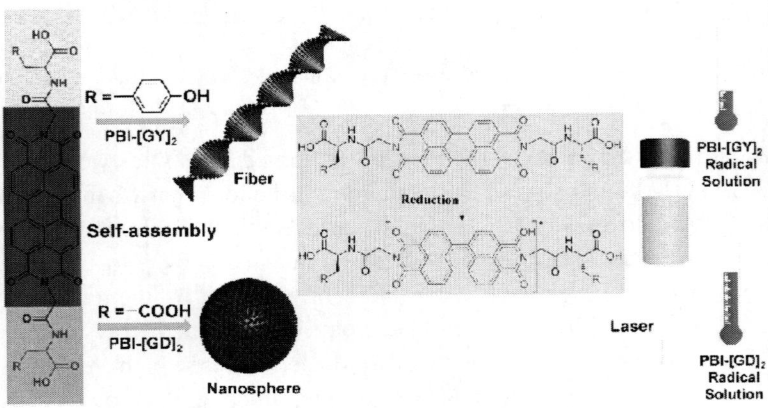

Figure 105. Representation of chemical self-assembly of sequence-dependent peptide using perylene bismide (Cui et al. 2018).

Octadecyltrichlorosilane (OTS) is the organosilane that is used the most often in the production of SAMs. This is mostly due to the fact that it is simple, readily available, and generates layers that are of a good, thick consistency. The formation of alkyltrichlorosilane monolayers is possible on wafers of pure silicon that have a surface composed of SiO_2 and contain around 5 1014 SiOH groups/cm2. Because the silicon-chloride bond is susceptible to hydrolysis, the solution must include a trace amount of water in order to form monolayers of a high quality. This is because the bond may be broken by water. Monolayers that are composed of methyl and vinyl-terminated alkyl silanes are autophobic to the hydrocarbon solution, which is why they emerge uniformly dry from the solution. This is in contrast to monolayers that are produced of ester-terminated alkyl silanes, which emerge wet from the solution that is used to make them. When the alkyl trichlorosilane in the solvent that is adhering to the substrate is exposed to water, a gel of polymeric siloxane is created. This gel causes a hazy layer to be produced on the surface of the substrate. The disadvantage of this approach is that it produces a hazy layer. One potential solution to this problem is the use of alkyl dimethyl chlorosilane, which has just a single anchoring site and is incapable of forming polymers. Chlorosilanes are sometimes preferred over alkoxysilanes due to the fact that they have a higher level of reactivity. The reactivity of chlorosilanes places significant restrictions on the spectrum of functional groups that may be placed at the end of the hydrocarbon tail. However, these restrictions can be overcome. On the other hand, groups such as amino, mercapto, epoxy, and thiocyanate are often necessary for the following binding of colloidal particles and biomolecules. On the other hand, methoxy silanes and ethoxy silanes are often accessible with a wide range of functional groups. Layers that are more consistent in appearance are created when the molecules in question are deposited in the gas phase as opposed to the liquid phase. Because of PDMS's intrinsic permeability to volatile chemicals, reactive silane designs have been generated by extracting them from the gas phase and casting them in moulds resembling soft lithography.

An important example of an organic SAM system is provided by gold's alkanethiols, which are denoted by the notation X(CH2)nSH. Utilizing gold as the material that makes up the substrate is particularly beneficial due to the absence of a stable oxide and the fact that it can be worked with under natural conditions. Submerging a brand-new, spotless, hydrophilic gold substrate for a short period of time (anywhere from a few minutes to several hours) in a diluted solution (103 M) of the organic sulphur compound (alkanethiols) in an inorganic solvent can produce a monolayer that is densely packed and oriented. This can be accomplished by creating a close-packed monolayer. When sulphur, which serves as the head group, interacts significantly (44 kcal/mol) with the gold substrate, a close-packed, ordered monolayer is produced. This monolayer is characterised by its high degree of packing density. It is possible to alter the end group of alkanethiol in order to adjust the degree to which the adsorbed layer is hydrophobic or hydrophilic. The deposition of alkanethiol SAM may also be done using the method of soft lithography. Before transferring it to substrates that are either planar or nonplanar, this approach involves putting alkanethiol ink on a PDMS stamp and then stamping it onto the surface of the substrate. Self-assembly of a wide range of intricate three-dimensional structures is possible when planar, nonplanar, or spherical surfaces that have been functionalized with alkanethiols are used.

Carboxylic acid derivatives undergo an acid-base reaction during their self-assembly onto surfaces (such as glass, Al_2O_3, and Ag_2O), which results in the formation of fatty acid monolayers. As the concentration of the solution decreases, the time required to create a complete monolayer increase. To generate a monolayer on gold, a greater concentration of the carboxylic acid is required as compared to that which is required for Al_2O_3. This is due to differences in the affinities of the COOH groups, which are greater for $Al{=}O_3$ and glass than they are for gold, as well as differences in the surface concentration of the salt-forming oxides that are present in the two substrates. Chemisorption of alkanoic acids on surfaces composed of amorphous metal oxide is not an uncommon occurrence. On Ag_2O, for example, the two oxygen atoms of the carboxylate connect to the substrate in a somewhat symmetrical fashion. This results in ordered monolayers with a chain tilt angle that is anywhere from 15 to 25 degrees from the surface normal. On the other hand, the oxygen atoms join symmetrically on CuO and Al_2O_3, and the chain tilt angle is extremely close to 0. As a result, the structure of the monolayers achieves a state of equilibrium that is independent of the various interactions that take place in the polymer chains.

Deoxyribonucleic acid (DNA), which is the fundamental component of all living things, may be used as a tool to facilitate the self-assembly of nanomaterials into macroscopic aggregates that have a range of advantageous physical properties. DNA is made up of two strands, which are then twisted around each other to form a double helix structure. After the two strands of nucleotides have been uncoiled, what is left behind are single strands of nucleotides. A nitrogenous base, a phosphate (PO4), and a sugar are the components that go into making these nucleotides (pentose ring). In order for a nucleotide to have the necessary structure, its component components need to be arranged in the correct sequence and have the appropriate architecture. Adenine (A), Guanine (G), Cytosine (C), and Thymine (T) are the four nucleotides that are often found in DNA (T). One of the most important aspects of the structure of DNA is that the nucleotides that are defined form a specific attachment to another nucleotide when the strands are wound together to form the double helix (A to T, and C to G). Using this special bonding capacity, nanophase materials and nanostructures may be assembled. For example, intricate three-dimensional structures built of nucleotide-functionalized nanogold particles have been assembled by affixing DNA strands to the gold by the use of an enabler or linker. These structures are made up of nucleotides. In a separate line of research, DNA was used in the process of assembling nanoparticles into macroscopic structures. An alkane thiol serves as the linker molecule in this process. It is responsible for linking the DNA template to the nanoparticle. The thiol groups at the ends of the linker molecule form covalent bonds with the colloidal particles, which results in the formation of the aggregate structure.

2.10. Template Manufacturing

A group of methods known as "template manufacturing" may be used to create 3-D organic or inorganic objects from a nano template. The materials, patterns, feature sizes, overall template sizes, and periodicities of these templates vary. E-beam lithography can be used to create nanotemplates, but its serial nature prevents it from being widely used. Self-assembly is the primary method for manufacturing massively parallel large-area nanotemplates. For use in creating templates, a number of nanotemplates have been studied. These include alumina membranes, nuclear track etched membranes, and polymer colloidal spheres. Using the strategies outlined in the preceding section, colloidal spheres may be deposited in a regular 3-D array. Aluminum may be anodized and oxidised to create porous aluminium oxide membranes. The oxidised film is made up of hexagonal close-packed pores arranged in columnar arrays with spacing that is similar to pore size. You may create variable pore sizes, densities, and heights by adjusting the electrolyte species, temperature, anodizing voltage, and time. By etching the oxide in the proper acid, the pore size and depth may be further altered. Nuclear-track etched membranes may be used to create templates for porous polycarbonate or mica membranes. This method relies on high-energy decay pieces from a radioactive source passing through a dielectric substance. The particles leave behind chemically reactive, damaged tracks that may be etched to produce gaps across the membrane's thickness. In contrast to previous techniques, the pore density and separation are independent of pore size. Only the irradiation procedure can determine the pore density. More recently, metallic nanowires with diameters as tiny as 15 nm have been produced via electrochemical deposition of metallic nanowires on the step edges of highly oriented pyrolytic graphite (HOPG) templates. Metals like copper and palladium have been successfully processed using this technique.

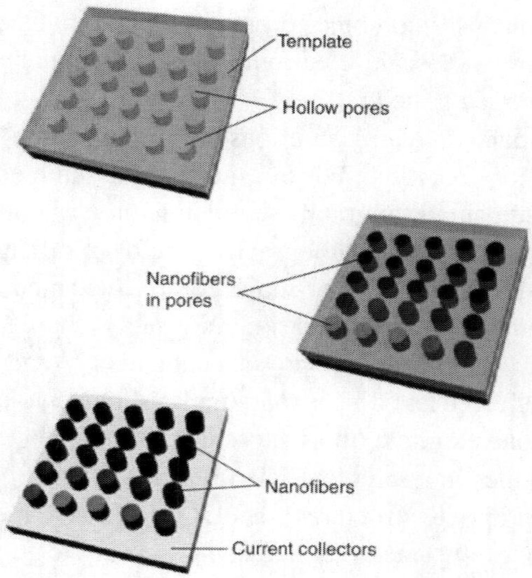

Figure 106. Template synthesis of nanofibers (Meyer and Croce 2009).

After the template has been created, the necessary substance is inserted into the template's interstitial spaces (in the case of colloidal spheres) or pores (in the case of alumina and polycarbonate membranes) as shown in Figure 106 (Meyer and Croce 2009). This may be accomplished via the use of a variety of different deposition techniques, such as electroplating and CVD. The ultimate structure may be a combination of the nano template and the substance that was deposited, or it could be an air-filled three-dimensionally complicated structure that was generated by selectively etching the template. For example, porous template matrices made from nickel, iron, and cobalt nanowires have been created by the electrochemical process. The formation of three-dimensional photonic crystals has been accomplished using the electrochemical deposition of CdSe and silicon onto polystyrene and silica colloidal assembly templates. An exciting example of template-assisted manufacturing is the production of nanometer-sized metallic barcodes. These nanobarcodes are produced by electrochemically reducing metallic ions inside the pores of an aluminium oxide membrane, and then releasing the ions by etching the template. This process creates the nanobarcodes.

Conclusion

In the realm of nanotechnology, nanosensors are an essential instrument that enable the detection of chemical and physical processes at the nanoscale. There are numerous ways that may be utilised to acquire the exact control over the size, shape, and composition of the materials used in the manufacture of nanosensors. In top-down methods like lithography and etching, a material is patterned to produce the desired structure. Lithography is a technique for creating intricately detailed patterns on a nanoscale by patterning a substance with light or electrons. It is possible to employ etching to produce structures with exact forms and dimensions by removing material to create a pattern. Self-assembly and chemical synthesis are examples of bottom-up processes that start with individual atoms and molecules and build things from the ground up. Self-assembly is a process in which discrete building elements, such proteins or lipids, spontaneously combine to create structures. Chemical synthesis, which is often used to manufacture nanoscale electronic devices, entails the controlled reaction of molecules to produce a desired structure.

Additionally, a number of hybrid solutions exist that blend top-down and bottom-up approaches to get the desired outcome. For instance, self-assembled nanostructures may have their characteristics altered using lithography or etching, or very complex devices can be made by combining chemical synthesis with lithography. The development of these approaches has led to the invention of devices with ever-more complicated and sophisticated capabilities, which has fueled the production of nanosensors. Nowadays, nanosensors are employed for a variety of purposes, such as sensing in physics, chemistry, and biology. The field-effect transistor (FET), a kind of electrical device that employs a gate voltage to regulate the flow of current via a channel, is one example of a nanosensor. Since they may be used to identify the presence of certain molecules in a sample, FETs are often utilised in the area of biosensors. In order to construct the appropriate structures for FETs, top-down methods like lithography are often used, as well as bottom-up methods like self-assembly to functionalize the device with the desired sensing material.

The optical nanosensor, a device that employs light to detect changes in a sample, is another example of a nanosensor. A variety of physical and chemical events, including as changes in temperature, pressure, and chemical concentration, may be detected using optical nanosensors. In order to build the needed structures for optical nanosensors, bottom-up methods like chemical synthesis are often used, together with top-down methods like lithography to produce the essential patterns and forms.

In conclusion, the building of nanosensors is a crucial field of study that requires exact control over the size, content, and form of the materials utilised. Nanosensors may be made using a variety of methods, including bottom-up methods like self-assembly and chemical synthesis as well as top-down methods like lithography and etching. The creation of nanosensors with distinctive qualities and capabilities has been made possible by improvements in these approaches, and the area is still developing as researchers discover new and better production procedures.

References

Cui, Lingyun, Yang Jiao, Anhe Wang, Luyang Zhao, Qianqian Dong, Xuehai Yan, and Shuo Bai. 2018. "Regulating Morphologies and Near-Infrared Photothermal Conversion of Perylene Bisimide via Sequence-Dependent Peptide Self-Assembly." *Chemical Communications* 54 (18): 2208–11. https://doi.org/10.1039/C8CC00177D.

Galeotti, F., M. Pisco, and A. Cusano. 2018. "Self-Assembly on Optical Fibers: A Powerful Nanofabrication Tool for next Generation 'Lab-on-Fiber' Optrodes." *Nanoscale* 10 (48): 22673–700. https://doi.org/10.1039/C8NR06002A.

Goyal, Amit Kumar, Amit Kumar Goyal, Hemant Sankar Dutta, Pankaj B. Agarwal, Pankaj B. Agarwal, Suchandan Pal, and Suchandan Pal. 2014. "Realization of Rod-Type Photonic Crystal Structure for Optical Sensing Using Dip-Pen Nano-Lithography." *12th International Conference on Fiber Optics and Photonics (2014), Paper M2B.4*, December, M2B.4. https://doi.org/10.1364/PHOTONICS.2014.M2B.4.

Howell, Samuel Tobias, Anya Grushina, Felix Holzner, and Juergen Brugger. 2020. "Thermal Scanning Probe Lithography—a Review." *Microsystems & Nanoengineering 2020 6:1* 6 (1): 1–24. https://doi.org/10.1038/s41378-019-0124-8.

Jain, I. P., and Garima Agarwal. 2011. "Ion Beam Induced Surface and Interface Engineering." *Surface Science Reports* 66 (3–4): 77–172. https://doi.org/10.1016/J.SURFREP.2010.11.001.

Kendre, P. N., and T. S. Satav. 2019. "Current Trends and Concepts in the Design and Development of Nanogel Carrier Systems." *Polymer Bulletin* 76 (3): 1595–1617. https://doi.org/10.1007/S00289-018-2430-Y/METRICS.

Kim, Jung Sik, and Jinho Ahn. 2018. "Mask Materials and Designs for Extreme Ultra Violet Lithography." *Electronic Materials Letters* 14 (5): 533–47. https://doi.org/10.1007/S13391-018-0058-6/METRICS.

Lang, Rossano, Alan de Menezes, Adenilson dos Santos, Shay Reboh, Eliermes Meneses, Livio Amaral, Lisandro Cardoso. 2012. "Ion-Beam-Induced Epitaxial Recrystallization Method and Its Recent Applications." *Recrystallization*, March. https://doi.org/10.5772/35355.

Li, Mingjie, Yulong Chen, Wenxin Luo, and Xing Cheng. 2021. "Interfacial Interactions during Demolding in Nanoimprint Lithography." *Micromachines 2021, Vol. 12, Page 349* 12 (4): 349. https://doi.org/10.3390/MI12040349.

Manouras, Theodore, and Panagiotis Argitis. 2020. "High Sensitivity Resists for EUV Lithography: A Review of Material Design Strategies and Performance Results." *Nanomaterials 2020, Vol. 10, Page 1593* 10 (8): 1593. https://doi.org/10.3390/NANO10081593.

Meyer, B., and F. Croce. 2009. "MATERIALS | Nanofibers." *Encyclopedia of Electrochemical Power Sources*, January, 607–12. https://doi.org/10.1016/B978-044452745-5.00058-7.

Nahal, Arashmid, Seyed Reza Hosseini, and Masoud Mahjour-Shafiei. 2021. "Ion-Beam Lithography for Fabrication of Diffractive Optical Phase Elements in Silver-Ion-Exchanged Glasses." *Journal of Materials Science: Materials in Electronics* 32 (18): 23349–62. https://doi.org/10.1007/S10854-021-06819-0/METRICS.

Patrick, Chris. 2018. "Ion Beam Sputter Deposition Allows Tailoring of Thin Film Properties." *Scilight* 2018 (51): 510013. https://doi.org/10.1063/1.5086197.

Peng, Zhiwei, Xiaolin Chen, Yulong Fan, David J. Srolovitz, and Dangyuan Lei. 2020. "Strain Engineering of 2D Semiconductors and Graphene: From Strain Fields to Band-Structure Tuning and Photonic Applications." *Light: Science & Applications 2020 9:1* 9 (1): 1–25. https://doi.org/10.1038/s41377-020-00421-5.

Peters, Robert F., Luis Gutierrez-Rivera, Steven K. Dew, and Maria Stepanova. 2015. "Surface Enhanced Raman Spectroscopy Detection of Biomolecules Using EBL Fabricated Nanostructured Substrates." *Journal of Visualized Experiments* 2015 (97). https://doi.org/10.3791/52712.

Pimpin, Alongkorn, and Werayut Srituravanich. 2012. "Reviews on Micro- and Nanolithography Techniques and Their Applications." *Engineering Journal* 16 (1): 37–55. https://doi.org/10.4186/EJ.2012.16.1.37.

"X-Ray Lithography-Integrated Circuit Fabrication." n.d. Accessed February 6, 2023. https://ebrary.net/184552/engineering/lithography.

Yilbas, Bekir Sami, Abdullah Al-Sharafi, and Haider Ali. 2019. "Surfaces for Self-Cleaning." *Self-Cleaning of Surfaces and Water Droplet Mobility*, January, 45–98. https://doi.org/10.1016/B978-0-12-814776-4.00003-3.

Chapter 6

Nanosensors: Characterization Techniques

Learning Outcomes

- The reader will have a deep insight into various characterization techniques used for analysing a nanoparticle.
- The significance of each characterization technique used for observing a particular behaviour of the nanoparticle can be clearly learned.
- The reader will have detailed knowledge about the working principle, instrumentation and the selection of appropriate characterization techniques to study the desired parameter or property or behaviour of the nanoparticle.
- The reader will be able to interpret the data obtained through various characterization techniques in a better way to add more value to their research.

Abstract

Tremendous interest has been shown in nanostructures, a rapidly expanding class of materials with numerous applications. One of the most important developments in nanotechnology over the past few decades has been the ability to design and create nano-objects with specific functionalities by incorporating functional molecules into a variety of host structures. Understanding how nanoparticles behave will reveal the material's enormous potential for use in a variety of application platforms. The interactions at the atomic and molecular levels must be carefully examined in order to use the nanoparticles as thinfilms and sensors in biological, clinical, and environmental fields, as well as in various automobile industries. After gaining a thorough understanding of the various characteristics and behaviours of the nanoparticles, the study can be expanded by engineering a variety of nanomaterial's parameters to suit the intended application. The various techniques for characterising the underlying physics and chemistry that affect the distinctive behaviour of many nanoparticles are covered in this chapter. Using the right characterization technique, characteristics such as morphology, size, shape, chemical composition, molecular vibrational patterns, thermal behaviour, lattice defects, etc., can be studied in detail.

1. Introduction

The purpose of this chapter is to provide a comprehensive collection of analytical methods that are commonly used to measure nanoparticles, providing information on one, or more, property of importance. The chapters provide up-to-date information and guidance on the use of these techniques, detailing the manner in which they may be reliably employed. The aim of the chapter is to encourage awareness of both the strengths and weaknesses of the various methods used to measure nanoparticles and raise awareness of the range of methods that are available.

Different forms of nanoparticles are used in a variety of applications. For instance, semiconducting nanoparticles are used in nano-coatings and biological or environmental sensing applications, whereas metal nanoparticles with metal behaviour are used as catalysts and conductive inks. Metal oxides are used as contrast agents. Using current or recently generated data, as well as newly developed predictive hazard assessment models, recent research activities have attempted to accurately predict the health and environmental risks associated with nanomaterials.

The characterization of nanoparticles has become increasingly important because they are synthesized from several mechanical or chemical routes (Koczkur, Mourdikoudis, Polavarapu, & Skrabalak, 2015), and they have huge potential technological applications and academic research interests across several fields.

Since nanoparticles have a large surface area-to-volume ratio, their properties are significantly altered with size, which makes the characterization an important step to understand their properties at different molecular levels. Other properties, such as texture, strain, shape anisotropy, crystalline phase, crystal defects, and crystal size impact the chemical, electronic, mechanical, and optical attributes of the nanoparticles. Without the proper characterization, the applicability of the specific nanoparticle would meet an immense challenge (Thanh, Maclean, & Mahiddine, 2014).

This chapter explains in detail about various characterization techniques used for analysing different properties and applications.

2. Classification of Characterization Techniques Based on Various Characteristics

2.1. Determination of the Morphological Characteristics

2.1.1. TEM

Transmission electron microscopy (TEM), a flexible technique that can identify a range of nanomaterials' size, shape, crystallographic structure, and chemical composition, may be used to study them (NM). The whole range of nanometre sizes, from 1 to 100 nm, may be consistently delivered by TEM analysis in the form of 2D pictures. It is feasible to quantitatively ascertain the physical properties (size, shape, and surface morphology) of certain nano-objects based on the specifics of their 2D projections when TEM imaging and image analysis are combined. Multiple attributes may be concurrently assessed for each individual particle, enabling the creation of descriptive statistics and related number-based distributions as needed by laws and regulations. To inspect and analyse the elements in nano-objects, the TEM can be used with the spectroscopic methods EDS and EELS.

Despite the abundance of data that may be collected, there are several factors that limit the use of TEM to describe nano-objects. The transfer of a representative portion of the sample having a sufficient number of particles to the specimen carrier is necessary for the accuracy of TEM analysis (TEM grid). The sample preparation, which includes the purification and concentration stages, as well as the preparation of the TEM specimen, have an impact on this. The main electrons' primary acceleration voltage determines the transmission electron microscope's spatial resolution. In order to achieve excellent spatial resolution, traditional

TEM commonly employs acceleration voltages of 100–300 kV. The beam damage restricts the analysis of "soft matter" that is susceptible to the beam, such as polymers, even though it is compatible with metallic or hard matter. The artefacts brought on by the high vacuum conditions further restrict the uses of conventional TEM. The dimension parallel to the electron beam is not normally assessed by TEM techniques, just the two dimensions of particles perpendicular to the electron beam. As a result, measurements of particles, such as platelets, that have a preferred orientation on the EM grid, may be skewed.

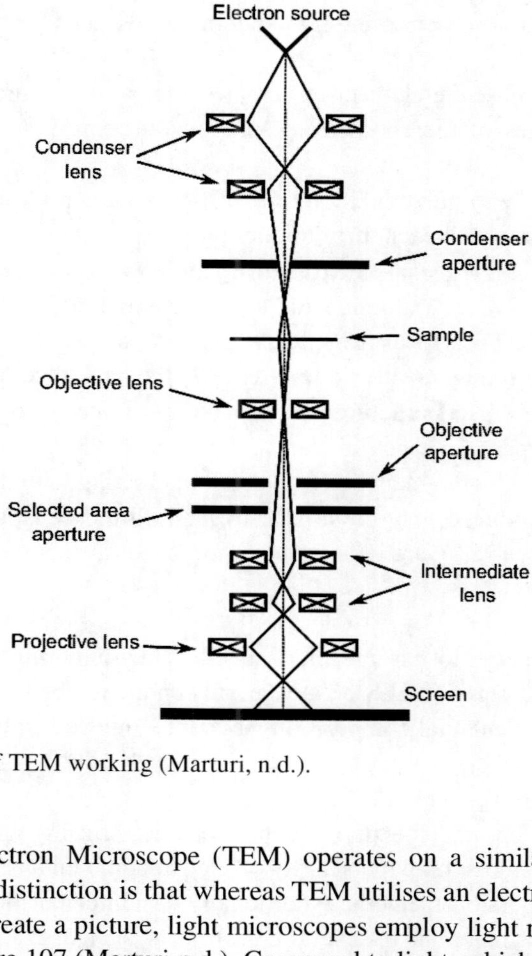

Figure 107. Schematic of TEM working (Marturi, n.d.).

2.1.1.1. Principle

The Transmission Electron Microscope (TEM) operates on a similar concept to the light microscope. The main distinction is that whereas TEM utilises an electron beam to concentrate on the specimen and create a picture, light microscopes employ light rays to focus and do the same as shown in Figure 107 (Marturi n.d.). Compared to light, which has a long wavelength, electrons have a shorter wavelength. In a light microscope, the resolution power rises as the wavelength of the light drops, whereas in a transmission electron microscope (TEM), when an electron lights the specimen, the resolution power grows as the wavelength of the electron transmission increases. Since the wavelength of electrons is around 0.005 nm, 100,000 times shorter than the wavelength of light, TEM offers about 1000 times greater resolution than a light microscope.

2.1.1.2. Working

- Electrons are emitted from an electron source at the top of the microscope and go through a vacuum in the column of the instrument.
- The electrons are focused using electromagnetic lenses to create a very tiny beam that is then guided into the target object.
- The detector is then affected by the electrons as they move through the object.
- Traditional bright field imaging depends on the sample's compositional density and crystal orientation to determine how incoming electrons scatter and vanish from the beam.
- The amount of unscattered electrons creates a "shadow picture" of the specimen, with various densities of the specimen's components represented in varying degrees of blackness.
- It is also feasible to create a 3D model of the specimen by rotating it and collecting several photographs at each turn (tomography).
- It is also possible to use electron diffraction to examine the crystal structure of materials having regular atomic structure (crystalline material).
- By mapping the back focal plane to the imaging device, discrete regions of electron localization caused by positive interference in the back focal plane may then be seen.
- The specimen's crystal structure may then be determined using the diffraction patterns.
- An energy-dispersive spectrometer (EDS) within the TEM may also find X-Ray emission caused by the main electron beam interacting with the material.
- The spectra produced may be utilised to identify the constituent elements since the resultant X-Ray energy are indicative of the atomic structure of the element they came from.

Additionally, the energy lost as a result of the inelastic scattering of electrons in specimen transmission may be measured (EELS). Valence and conduction band electrical characteristics, chemical bonding, and elemental composition may all be inferred from this knowledge.

2.1.1.3. Specimen Preparation

Nanoparticles found in liquid dispersions can be prepared for using the mentioned procedures. For dry powders, a number of papers have offered recommendations for formulating and disclosing suitable dispersion methods. Additionally, a number of dispersion protocols have been created for various materials in several EU projects, such as NanoDefine and NanoREG, and the protocols may be found in the corresponding public reports.

Achieving an equal distribution of the target particles on an appropriate carrier, in this case a TEM grid, is the aim of the specimen preparation process. In order for automated image analysis processes to properly extract the measurands of interest from captured TEM pictures, a sufficient number of particles must be put on the TEM grid, ideally as single particles. However, caution must be exercised to avoid overloading the grid with particles. For subsequent (automatic) image analysis, it is often recommended to coat a few percent of the TEM grid surface with nanoparticles.

The particles must be evenly distributed on a special TEM grid for TEM studies. These are uniform grids with varying mesh sizes, typically 3.05 mm in diameter, manufactured of different metals. Although copper (Cu) is the most often utilised material for this purpose, it is also commercially possible to find gold, nickel, and molybdenum. These grids can be found with silicon oxide coating in addition to foils that are primarily carbon and/or formvar and are transparent to electrons.

In a dispersion, particles have a surface charge known as zeta potential. The surface of the TEM grid must be opposing the charge of the particles for an effective and dependable deposition of the particles on the grid. The charge of the particles in the particular dispersion is frequently known. If this is not the case, the charge can be determined by determining how mobile the particles are when they are dispersed electrophoretically, for instance, using phase analysis light scattering. Freshly made carbon-coated TEM grids often have a little negative charge and are hydrophobic. The TEM grids' surface steadily increases in hydrophilicity over time. The TEM grids should be hydrophilic since nanoparticles are often disseminated in aqueous (polar) solutions. This can be accomplished via a glow discharge process. If the dispersion's particles have a positive charge, the grid doesn't need to be further treated. However, the TEM grids should also be handled if the particles have a negative charge.

2.1.1.4. Sample Preparation

A TEM specimen has to be thin enough to transmit enough electrons for an image to form with the least amount of energy loss. Therefore, a key component of the TEM study is specimen preparation. Ultrasonic disc cutting, dimpling, and ion milling are frequently used preparation processes for the majority of electronic materials. Dimpling is a specimen preparation method that results in a specimen with a thinned centre area and an outside rim thick enough to allow for easy handling. Traditionally, ion milling has been the last step in specimen preparation. High voltage is used in this procedure to accelerate charged argon ions to the specimen surface. Material is removed from the specimen surface by the ion impingement due to momentum transfer.

The drop-on-grid or grid-on-drop technique is one of the easiest ways to transfer particles from a liquid dispersion to the TEM grid. The TEM grid is either floated on a drop of nanoparticle dispersion (grid on drop) after functionalization, such as glow discharge followed by Alcian blue treatment, or a drop of dispersion is deposited on the TEM grid (drop on grid). TEM grids are taken out of the liquid and cleaned twice to three times in a drop of deionized water after a predetermined amount of time, ranging from a few seconds to a few minutes. This washing phase stops drying artefacts from forming, such as solid salt precipitates, which might cover up the particles of interest on the TEM grid.

The (ultra)centrifugation of the particles directly on the TEM grids is an alternate method for depositing particles from liquid suspension on the TEM grid. A centrifuge equipped with special rotors for TEM grids can be used to do this. All of the water column's particles are deposited above the TEM grid as a result of the centrifugation technique. As a result, using the particles found on the TEM grid, it is possible to determine not only the particle size distribution but also the absolute number concentration of the particles in the dispersion. The ideal particle concentration in the liquid dispersion, the centrifugation speed, and the duration may all be computed using the predicted particle size and density on the grid. The schematic of nickel chromium thin wire sample preparation for TEM analysis is shown in Figure 108 (Zhou et al. 2015).

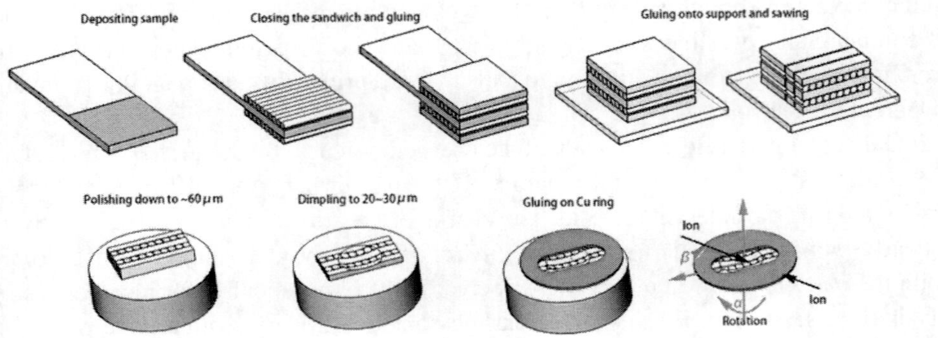

Figure 108. Schematic of NiCr thin wire sample preparation for TEM analysis (Zhou et al. 2015).

Most recently, an intriguing method that aerosolizes the dispersions with an electrospray before depositing the particles electrostatically onto electrically conductive TEM grids has been reported. This method uses an electrospray instrument to produce tiny droplets. A thin jet of liquid suspension is driven to the TEM grid as the collector and expelled from a conically formed meniscus as the end part of the liquid suspension in a capillary tip. The droplets explode as a result of electrostatic repulsion, which causes drying and ultimately results in direct deposition of particles (without solvent) on the TEM grid. Initial findings demonstrated the electrospray's relative tolerance for diverse liquid dispersions, and it was possible to get excellent particle coverage for a variety of different materials. It may be shown that no appreciable particle losses occur when electrospraying on substrate. This method would therefore allow for an approximate estimation of the particle number concentration in a liquid solution. Although this is a highly promising approach, it calls for advanced equipment that, however, is not yet offered on the market.

2.1.1.5. Advantages and Limitations of TEM

Most recently, an intriguing method that aerosolizes the dispersions with an electrospray before depositing the particles electrostatically onto electrically conductive TEM grids has been reported. This method uses an electrospray instrument to produce tiny droplets. A thin jet of liquid suspension is driven to the TEM grid as the collector and expelled from a conically formed meniscus as the end part of the liquid suspension in a capillary tip. The droplets explode as a result of electrostatic repulsion, which causes drying and ultimately results in direct deposition of particles (without solvent) on the TEM grid. Initial findings demonstrated the electrospray's relative tolerance for diverse liquid dispersions, and it was possible to get excellent particle coverage for a variety of different materials. It may be shown that no appreciable particle losses occur when electrospraying on substrate. This method would therefore allow for an approximate estimation of the particle number concentration in a liquid solution. Although this is a highly promising approach, it calls for advanced equipment that, however, is not yet offered on the market.

TEMs are often highly expensive to buy. It takes a lot of time and effort to prepare specimens for TEM viewing, and using dehydrators and chemical fixations might result in the risk of artefacts. They are employed in secluded locations where they are not exposed because of how sensitive they are to vibrations and electro-magnetic motions. Unless they employ a fluorescent screen at the conclusion of viewing, it generates monochromatic visuals. Additionally, it uses more energy.

2.1.2. SEM

To outperform the transmission electron microscope, Mafred von Ardenne created the first scanning electron microscope in 1937. He focused an electron beam on the raster and utilised high-resolution power to scan the area. He also tried to lessen the issues caused by the chromatic aberrations in pictures taken using transmission electron microscopes. Scientists and research organisations, such Cambridge Scientific Instrument Company, continued their researches and in 1965 created the Stereoscan, a fully functional scanning electron microscope.

SEMs with high resolution are very effective at characterising the size and form of nanoparticles. Sample setup and picture capture are both rather rapid and easy processes. A concentrated, high-energy electron beam is utilised by a scanning electron microscope (SEM) to produce high resolution, magnified, two-dimensional pictures of the material. An area of the solid sample's surface is chosen to receive the electron beam for this purpose. Different signals are generated as a result of the interaction between the beam's electrons and the sample. These signals are captured and put through further processing to create digital pictures. A beam of low-energy electrons is used in SEM to focus and scan specimens. In order to achieve superior resolution than optical microscopes, electron microscopes employ a very small wavelength. A scanning electron microscope's magnification generally varies from 20X to over 30,000X. A scanning electron microscope has a spatial resolution range of 50 to 100 nm.

2.1.2.1. Principle

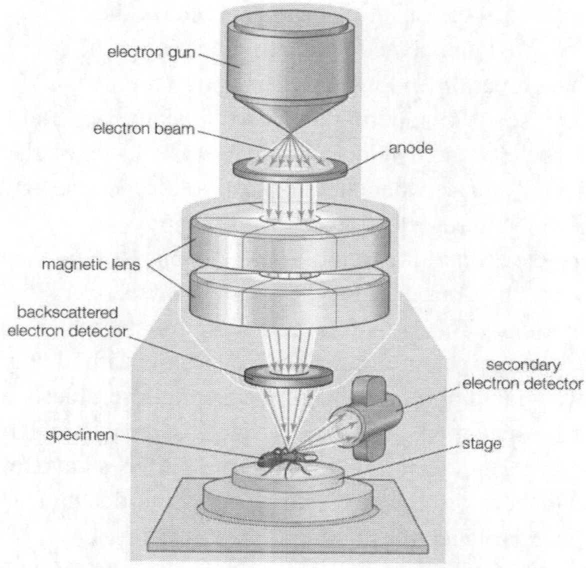

Figure 109. Schematic of a SEM working principle ("Scanning Electron Microscope | Definition, Images, Uses, Advantages, & Facts | Britannica" n.d.).

The scanning electron microscope employs emitted electrons as contrasted to transmitted electrons, which are used by the transmission electron microscope as shown in Figure 109

("Scanning Electron Microscope | Definition, Images, Uses, Advantages, & Facts | Britannica" n.d.). According to the scanning electron microscope's operating theory, kinetic energy is used to generate signals from electron interactions. These electrons, which are utilised to observe crystalline elements and photons, are secondary electrons (SE), backscattered electrons (BSE), and diffracted backscattered electrons (DBSE). A picture is created using BSE and SE. While the backscattered electrons reveal contrast in the composition of the specimen's components, the secondary electrons that are released by the specimen serve as the principal means of identifying the morphology and topography of the specimen.

2.1.2.2. Working

1 The tungsten filament lamps at the top of the column, which are equivalent to those used in TEM, are the source of the electrons
2 After thermal energy is provided to the electron source, the electrons are released and permitted to flow quickly to the anode, which is charged positively.
3 Primary electrons with high energy levels and secondary electrons with low energy are emitted from the specimen surface as a result of the electron beam.
4 By interacting with the specimen, the electron beam generates signals that provide details about the specimen's surface topography and composition.
5 Without any preparation, even air-dried samples may be inspected to see how the specimen appears under the scanning electron microscope. However, microbiological specimens need to be fixed, dehydrated, and dried in order to maintain the structural properties of the cells and prevent cell collapse when exposed to the high vacuum of the microscope.
6 To enable spatial dispersion of electric charges on the surface of the specimen, the samples are mounted and coated with a thin coating of heavy metal components. This improves picture creation and increases image quality.
7 This microscope scans by tapering an electron beam back and forth over a small area of the microscope. Secondary electrons are a small cluster of electrons that are released by the specimen's surface when electrons strike it. These secondary electrons are then captured by a specialised detection device.
8 The secondary electrons hit a scintillator as soon as they enter the detector and get there (a luminescence material that fluoresces when struck by a charged particle or high-energy photon). This produces brief light bursts, which are transformed into an electric current via a photomultiplier and sent to the cathode ray tube as a signal. This creates a viewable and photographable image that resembles a television image.
9 The nature of the specimen, i.e., raised surfaces to receive high quantities of secondary electrons entering the detector while depressed surfaces have fewer electrons reaching the surface and therefore fewer electrons enter the detector, strongly influences the quantity of secondary electrons that enter the detector.
10 As a result, on the screen, elevated objects will seem brighter and depressed areas darker.

2.1.2.3. Sample Preparation

The sample preparation process has a significant impact on the accuracy of SEM nanoparticle size and shape measurement results. The properties of the nanoparticles and the measurand, the quantity or quality to be measured, dictate more or less great procedures rather than a single ideal preparation technique for all nanoparticles. Making sure that the test or analytical samples are representative and pertinent to the measurand is essential. The equipment required to prepare and handle nanoparticles must be maintained sanitised and stored dry in a clean enclosure or bench, ideally in high-efficiency particulate air (HEPA) filtered air where the sample preparations are carried out.

Typically, nanoparticle samples are powders or solutions that frequently need to be diluted in order to photograph discrete particles or to reduce particle accumulation. Smaller particles that are "hidden" behind bigger particles cannot be seen if the presence of smaller particles on larger particles may be seen using high-resolution SEM. It is feasible to measure contacting particles, but to prevent unacceptably large measurement error, particle separation may need to be helped by an operator. In order to create homogenous, representative test or analytical samples, fewer nanoparticles from the original nanoparticle material must be separated and deposited on a substrate surface. For a large number of nanoparticles to be measured across, the contrast between the particles and substrate needs to be strong and the particles close enough to one another.

Numerous nanoparticles may be in liquid colloidal phase, or suspended within another substance, where agglomeration is prevented by electrostatic or steric stabilising techniques. These need to be diluted in order to reduce the quantity of contacting nanoparticles on the measuring substrate or grid. Results compared to simple deposition are better when nanoparticles are captured using charge-based methods and the suspension media is carefully removed.

For the fabrication of nanoparticle sample, TEM grids are commercially available. To promote effective particle dispersion and capture, these surfaces may contain surface coatings or functionalized surfaces. To the raw liquid nanoparticle material, dilution fluids and surfactants may be added to aid in the creation of usable test samples. Both of these might change the apparent size of the nanoparticles depending on imaging settings, therefore due precautions must be taken and, if required, data must be corrected for these effects.

2.1.2.4. Applications of SEM

- The information obtained from SEM images are used for spot chemical analysis in the energy dispersive X-Ray spectroscopy.
- It can be used to study the surface of solid objects and analyse the distribution of atoms in various elements.
- Cosmetologists make use of a scanning electron microscope to analyse fine details of tiny cosmetic components.
- Manufacturing industries employ scanning electron microscopes to look for the contaminants and impurities in manufactured items.
- The pharmaceutical industries use them for good-bad testing of drugs, medicines, and other products.
- Scanning electron microscope is also used in qualitative chemical analysis of elements by proving a clear and magnified image of the crystalline structures.

- A scanning electron microscope is quite advantageous in nanotechnology and other related fields. It can provide precise measurements and detailed images of objects that are over 50nm in size.
- It can be used to distinguish different phases of a multiphase sample.
- Some of the scanning electron microscopes are equipped with diffracted backscattered electron detectors, which helps to examine and determine the microfabric and crystallographic orientation of substances.
- A scanning electron microscope is typically used to produce high definition images of objects that can display spatial variation of chemical compounds.
- It is generally used in the medicine domain to observe the bacterial interaction with the skin and body organs. This helps the doctors determine the nature of the bacterial disease and to find the cure.

2.1.2.5. Limitations of SEM

- Scanning electron microscopes are comparatively expensive.
- Some microscopes must fulfil certain special conditions before their use. For instance, the room must be free of vibrations and electromagnetic radiation.
- Scanning electron microscopes have a bulky structure.
- A consistent voltage level must be maintained for the proper operation of a scanning electron microscope. This may require additional electronic circuitry or voltage regulators to fix the voltage magnitude to a constant value.
- A cooling system should be attached with such microscopes.
- The sample should be small enough to fit within the chamber of the microscope. The horizontal dimensions of the sample should not exceed 10 cm, while the vertical dimension is much more constrained and must be less than 40 mm.
- The sample to be examined with the help of a scanning electron microscope needs to be solid in nature. Wet samples are unsuitable and are required to be decrepitated first.
- A scanning electron microscope cannot be used for light materials such as hydrogen, helium, lithium, etc.
- To study samples that are insulators with the help of a scanning electron microscope, an electrically conductive coating is required to be applied on their surface. However, it can be neglected if the device is able to work in low vacuum mode.
- Scanning of living samples with the help of a scanning electron microscope is not possible.

2.2. Determination of the Structural Characteristics

2.2.1. BET

Research on nanotechnology has recently branched out into areas including medicine, the environment, space exploration, and energy. When it comes to the number of atoms in a bulk material, the surface area to volume ratio is negligible. However, when the particles are just 1 to 100 nm wide, other features start to emerge. For instance, nanoparticle zinc oxide can have surface areas as high as 54 m2/g, whereas commercial grade zinc oxide has a surface area range of 2.5 to 12 m2/g. Since the nanoparticles outperform the bulk material at blocking UV rays, they are beneficial in products like sunscreen. Knowing a nanoparticle's surface area is crucial since many valuable aspects of nanoparticles result from their small size. In 1938, Stephen Brunauer, Paul Emmett, and Edward Teller created the BET hypothesis. The Langmuir hypothesis, created by Irving Langmuir in 1916, was expanded upon by the BET theory. The Langmuir hypothesis connects the gas pressure of a medium above a solid surface at a certain temperature to the monolayer adsorption of gas molecules, also known as adsorbates, onto a solid surface.

For the purpose of calculating the surface area of solid or porous materials, the Brunauer-Emmett-Teller (BET) theory is applied. Given that the size of a material's surface has an impact on how that solid will interact with its surroundings, it provides crucial information about their physical structure. The surface area of a substance is frequently connected to several aspects, including dissolving rates, catalytic activity, moisture retention, and shelf life. Surface area analysis is one of the most frequently used techniques in material characterisation and is essential to the design and production of solids.

Although this approach is frequently used for most materials, it is most accurate for materials having a Type II or Type IV isotherm and a high enough amount of surface-gas contact. The BET hypothesis might not apply to materials with different types of isotherms for a variety of reasons, thus care should be taken.

2.2.1.1. Principle

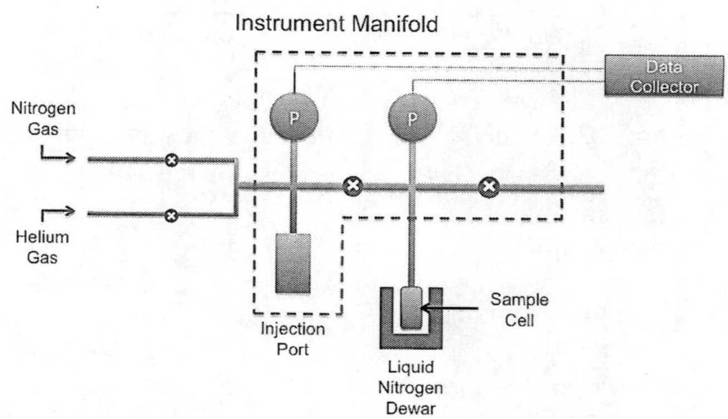

Figure 110. Schematic representation of BET working ("2.3: BET Surface Area Analysis of Nanoparticles - Chemistry LibreTexts" n.d.).

Adsorption is the process by which gas molecules or atoms attach to a surface. Absorption, which happens when a fluid permeates a liquid or solid, should not be confused with adsorption. The exposed surface area, temperature, gas pressure, and the degree of gas-solid interaction all have an impact on how much gas is adsorbed. Because it is widely accessible in high purity and interacts strongly with most materials, nitrogen is routinely used in BET surface area analysis. Due to the normally weak contact between gaseous and solid phases, the surface must be cooled using liquid N_2 in order to produce significant amounts of adsorption. The sample cell is then filled with known quantities of nitrogen gas. Relative pressures lower than atmospheric pressure can be achieved by creating conditions of partial vacuum. No matter how high the pressure is raised, no further adsorption can occur after the saturation pressure. Pressure transducers that are extremely accurate and precise are used to measure pressure variations caused by the adsorption process. Once the adsorption layers have formed, the sample is removed from the nitrogen atmosphere, which allows the adsorbed nitrogen to be released from the substance and quantified. The acquired information is shown using a BET isotherm, which displays the amount of gas adsorbed as a function of relative pressure. There are five distinct types of adsorption isotherms that might exist.

2.2.1.2. Working

The solid sample is chilled to cryogenic temperature under vacuum in order to calculate the surface area (using liquid nitrogen). The solid sample (adsorbent) is dosed with nitrogen gas in regulated increments (as is normal for an adsorbate) as shown in Figure 110 ("2.3: BET Surface Area Analysis of Nanoparticles - Chemistry LibreTexts" n.d.). The relative pressure (P/P_0) is allowed to equalise following each dosage of nitrogen gas before calculating the weight (W) of nitrogen adsorbed. In a precise sense, the BET equation describes a linear plot of $1/((P_0/P)-1)$ versus. When nitrogen is used as the adsorbate, P/P_0 is often constrained to a narrow area of the adsorption isotherm, typically in the P/P0 range of 0.05 to 0.35. This figure allows one to calculate the weight of nitrogen (W_m) that makes up a monolayer of surface coverage. Using the BET equation and the known molecular cross-sectional area of the nitrogen molecule, the slope and intercept of the BET plot may be used to determine the sample's total surface area.

$$\frac{1}{W\left(\frac{P_0}{P-1}\right)} = \left(\frac{1}{W_m C} + \frac{C-1}{W_m C}\right)\left(\frac{P}{P_0}\right)$$

The value of the term C, or the BET C constant, which is connected to the energy of adsorption in the first adsorbed layer, provides insight into the strength of the adsorbent/adsorbate interactions.

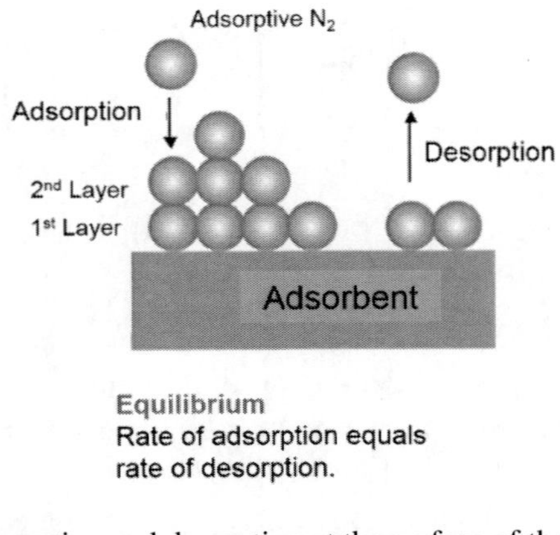

Figure 111. Illustration of adsorption and desorption at the surface of the adsorbent ("Surface Characterization: BET Surface/Specific Surface | 3P Instruments" n.d.).

The gas pressure is steadily expanded until all pores are loaded up with nitrogen particles, so, all in all estimations of the pore volume and size dissemination are performed. The nitrogen gas that has collected in the framework is then vanished by step by step bringing down the gas pressure. Data with respect to the pore volume and pore size dispersion might be acquired by examining the adsorption and desorption isotherms as illustrated in Figure 111 ("Surface Characterization: BET Surface/Specific Surface | 3P Instruments" n.d.).

2.2.1.3. Sample Preparation

To unequivocally assess the surface region, the sample should initially be degassed to take out water and different debasements. Tests are warmed to a high temperature and degassed in a vacuum. To diminish the degassing time, the greatest temperature that might be utilized without hurting the sample's construction is for the most part chosen. To ensure that undesired fumes and gases are killed from the sample's surface, the IUPAC prompts degassing tests for something like 16 hours. More modest degassing periods are frequently noticed for tests that can support more noteworthy temperatures without underlying changes. The BET needs no less than 0.5 g of test to compute the surface region precisely.

Glass cells are utilized to hold tests while they are degassed and exposed to Wager investigation. To lessen how much void space inside the cell, glass poles are situated there. The machine's outgas port is associated with the cells, which are set inside warming mantles. The cell is shipped to the investigation port once the sample has been degassed. The sample is cooled and kept at a reliable temperature utilizing fluid nitrogen dewars. For quantifiable degrees of adsorption to happen, a low temperature should be kept up with between the gas particles and the sample's surface. An adjusted cylinder is utilized to infuse the adsorbate, which in this case is nitrogen gas, into the sample cell. When each estimation, the sample cell's dead volume should be aligned. Since helium doesn't adsorb onto the sample, helium gas is used for a clear disagreement request to achieve that.

2.2.1.4. Limitations

Since estimations must be utilized to compute the surface area of dry powders, the BET technique has a few downsides. The adsorption of gas particles requests a great deal of time in this strategy. There should be a great deal of manual readiness.

2.2.2. XRD

Scientists are ceaselessly taking a stab at a more exact instrument for their nanoscale characterisation as new strategies for making nanomaterials are created. Since it might associate with the electrons in an iota's inward shell, X-beam diffraction (XRD) is rapidly turning into the most famous strategy for depicting nanoparticles. The XRD procedure is the most ideal for powdered samples that have quite recently been made by drying the comparing colloidal answers for the samples. Measuring the creation of nanoparticles is made more straightforward by looking at the sample's pinnacle area and power with designs from a few diffraction data set references utilizing XRD gear.

2.2.2.1. Principle

By warming a fiber in a cathode beam tube, one method for creating X-beams (energies fluctuate from 100 eV to 100 keV) is to utilize that gadget as shown in Figure 112 (Zhang et al. 2017). At the point when a X-beam test fulfills the Bragg Condition, $n\lambda=2d\sin\theta$, the occurrence monochromatic light emission causes a top in force (helpful impedance).

Here, n is any integer, d is the distance between diffracting planes, and is the incidence angle. λ is the wavelength of the beam. The diffracted X-Ray radiation is measured and examined by a detector, which then transforms it into a count rate that is shown on a computer screen.

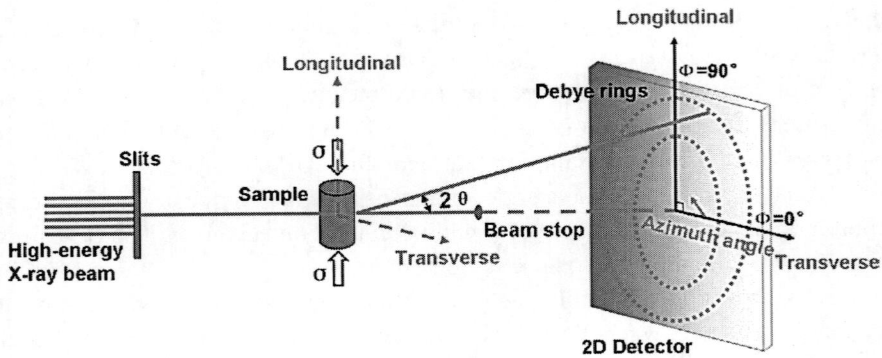

Figure 112. Illustration of XRD working (Zhang et al. 2017).

The atomic location inside the lattice planes affects the XRD peak intensity. As a result, XRD may be effectively employed to get structural data for nanomaterials such lattice parameters, phase nature, and crystalline structure. Data collection is straightforward, making it a natural choice for characterisation of nanomaterials.

2.2.2.2. Crystal Structure and Size Dependent Nature of XRD Peaks

Quantum dab nanoparticles made of cadmium sulfur (Discs) are generally utilized in an assortment of nanoscience and nanotechnology areas. For sample, XRD investigation might be utilized to gauge the molecule size of these quantum specks. XRD tops extend as crystallite size approaches the nanoscale. Top widening forcefully increments when the glasslike size diminishes from 50 to 25 nm. Moreover, when size diminishes to 10 nm, the pinnacle enlarging turns out to be significant to such an extent that top strength starts to debilitate; tops much of the time cross-over and are trying to observe.

Thus, size-subordinate XRD top widening significantly affects how nanomaterials are portrayed. For instance, assuming that the XRD design shows sharp pinnacles that are more steady with particles of a lot bigger size, yet the TEM examination uncovers round particles with a typical measurement of 10 nm, then, at that point, most of the mass sample isn't comprised of 10 nm particles; rather, they address just a minority part of the populace. Also, the hexagonal wurtzite structure and cubic zinc blend polymorphs can every so often exist together in the Cds nanoparticles. The XRD sample of the mix appears to have bigger relative forces than wurtzite in this case since a portion of the wurtzite tops cross-over with the zinc blende tops.

A successful strategy for grouping materials and recognizing different glasslike stages is X-beam diffraction. Nonetheless, it very well may be hard to infer cross section constants with decimal accuracy for nanocrystalline materials that have observable pinnacles. Established researchers faces difficulties in further improving the accuracy and goal of the XRD strategy for nanoanalysis. We can't, notwithstanding, excuse the likelihood that a cautious examination of the XRD information could yield canny information and assist with connecting minuscule outcomes to the mass material. Also, flow research inspects the production of elective systems to expand the use of X-beam diffraction.

2.2.3. IR

The infrared section of the electromagnetic spectrum, or light with a longer wavelength and a lower frequency than visible light, is the subject of IR spectroscopy, sometimes known as infrared spectroscopy. The study of a molecule's interaction with infrared light is known as infrared spectroscopy. Three methods may often be used to investigate the idea of IR spectroscopy: reflection, emission, and absorption measurements. Finding the functional groups of molecules—relevant to both organic and inorganic chemistry—is the main use of infrared spectroscopy.

An IR spectrum is basically a graph with the frequency or wavelength on the X-axis and the amount of absorbed infrared light on the Y-axis. IR Spectroscopy finds infrared light frequencies that are absorbed by molecules. Due to the fact that these particular light frequencies match the frequency of the bonds in molecules, molecules have a tendency to absorb them.

Infrared radiation contains the energy expected to invigorate atomic securities and influence them to vibrate all the more brutally. In any case, just polar bonds will collaborate with electromagnetic infrared radiation. The electric field part of an electromagnetic wave can invigorate the vibrational energy of an atom when there are unmistakable districts of halfway sure and negative charge present in the particle. The dipole snapshot of the specific particle changes couple with the adjustment of vibrational energy. The extremity of the connection influences how much energy is assimilated. Balanced non-polar bonds in N≡N and O=O can't associate with an electric field; thus they can't retain radiation.

2.2.3.1. Principle

The IR spectroscopy hypothesis is based on the idea that molecules have a tendency to absorb particular light frequencies that are unique to the corresponding structure of the molecules. The energies depend on the atomic masses, the related vibronic coupling, and the geometry of the molecular surfaces. It is commonly known that molecules experience a variety of vibrational movements that are unique to their atoms, in addition to the simple rotation of groups around a single bond. All organic substances then absorb infrared light, which has an energy similar to the vibrations. Similar in principle to other spectrometers, infrared spectrometers enable chemists to get absorption spectra for compounds that only reflect the molecular structures.

The infrared spectrum is the most basic measurement that is made using infrared spectroscopy. This graph shows the observed infrared intensity as a function of light frequency. Using IR Spectroscopy, one may determine which functional groups exist by looking at the atoms' vibratory patterns. Higher stretch frequencies are oscillated by lighter atoms and stronger bonds (wavenumber).

2.2.3.2. Working

The Figure 113 shows the infrared spectroscopy equipment. First, a single IR beam is divided in two and directed at the reference and the sample, respectively. Both of these beams are now reflected in order to travel via a splitter and a detector. After the CPU decodes the data that was passed by the detector, the necessary reading is printed out ("IR Spectroscopy - Principle and Instrumentation of Infrared Spectroscopy" n.d.).

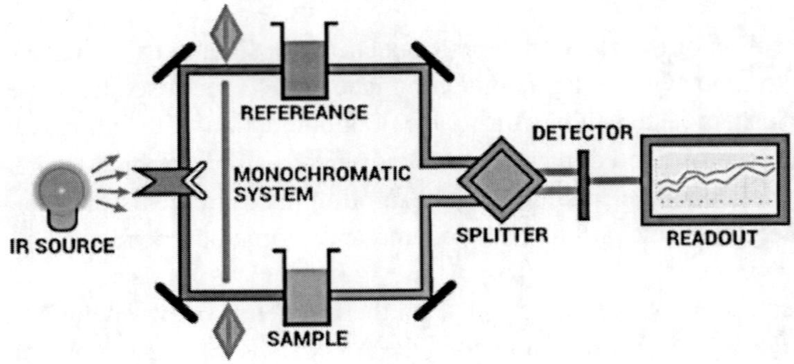

Figure 113. Schematic of IR spectroscopy working ("IR Spectroscopy - Principle and Instrumentation of Infrared Spectroscopy" n.d.).

In order to operate, IR instruments need an energy source that continuously generates IR radiation that is strong enough to be detected and reach the required wavelength. It is possible to gauge the intensity of infrared light using detectors. They use pyroelectric detectors as well as thermocouples, bolometers, thermisters, and golay cells.

2.2.3.3. Sample Types

- *Solid* – Strong samples are arranged utilizing various procedures, including as the squeezed pellet strategy, strong altercation arrangement reflect on process, strong movies, etc.
- *Liquid* – The fluid sample cell, which is made of salt halides, can hold fluid samples. Watery solvents are incapable in light of the fact that they break down salt halids. Just natural solvents, similar to chloroform, are permitted to be utilized.
- *Gas*– It is similar to the sampling of liquids.

2.2.3.4. Advantages of IR Spectroscopy

High Sweep Speed: In one moment, infrared spectroscopy can accumulate information for the entire recurrence range. To finish the output before the substance starts to corrupt, IR can be used to assess a synthetic that isn't especially steady.

High Goal: The goal of an infrared spectrometer is considerably higher than that of a common crystal spectrometer, which has a goal of just around 3 cm^{-1}. Grinding's infrared spectrometer, for sample, has a goal of 0.2 cm^{-1}, while the FT's goal goes from 0.1 to 0.005 cm^{-1}.

High Responsiveness: The Fourier Change wipes out the prerequisite for the cut and monochromator in the infrared spectrometer. This will bring about a more noteworthy reflection specularity and less energy being lost all through the investigation interaction. Since of this, even incredibly minute measures of analytes may be distinguished on the grounds that the energy that arrives at the finder is adequate. Today, a sample just 1 to 10 grams may be distinguished utilizing infrared spectroscopy.

Extensive variety of Utilizations: Practically all natural synthetics and certain inorganic mixtures might be concentrated on utilizing infrared spectroscopy. Both subjective and quantitative examination can involve it in different ways. Also, there are no stage limitations on the infrared spectroscopy test. It very well may be a gas, a fluid, or a strong, significantly growing the assortment of analytes.

Adequate Data: Infrared spectra may give us an abundance of primary information about the analytes, including the sort of particle, the utilitarian gathering, the stereoscopic design, the sum and area of substituent gatherings, and that's just the beginning. Infrared spectroscopy has formed into an incredible way to deal with recognize different sorts of synthetic compounds, contingent upon the data that is open from the useful piece and the unique finger impression part.

Non-Horrendous: The sample isn't harmed by infrared spectroscopy.

2.2.3.5. Limitations of IR Spectroscopy

Test Limitation: Since water incredibly assimilates IR light, infrared spectroscopy isn't fitting to tests that contain water.

Range intricacy: The understanding of the IR range requires broad ability because of its intricacy. Once in a while, a compound's design not entirely settled unhesitatingly from a solitary IR spectra. To completely comprehend the novel design, more spectroscopic strategies like atomic attractive reverberation (NMR) and mass spectrometry (MS) are as yet required.

Quantitative examination: Infrared spectroscopy is compelling for breaking down a large number of tests subjectively, however it has constraints with regards to quantitative examination, like in circumstances with very high and low focuses.

2.3. Estimation of the Nanoparticle Size

2.3.1. AFM

One of the filtering test microscopy (SPM) techniques called nuclear power microscopy (AFM) regularly utilizes a tip to truly communicate with a sample surface as shown in Figure 114 ("Atomic Force Microscopy: General Principles and Applications" n.d.). By connecting to the actual amount recognized during the contact of the tip with the sample, it is feasible to produce an image of the sample surface. A high-goal 3D profile on the request for parts of a nanometer is given by AFM. AFM normally gauges the power between the fine tip and the sample, which gives level data about the sample being estimated (van der Waals powers, dipole collaborations, or electrostatic power).

One of the scanning probe microscopy (SPM) methods, atomic force microscopy (AFM), shares with SPM the use of a tip that interacts physically with the surface of a sample. In relation to the physical quantity measured during the contact of the tip with the sample, it is possible to generate a picture of the sample's surface. AFM offers a high resolution 3D profile on the order of nanometer-sized fractions. AFM typically measures the force between the fine tip and the sample, which gives height information about the sample being measured (van der Waals force, dipole-dipole interactions, or electrostatic force).

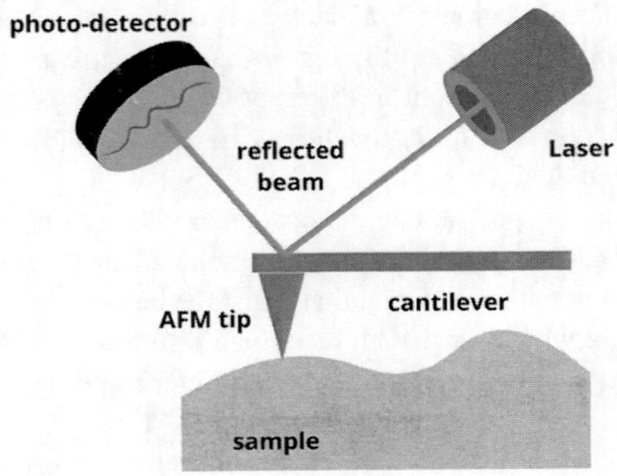

Figure 114. Schematic of AFM working ("Atomic Force Microscopy: General Principles and Applications" n.d.).

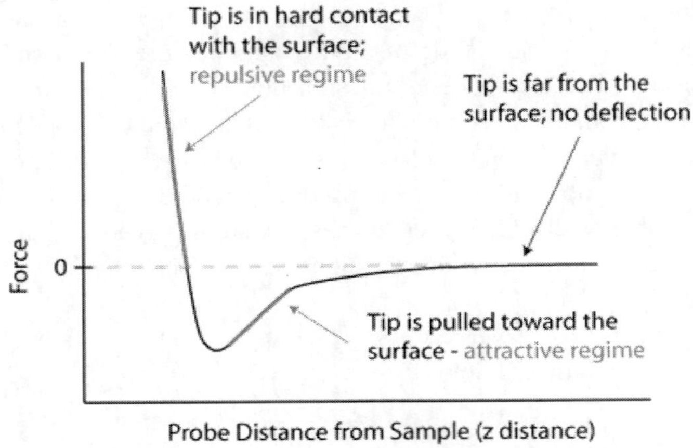

Figure 115. Relation between force and distance in AFM ("Atomic Force Microscopy - Nanoscience Instruments" n.d.).

2.3.1.1. Working

The construction and working principle of the AFM is discussed in the steps below:

- The piezoelectric crystals are made of ceramic materials which have the special ability to expand and contract when a voltage gradient is present or vice versa, i.e. it will generate an electric potential on application of a mechanical force or pressure.
- The AFM consists of a sharp tip, sometimes called as a probe, attached to a cantilever. The end of the sharp tip is used to scan the surface of the sample.
- The cantilever generally will be made of silicon (Si) or a silicon nitride (Si_3N_4) and the tip will have the radius of curvature in the order of nanometers.
- When the tip is brought in close proximity to the surface of the sample, then according to Hooke's law a deflection in the cantilever will be generated due to the force between the tip and the sample surface.

- A laser light is made to fall on the cantilever vertically, as the cantilever experiences a displacement there will be an equivalent change in the laser light being reflected from the backside of the cantilever surface.
- This deflection in the cantilever is recorded using the photodetector and is converted into a corresponding electrical signal.
- The intensity of the electrical signal generated will be proportional to the displacement of the cantilever.
- The electrical signal thus generated is sent to the feedback loop to maintain the distance between the tip and the sample along z-axis, whereas the sample is scanned along the x-y plane.
- Until the tip and the sample are maintained at a constant distance, the height of the sample along z-axis can be obtained from the deflection of the cantilever. Hence, the feedback loop plays a vital role in maintain that constant distance.
- The obtained electric signals are recorderd as a function of x-y plane which is simply the relation between force and distance as shown in Figure 115 ("Atomic Force Microscopy - Nanoscience Instruments" n.d.). Thus the 3D image of the sample containing the height information is obtained.

Others characterization techniques that can be used: XRD, SEM, TEM

2.4. Determination of the Elemental Studies

2.4.1. EDX

Energy-dispersive X-Ray spectroscopy (EDS), one of the usually utilized scientific procedures for deciding the basic organization of strong matter, has as of late achieved huge unmistakable quality as to its utilization in the substance examination of nanoparticles, especially when joined with the utilization of a SEM and the transmission activity method of SEM (STEM-in-SEM), where the imaging capacity of the magnifying instrument recognizes the species. The data delivered by EDX investigation comprises of spectra with tops relating to the constituent parts of the material under study. Moreover plausible are picture investigation and basic planning of a sample. Since individual nanoparticles can now be promptly examined with EDS at a SEM thanks to technology advancements with very sensitive EDS detectors, this development was primarily driven by them. Within the same scanned region as the high-resolution information of the sample surface morphology supplied by the electron microscope, qualitative information on elemental composition with a spatial resolution of around 10 nm may be obtained. EDX becomes extremely effective when used in a multi-technique approach, especially in contamination analysis and industrial forensic science investigations. The method can map out the spatial distribution of the elements and be qualitative, semi-quantitative, or quantitative. The EDX technique is non-destructive, allowing for minimal to no sample preparation and in situ examination of specimens of interest.

2.4.1.1. Principle

EDS, EDX, or EDXA, commonly known as energy-dispersive X-Ray spectroscopy, is a potent method that enables the user to examine the elemental makeup of a chosen sample. The main mechanism by which EDS works is the ability of high intensity electromagnetic radiation (X-Rays) to expel "core" electrons from an atom (electron that is not in the outermost shell) as shown in Figure 116. This theory, sometimes referred to as Moseley's Law, established a direct connection between the atomic number of an object and the frequency of light it emits. A higher energy electron can fill the hole left by the removal of these electrons, and when it relaxes, it will release energy. By exposing a sample to X-Rays, one may determine which elements are present and in what proportion by measuring the energy released during this relaxation process, which is specific to each element on the periodic table.

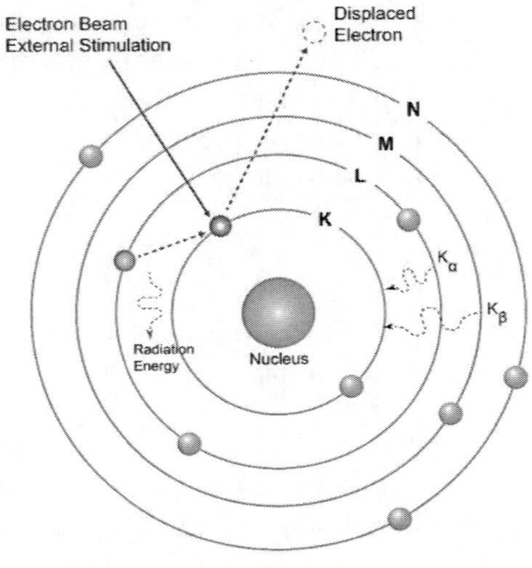

Figure 116. Illustration to demonstrate the principle of EDX.

2.4.1.2. Working

The underpinning of the natural examination strategy known as EDX microanalysis is the production of particular X-Rays in the sample's particles by the approaching pillar electrons as shown in Figure 117 (Colpan, Nalbant, and Ercelik 2018). Flexible dissipating and an inelastic dispersing are two fundamental actual cycles that follow the impact with the iotas. Inelastic dissipating is the deficiency of energy with no recognizable shift in course and is regularly created by connections with both the bound electrons and the core in the iotas. Flexible dispersing is the course adjustment of the electron with no recognizable energy misfortune, which is commonly brought about by cooperations with the core involving the materials. The energy of the X-beam photon is the potential energy coming from the distinction between the two orbitals participated in the change, making the X-beams normal for the component. The iotas are ionized, and when they return to their ground state, they will create trademark X-beams. A photon-energy delicate finder may then be utilized to screen the X-beam discharge at different frequencies. These X-beams are characteristic of the component from which they come and uncover the components that are available in the material.

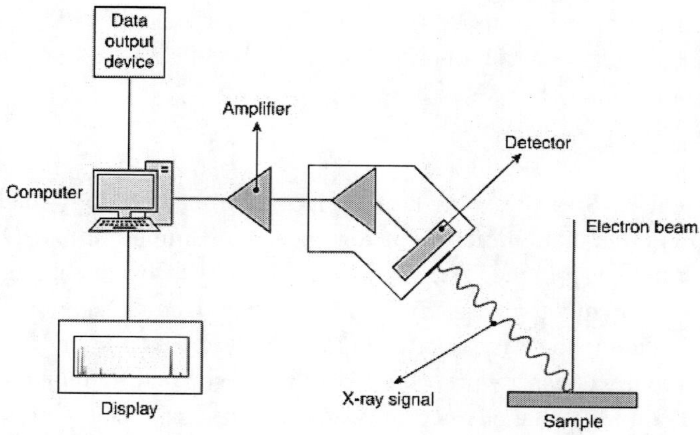

Figure 117. Schematic of EDX working (Colpan, Nalbant, and Ercelik 2018).

The mid-energy (1–20 keV) X-Rays gathered during any given analysis period are all simultaneously shown by the EDX detector system, and the energy of the X-Rays is reproduced as a spectrum, which is a histogram plot of the number of counts versus the X-Ray energy. Semiqualitative and semiquantitative data are both included in the spectrum. The element is identified by the position of a peak in the spectrum and its energy; the area under the peak is proportional to the number of atoms of the element in the region that was exposed to radiation. When the electron beam is slowed by the electric fields of the atomic nuclei of the elements present in the material, X-Rays are also generated. Below the peaks of that spectrum, there is a continuous emission made up of these X-Rays.

2.4.1.3. EDX and Nanomaterials

The far and wide worry over the likely gamble of openness to nanoparticles is regularly connected to the change in harmfulness that happens when nanoscale is reached. For instance, synthetics like TiO_2 and undefined SiO_2 that are frequently viewed as adequate for individuals to be around in the micrometric reach might adversely affect individuals' wellbeing when they are around at the nanometric scale. To address the security of nanoscale materials, it is pivotal to comprehend natural responses to nanoparticles.

Since nanotechnologies have the ability to safeguard payloads from both metabolic and off-target impacts as well as permit designated conveyance of freight to resistant cells, inoculation and medicine conveyance have as of late been gotten to the next level. The comprehension of NPs' potential poisonousness is critical for organic applications and targets to accomplish this objective. Human monocyte-macrophages sequester HS NPs in a compartment that empowers free dispersion with the extracellular climate utilizing the EDX. Since calcium phosphate and hydroxyapatite nanoparticles are many times utilized in clinical applications, like bone embed materials, DNA and SiRNA conveyance vectors, and slow-discharge immunizations, they explored this nanoparticle. In order to clarify the processes by which cells absorb this NP, investigators utilised SEM and EDX. The scientists showed that surface-connected compartment (SCC), a complex labyrinth-like structure immediately connected to the extracellular space, is where hydrophilic HA NPs are sequestered. These findings demonstrate that this specialised compartment has the capacity to remove significant amounts of foreign material from the extracellular space, followed by gradual disintegration, while minimising harm to cells or tissues nearby. The research on the potential effects of dietary Ag nanoparticles on the gut microbiota showed that oral exposure to human relevant levels of AgNPs can cause microbial changes in the gut.

2.4.1.4. Limitations

Despite the fact that EDS is an exceptionally supportive methodology, there are a couple of issues with the technique that cutoff its convenience. Most importantly, EDS isn't commonly an exceptionally touchy technique. The energy delivered by X-beams in the wake of striking a sample won't be adequate to precisely gauge a component's rate in the event that the centralization of that component is excessively low. Second, for components with a low nuclear number, EDS frequently doesn't work. Since both hydrogen and helium just have a n=1 shell, no center electrons should be taken out to deliver X-beam emanation. Conversely, lithium and beryllium have such low nuclear numbers that the energy of the X-beams transmitted by Li or Be tests is lacking for estimation and, subsequently, they are much of the time unfit to be inspected.

The thickness of the sample is an extra test with the strategy. Because of electrons having the option to relocate all the more effectively to external energy levels because of test thickness, the outcomes might digress from those normal. Covering x-beam outflows are one more type of mistake that can change KeV values. Moreover, just top layers might be precisely surveyed by the technique since X-beams are not particularly great at infiltrating materials more profound than a couple of nanometers. In this way, a crisscross between the outside and internal material layers will not necessarily in every case be found in EDS.

2.4.2. XPS

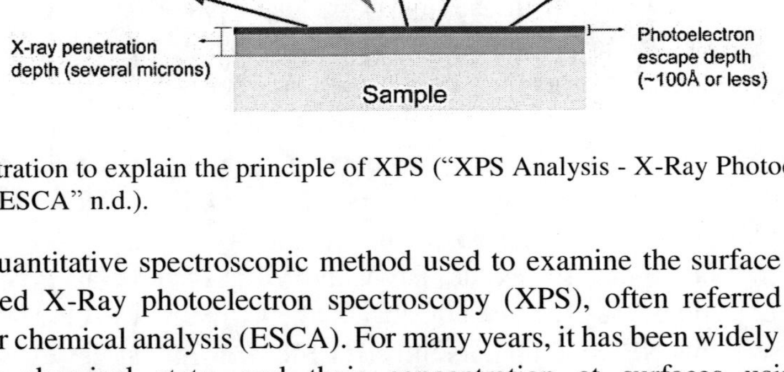

Figure 118. Illustration to explain the principle of XPS ("XPS Analysis - X-Ray Photoelectron Spectroscopy, or ESCA" n.d.).

The sensitive quantitative spectroscopic method used to examine the surface chemistry of a material is called X-Ray photoelectron spectroscopy (XPS), often referred to as electron spectroscopy for chemical analysis (ESCA). For many years, it has been widely used to identify elements, their chemical state, and their concentration at surfaces using the X-Ray photoelectron spectroscopy (XPS) technique. For determining the makeup and thickness of thin (less than 10 nm) surface coatings, it is the most precise technique. The restricted distance that electrons can go through materials without losing energy brings forth the technique's data profundity, and these aspects are similar to those that are critical in nanoparticle research. The best method for using XPS is related to different strategies that can measure the size and type of the particles in a nanomaterial. Surfaces play a vital part in the usefulness, poisonousness, and conduct of nanoparticles, consequently XPS is a critical strategy to ensure that particles are very much portrayed.

2.4.2.1. Working

Electron bombardment of a water-cooled target produces X-Rays as shown in Figure 118 ("XPS Analysis - X-Ray Photoelectron Spectroscopy, or ESCA" n.d.). This target, which is positively charged to speed up the electrons, is known as the anode. The target produces peaks from X-Ray fluorescence as well as a wide range of X-Ray energies from Bremsstrahlung radiation. The monochromator, which is typically a single crystal utilising Bragg reflection, is used to concentrate the X-Rays back onto the sample by choosing a specific range of X-Ray energies at the location of one of these peaks. The majority of laboratory-based equipment uses aluminium as the anode material, and quartz crystals are used to monochromatize the Al Kα radiation to produce X-Rays with an energy of 1486.6 eV. At the point when electrons are removed from the sample, they go through a focal point, which modifies their dynamic energy and focuses them on the analyser's entrance. The analyzer has two concentric sides of the equator, with the inward bearing a positive charge in contrast with the external. With the legitimate dynamic energy, or pass energy, electrons entering the analyzer through the section gap will go through the analyzer to the leave cut and strike the identifier. Because of some precise deviation, electrons with pass energy will likewise go through the leave gap. Diversely dynamically vigorous electrons will influence the metal sides of the equator or the gap and go undetected. An XPS equipment may also have an electron source to flood the sample with low-energy electrons and an ion source to pelt it with high-energy ions, among other characteristics. Insulating or weakly conductive materials are utilised as the electron source because they will charge positively when exposed to X-Rays, changing the electrons' energy. In order to perform examination on buried layers and interfaces, the ion source is employed to clea samples or remove material from them. Figure 119 shows the schematic of XPS working (Seyama, Soma, and Theng 2013).

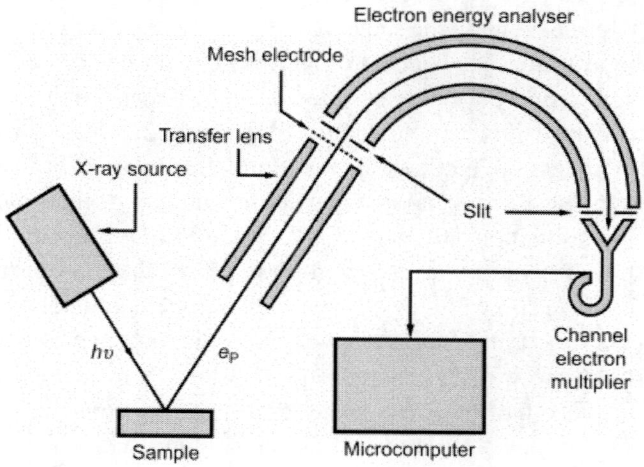

Figure 119. Schematic of XPS working (Seyama, Soma, and Theng 2013).

Given the multifaceted design of these instruments, it's critical to do routine tests to ensure they convey reproducible outcomes. Consistently getting spectra from a reference material, for sample, falter cleaned gold, and contrasting them with spectra got after establishment or adjustment could assist with guaranteeing this.

2.4.2.2. Data Interpretation

The most direct and regular method for addressing XPS information from tests with a few perceivable components is to change over the powers into an equivalent homogenous sythesis that is then expressed as a nuclear portion or rate. For every component, a pinnacle is decided explicitly for this. Tops in the XPS range demonstrate the extent of electrons with a specific restricting energy. Less electrons are demonstrated by more limited tops. For example, in the event that one pinnacle is around 50% of the level of another pinnacle, it implies that half as numerous electrons were related to the limiting energy at the higher top as there were at the more limited top. Thus, the pinnacle powers uncover a material's rate cosmetics. The seriously restricting energy there is, the more attracted an electron is to the core. For instance, tops from electrons in the main state will have more energy than tops from electrons in the subsequent state. The energy of electrons in 2s will be higher than that of 2p. While certain instruments offer pinnacle ID attributes, it is as yet conceivable to distinguish pinnacles and lines on spectra by contrasting them with norms made of different materials. For help with deciphering the spectra, Moulder (1992) gives instances of these guidelines.

2.4.2.3. Limitations

At the point when lines or pinnacles cross-over, it very well may be trying to tell which particles' electrons are being addressed. Drill lines are impressively designed lines in the XPS information yield range that are irrelevant to the dynamic energy from electron ionization. Little lines or pinnacles known as X-Ray apparition lines are somewhat askew from what might be expected to owe to X-Ray sources other than the actual sensor. The way of behaving at the surface and the mass might vary altogether since only the surface could be examined.

Since nanoparticles must be analysed collectively, XPS will produce average attributes for the population under study. The particles should be displayed as a monolayer or sub-monolayer on a substrate that does not contain the same materials as the particle in order to guarantee that studies based on single particles are accurate. The alternate strategy is to analyse a thick layer of nanoparticles that obscures the substrate since many coatings include carbon and almost all substrates have accidental carbon contamination. Since the signal from particles below the first layer may alter the findings and, particularly at low emission angles, shadowing of the particles by the nearby particles must be taken into account, the validity of the single particle model is called into doubt in this situation.

Since nanoparticles should be investigated all in all, XPS will deliver normal credits for the populace under study. The particles ought to be shown as a monolayer or sub-monolayer on a substrate that doesn't contain similar materials as the molecule to ensure that reviews in light of single particles are precise. The substitute system is to investigate a thick layer of nanoparticles that clouds the substrate since numerous coatings incorporate carbon and practically all substrates have inadvertent carbon tainting. Since the sign from particles underneath the principal layer might change the discoveries and, especially at low outflow points, shadowing of the particles by the close by particles should be considered, the legitimacy of the single molecule model is called into uncertainty in this present circumstance. While playing out a total recreation of electron transport, precise outcomes might be accomplished for circular particles with uniform coatings. This study said that XPS was more precise than TEM, which was compelled by factual significance and picture contrast concerns and missing the mark on characterized process for laying out stage limits, in estimating the thickness of silicon oxide on silicon particles.

For nanomaterials with the vital data profundity, XPS offers helpful substance information that can be utilized to assess the circulation of components and their mixtures at the surface as well as their surface arrangement. It has the advantage of being viable on practically any strong substance and being less hurtful to radiation-delicate materials. Except for hydrogen and helium, XPS can recognize all components, distinguish their oxidation states by taking a gander at the pinnacle areas and shapes, and measure how much every part. The qualities procured from XPS forces might be dependably and promptly associated with synthetic pieces and actual aspects if the design of the nanoparticles is known. Utilizing the methodology on contemporary instruments is very straightforward and available.

2.4.3. Raman spectroscopy

One of the vibrational spectroscopic methods used to learn about molecule rotations, vibrations, crystal structures, and low-frequency modes in a sample is Raman spectroscopy. This spectroscopic method irradiates the sample using a monochromatic laser light source that emits light in the visible, near infrared, or near ultraviolet ranges. The molecular vibrations, phonons, or other excitations of the sample interact with this monochromatic light, shifting the energy of the laser photons up or down and producing Raman scattering, an inelastic scattering of the laser light that is recorded by a CCD camera. These changes in laser intensity provide researchers information about the sample's vibrational modes, making it feasible to identify molecules by their distinctive fingerprinting patterns.

2.4.3.1. Raman Scattering

At the point when light communicates with a substance, it is scattered, yet for all intents and purposes every last bit of it holds its unique energy. Flexible dissipating or Rayleigh dispersing is what this is. Inelastic dissipating, once in a while alluded to as Raman dissipating, happens when simply a minuscule level of the scattered light encounters an energy gain or misfortune as shown in Figure 120.

The polarizability of the electrons in a sample particle decides major areas of strength for how Raman Impact will be. The photon invigorates the sample in this sort of inelastic dissipating, making the atom momentarily occupy the virtual energy express, an unsteady condition, before the photon is removed. Stirs up and Hostile to Stirs up Raman dissipating are the two types of Raman scattering.

The electron in the ground state is stimulated and returns to the fundamental vibrational state when incoming light or a photon with energy E_0 interacts with a molecule of the sample. The molecule absorbs energy from the incoming photon during this process. Therefore, the dispersed light that results will have an energy E that is less than the incident light, i.e, $E<E_0$. Stokes Raman scattering is the name of this type of scattering a shown in Figure 121.

The electron that is already in the excited vibrational state will revert to the ground state when incoming light or a photon of energy E_0 interacts with a molecule of the sample. The molecule releases extra energy to the incident photon during this process. Therefore, the dispersed light that results will have an energy E greater than the incident light, or $E>E_0$. Anti-Stokes Raman scattering is the name of this type of scattering.

Around 99.999% of the episode light go through versatile or Rayleigh dispersing, which can't be utilized for sub-atomic portrayal. While the occurrence light of about just 0.001% goes through inelastic or Raman dispersing with an energy contrast of $E_0 \pm E$.

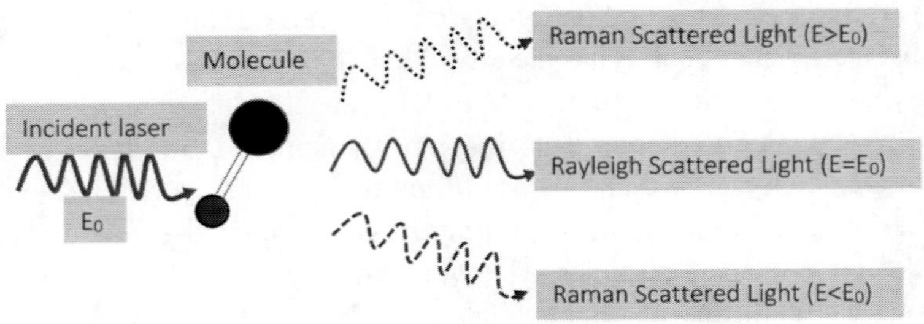

Figure 120. Scattering of light by molecules.

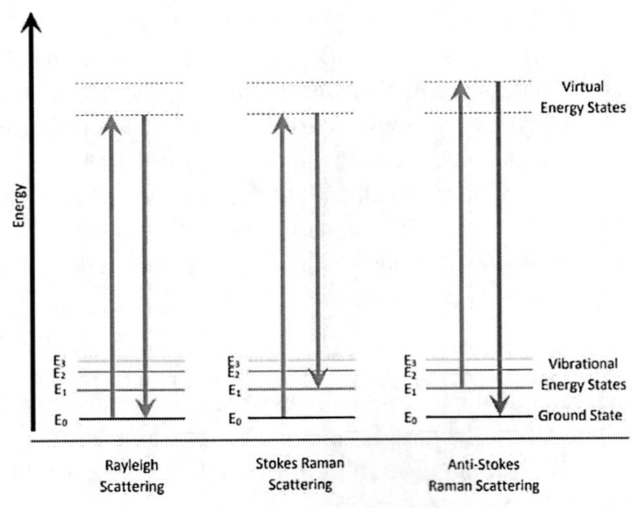

Figure 121. Rayleigh and Raman scattering process.

2.4.3.2. Working

The schematic of the working of Raman spectroscopy is shown in Figure 122 ("The Raman Spectrophotometer" n.d.). The Raman spectroscopy consists of four major components:

1. Excitation Source (Laser)
2. Sample illumination system and Light collection optics
3. Wavelength selector (Filter or Spectrophotometer)
4. Detector (Photodiode array, CCD or PMT)

A sample is by and large enlightened with a laser beam in the bright ultra-violet (UV), visible (Vis) or near infrared (NIR) range. The dissipated light is then gathered with a focal point and is sent through an obstruction channel or spectrophotometer to accomplish the Raman range of the expected sample.

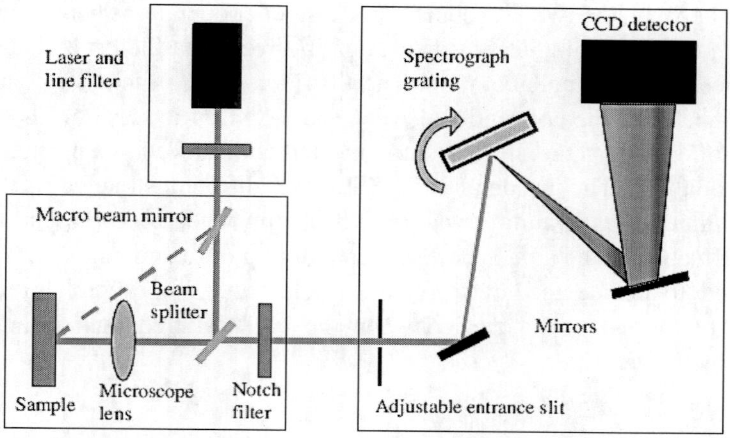

Figure 122. Schematic of Raman spectroscopy working ("The Raman Spectrophotometer" n.d.).

These signs should be switched off since Raman dispersing is a lot more vulnerable than Rayleigh dissipating. The force of the Rayleigh dispersed sign will regularly be close to the power of the laser energy and frequency since it will normally have forces that are a lot higher than those of the Raman dissipated signal. Basically removing the phantom band close to the laser where the Rayleigh dissipated sign will have significant effect tackles this issue. To do this, the bothersome ghastly district is removed utilizing a frequency choice or score channel. The Rayleigh scatted range impacts can likewise be limited by utilizing a few scattering stages.

Finally, the Raman scattered light spectrum is picked up by a detector. The Raman scattered light can be detected by single-point detectors like photon-counting Photomultiplier Tubes (PMT), multi-channel detectors like Photodiode Arrays (PDA), or Charge-Coupled Devices (CCD).

2.4.4. Nano FTIR

The infrared part of the electromagnetic radiation spectrum, which has a longer wavelength and a lower frequency than visible light, is measured using Fourier Transform Infrared Spectroscopy (FTIR) Analysis. When exposed to infrared light (IR), a sample contains this spectrum that may be measured as shown in Figure 123 (Ojeda and Dittrich 2012). The fundamental idea at play is that different elements' bonds absorb light at various frequencies. The fusion of several methodologies results in new developments continually. AFM, scattering-type scanning near-field optical microscopy, and FTIR spectroscopy are all used to inform one such technique, known as nanoscale FTIR (s-SNOM). As we begin to create some distance from utilizing mass materials and toward nano-and micron-sized materials, the ability to assess a material or substance with a nanoscale goal is turning out to be progressively more pivotal. Be that as it may, to accomplish exact portrayal and productivity, it is important to know about the geographical design and characteristics of these materials.

2.4.4.1. Principle and Working of Nano FTIR

Utilizing a metal-covered AFM tip, Nano-FTIR focuses an IR light bar from a light source through a pillar splitter and an explanatory mirror and onto the tip to check the outer layer of a sample. The laser pillar utilized by nano-FTIR, instead of s-SNOM, is the essential differentiation between the two procedures. Because of cooperations between the light and the metal covering, the tip is enlightened when a light source is sparkled at it. A huge close field is likewise delivered by this collaboration. The tip filters the surface and cooperates with the sample, very much like the conventional AFM and s-SNOM modes do. These collaborations give the place of every iota the tip has communicated with as well as a portion of its properties or underlying qualities. The upside of nano-FTIR over different strategies is that the produced close field communicates straightforwardly with the tip when it comes into contact with the sample. This affects how the light is backscattered during the examination, as the backscattered light is utilized to decide the area of the particles on a superficial level. A close field interferogram of the surface is created when the backscattered light is interferometrically distinguished by the identifier.

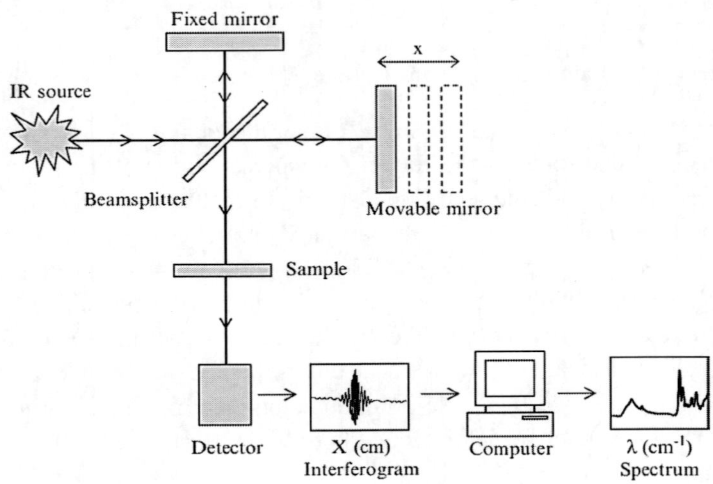

Figure 123. Schematic of FTIR spectroscopy (Ojeda and Dittrich 2012).

The close field signal is upgraded, lessening the foundation commotion while catching the adequacy and period of the light since the sample is placed in one of the interferometer arms (rather than outside the interferometer as in FTIR). Thus, a FTIR range is produced, which might be contrasted with existing data sets (however at a more noteworthy phantom goal than customary FTIR).

2.4.4.2. Application and Advantages of Nano FTIR in Nanoscale Operation

The ability to scan the surface with nanoscale resolution is of great benefit to anyone performing a surface analysis on a material or substance. The nanoscale resolution arises from the instrument performing hyperspectral imaging on the material by analyzing both the phase and amplitude of the detected light. This differs from FTIR which doesn't analyze the phase information of the detected light. This enables a spectrum to be produced for every pixel of the image/scanned area, and because an atomically sharp tip is used to analyze the surface, each of these pixels has a nanoscale resolution that can provide a lot of information about the surface of the material and its properties. The average spatial resolution that is achieved using nano-FTIR is between 10 and 20 nm.

The utilization of a broadband IR light source, which recognizes nano-FTIR from FTIR and related procedures, permits it to give IR spectroscopy investigation on exceptionally minuscule measures of materials, remembering for single particles and monolayer nanomaterials.

Different materials can likewise be utilized with nano-FTIR. The spectra of numerous substance mixtures, polymers, and organic materials might be rapidly and precisely recognized by contrasting them with a known information base. Be that as it may, a portion of these milder materials must be dissected since the actual procedure is non-disastrous and doesn't need test readiness, not at all like different strategies whose functioning rule or test planning techniques could harm gentler samples. A technique is often joined with different strategies to give beneficial information, however in spite of what its name proposes, it can give information from the terahertz (THz) to the noticeable ghostly reach, making it an extremely flexible examination strategy. To this end it has been utilized all the more often as of late on the slender materials and nanomaterials that are filling in notoriety.

2.5. Determination of the Optical Properties

2.5.1. UV-Vis/ Absorption spectroscopy

The amount of radiation absorbed by the sample is determined using the wavelength-dependent molecular spectroscopic technique known as absorption spectroscopy. The presence and concentration of a certain drug in the sample may be determined using absorbance spectroscopy. The most popular types of absorption spectroscopy are those that employ ultraviolet, visible, and infrared light.

2.5.1.1. Absorption Spectrum

The level of radiation from the approaching radiation that is consumed by the material over a scope of frequencies is known as the ingestion range of a material. The nuclear and atomic cosmetics of the particular material administers this reach. At the point when a sample is presented to radiation, the electrons inside it retain photons with energy that roughly matches the energy hole of the particles. This excitation makes the electrons move from a lower energy state orbital to a higher energy state orbital. The electromagnetic range displays a recurrence subordinate variety in the strength of retained radiation, which is alluded to as the retention range.

2.5.1.2. Working

Bright Noticeable spectroscopy is the type of retention spectroscopy that we employ here. The assimilation range of the sample is focused on in this spectroscopic technique in the brilliant and perhaps ghostly location. In the electromagnetic range, the iotas and molecules in the sample will run into an electronic advancement. The electronic transition will involve the σ, π and n electrons. In general, there are four possible electronic transitions that can occur, namely, $\sigma \rightarrow \sigma^*$, $n \rightarrow \sigma^*$, $n \rightarrow \pi^*$ and $\pi \rightarrow \pi^*$. Whereas the σ and π electrons are called as bonding electrons in the ground state, the n electrons are called as non-bonding electrons and the σ^* and π^* electrons are called as anti-bonding electrons in the excited state. Figure 124 illustrates the transition of electronic spectra utilized in UV-Vis spectroscopy ("14.11: Electronic Spectra: Ultraviolet and Visible Spectroscopy - Chemistry LibreTexts" n.d.).

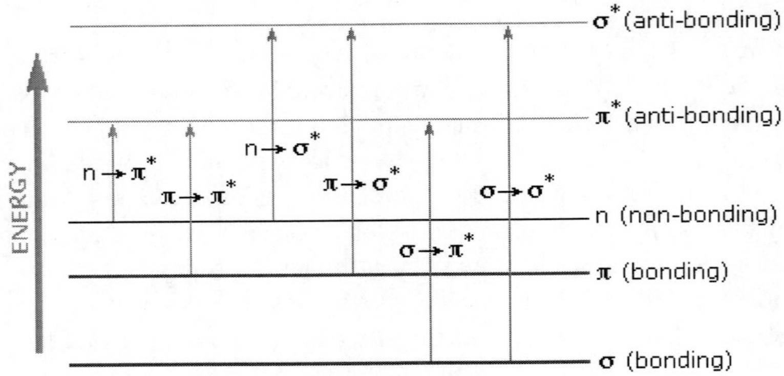

Figure 124. Illustration of the electronic spectra of UV-Vis spectroscopy ("14.11: Electronic Spectra: Ultraviolet and Visible Spectroscopy - Chemistry LibreTexts" n.d.).

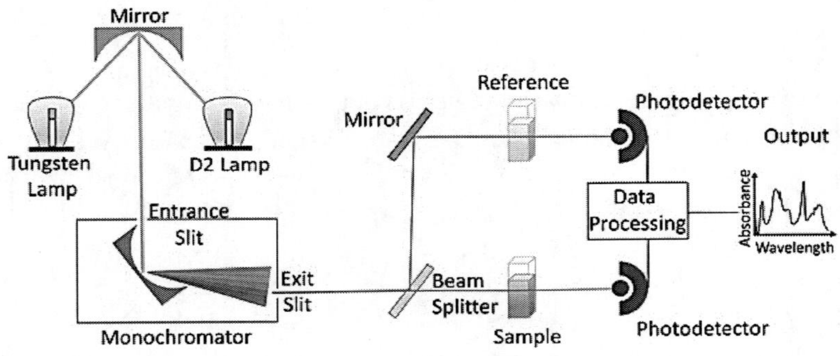

Figure 125. Schematic of UV-Vis Spectrophotometer working (Rocha et al. 2018).

The Figure 125 shows the schematic of the working of a UV-Vis spectrophotometer (Rocha et al. 2018). Following are the means engaged with using bright noticeable spectroscopy to decide a sample's retention spectra.

- A light emission estimating m is aimed at the sample by the brilliant source.
- A frequency selector or a monochromator is utilized to pick the particular locale of the electromagnetic range, for this situation the noticeable and bright ghostly reach.
- The required sample is put inside a reasonable sample compartment, and the picked scope of the produced radiation is allowed to go through. The most widely recognized sample compartments are made out of quartz glass or combined silica.
- The discovery gadget gets the light that is as yet coursing through the sample. This photodetector is utilized to quantify the got radiation's frequency.
- The photomultiplier tube, which is many times a variety of photodiodes and electron multipliers, is utilized to enhance the incredibly low radiation that the identifier will actually want to distinguish.

- The discoveries are then contrasted with the radiation data from the occurrence source to decide the ingestion spectra, which not entirely settled and introduced utilizing a read-out.

2.5.2. PL

A methodology habitually used to depict the optoelectronic qualities of semiconductors and different materials is photoluminescence. Its functioning idea is clear: a laser with energy more noteworthy than the bandgap energizes electrons in the material's valence band to move them into the conductance band. Accordingly, the photoexcited transporters loosen up before precipitously recombining with conduction band openings. With regards to coordinate semiconductors, the additional energy is delivered as light (unconstrained discharge). Deciding the material's response as far as power as a component of frequency by taking a gander at the light's spectrum is possible. Thusly, you might acquire insights concerning the band structure, for sample, the bandgap width, the general light creation productivity, the material's quality (inhomogeneous widening), and so on. Controlling the sample's environmental factors, for example, by presenting an attractive field or changing the sample's temperature, could yield extra data.

On a solitary seat top stage, Raman examination and PL identification might be utilized to evaluate a material's vibrational and electrical qualities. Confocal planning abilities with sub-micron spatial goal are made conceivable by joined Raman-PL frameworks. From the UV to the NIR, an extensive variety of excitation frequencies are possible, empowering control of the volume examined as well as the profundity of entrance into the material.

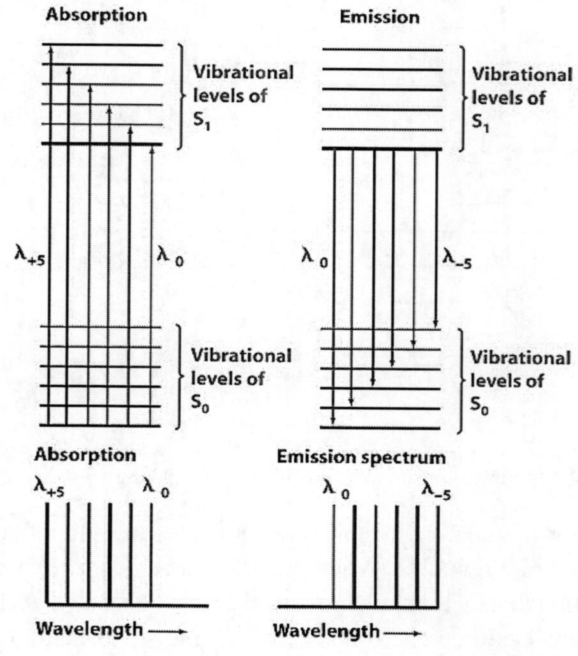

Figure 126. Energy level diagram of absorption end emission spectrum ("Photoluminescence Spectroscopy and Its Applications - Physical Methods in Chemistry and Nano Science - OpenStax CNX" n.d.).

Fluorescence and brightness are two results of photoluminescence used in fluorescence spectroscopy. The Photoluminescence quantum yield or PLQY of a particle or material is characterized as the quantity of photons discharged, as a negligible portion of the quantity of photons consumed is one of the normal procedures for Fluorescence Spectroscopy. The Figure 126 shows the energy level diagram of the absorption and emission spectrum of Photoluminescence spectroscopy ("Photoluminescence Spectroscopy and Its Applications - Physical Methods in Chemistry and Nano Science - OpenStax CNX" n.d.).

Chromophore accumulation frequently brings about total caused extinguishing, which lessens light discharge in most photoluminescent frameworks (ACQ). This suggests that fluorophores should be utilized and concentrated on in weakened arrangements or as single particles. This prompts diminished awareness of fluorescence-based gadgets, for example, biosensors and bioassays. Nonetheless, there have recently been situations when luminogen accumulation contributed emphatically to the light-producing process instead of adversely. This collected prompted emanation (AIE), especially comparable to strong state gadgets, has a great deal of likely significance. A helpful procedure for inspecting a fluorophore's brilliant qualities is photoluminescence spectroscopy.

2.5.2.1. Working

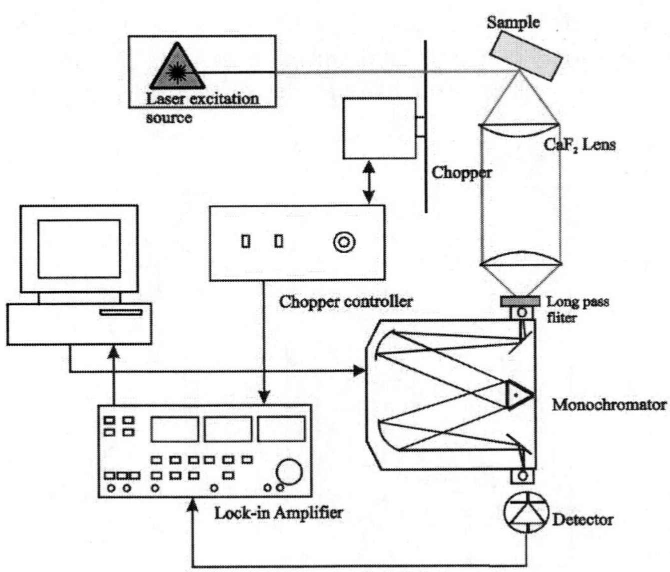

Figure 127. Schematic of Photoluminescence Spectroscopy working (Hughes 2007).

An optical chopper (or another light-modulating device) may modulate a continuous-wave (CW) laser's light at up to a few kHz. When the modulated beam strikes the sample, it excites the electrons in the conductance band, moving them from the valence band. The sample's spontaneous emission is captured and transferred to a monochromator or spectrometer, where the light's intensity versus wavelength is determined. Using optical filters to prevent laser light is a good idea because it is also gathered and typically has a much higher intensity.

Particularly in open table-top installations, ambient light can substantially affect the measurement. To optimise the rejection of spurious light components, the laser light and the output light must be modified and monitored with a lock-in amplifier as shown in Figure 127 (Hughes 2007).

2.5.2.2. Bandgap Determination of Nanoparticles Using PL

The band gap, or radiative progress in semiconductors, is the energy contrast between states in the conduction and valence groups. It is feasible to assess the range dissemination of PL from a semiconductor to nondestructively sort out the electronic band hole. This makes it conceivable to gauge the natural cosmetics of compound semiconductors, which is a significant material trademark influencing the viability of sun powered cell gadgets.

2.5.2.3. Surface Area Defects and Shape Determination of Nanoparticles Using PL

Conventional techniques that are often utilised, such XRD, IR, and Raman spectroscopy, are frequently insufficiently sensitive for supported oxide catalysts with low metal oxide concentrations. However, photoluminescence may be utilised as a probe of electron-hole surface processes since it is highly sensitive to surface effects or adsorbed species of semiconductor particles.

2.5.2.4. Understanding the Recombination of Electrons in a Nanomaterial Using PL

Recombination, the process of returning to equilibrium, can entail both radiative and nonradiative processes. The ratio of radiative and nonradiative recombination rates and the amount of PL that a material emits are closely correlated. Nonradiative rates are often linked to impurities, and photoluminescence intensity, as well as how it depends on temperature and photo-excitation intensity, is directly linked to the primary recombination mechanism. In order to understand the underlying physics of the recombination mechanism, study of photoluminescence may be used to qualitatively monitor changes in material quality as a function of growth and processing conditions.

2.5.2.5. Limitation of PL

- Even though photoluminescence is not frequently a quantitative approach, it may be utilised to locate optical centres at extremely low concentrations. This is the main scientific limitation of photoluminescence because many optical centres may have several excited states that are not filled at low temperatures.
- The signal's extinction is yet another shortcoming of photoluminescence spectroscopy. For instance, no sharp-line photoluminescence was seen when silicon photoluminescence centres at 969 meV were characterised after the centres had captured self-interstitials.

2.6. Determination of the Surface Charge

2.6.1. Zeta Potential

The net surface charge of nanoparticles is connected with the zeta potential. It's fundamental for sorting out the colloidal soundness of charged particles and fathoming how well your framework functions under different conditions. Because of the breakdown of particles at the fluid strong connection point or the adsorption of surface-dynamic substances, nanoparticles in fluids are fundamentally electrically charged. This electric interfacial charge's component known as the zeta potential is urgent to the aggregative dependability of nanodispersions. Zeta potential still up in the air to explore or screen surface adsorption of nanoparticles and different mixtures from the fluid stage as well as to recognize the zone of strength for nanodispersions. Zeta potential is an idea that must be laid out by means of the use of the right hypothesis; it can't be estimated straightforwardly.

Zeta potential relies upon both molecule and medium qualities as well as how they communicate at the connection point; it isn't just an element of suspended particles. This interfacial harmony and, consequently, the zeta potential are both affected by changes in the fluid's substance and ionic organization. Subsequently, both sample readiness and estimation procedures could impact the result of the estimation. Unseemly sample readiness and estimation procedures or the ill-advised utilization of hypothetical models for processing zeta potential from estimation information lead to mistaken discoveries much of the time.

2.6.1.1. Principle

All solid-liquid and liquid-liquid colloidal systems display the physical characteristic known as zeta potential. The Stern layer, a thin layer of ions with the anti-particle surface charge, surrounds the surface of all dispersed particles. The double layer is a layer of more loosely-associated ions of opposing charge to the surface that is farther from the surface and moves with the particle as it passes through a medium as a result of Brownian motion or sedimentation. The voltage at the edge of the sliding (shear) plane in relation to the bulk dispersion media, where ions, molecules, and other agents are no longer connected to a particle's surface, is known as the zeta potential. Due to repellent electrostatic interactions between particles with similar charges, two neighbouring particles will not aggregate if they have sufficiently strong zeta potentials of the same sign (Pate and Safier 2016).

2.6.1.2. Working

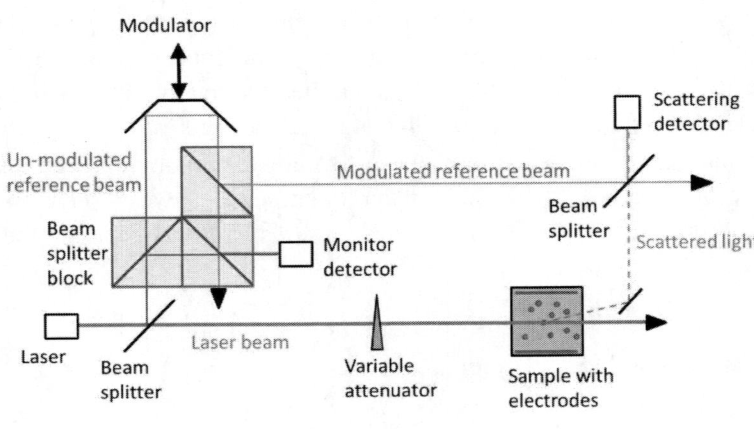

Figure 128. Schematic of Zeta potential measurement working ("Zeta Potential :: Anton Paar Wiki" n.d.).

A sample is placed into a single-use, foldable capillary cell to test the zeta potential. Two conductive electrodes are included in the cells, which fold inward to establish contact with the liquid sample on the inside and the instrument's applied voltage on the outside. Charged particles' zeta potential inside the cell controls how rapidly they move through the medium. Particles travelling more quickly have a bigger zeta magnitude, whereas those moving more slowly have a smaller zeta potential. The particles are illuminated by a laser, and the particle speed is inferred indirectly from a Doppler frequency shift of the scattered light as shown in Figure 128 ("Zeta Potential : Anton Paar Wiki" n.d.). This frequency shift can be converted into an electrophoretic mobility value. The Henry Equation is used to determine the zeta potential.

2.6.1.3. Stability Determination of Nanoparticles Using Zeta Potential

Zeta can be used to forecast a particle's long-term stability. For instance, whereas particles with zeta potentials between -10 mV and +10 mV may undergo quick agglomeration unless they are sterically protected, particles with zeta potentials bigger than 60 mV have good stability.

2.6.1.4. Limitations

Zeta potential relies upon both molecule and medium qualities as well as how they cooperate at the connection point; it isn't just an element of suspended particles. This interfacial balance and, thusly, the zeta potential are both affected by changes in the fluid's substance and ionic organization. Subsequently, both sample planning and estimation procedures could impact the result of the estimation. Unseemly sample planning and estimation strategies or the ill-advised utilization of hypothetical models for registering zeta potential from estimation information lead to inaccurate discoveries much of the time. After some time, the collapsed slim cells erode, particularly when estimations are made in exceptionally conductive or pungent circumstances. Consequently, it's pivotal to affirm that the zeta cells are estimating accurately by contrasting them with a typical reference material.

2.7. Visualization in the 3-Dimensional Regime

2.7.1. 3D-Tomography

Understanding the association between an article's design and capability requires 3D underlying investigation. XRD, SEM, and TEM are only a couple of the insightful strategies that are used to offer underlying data. For a long time, underlying examination in the physical and organic sciences has been a significant utilization of TEM, quite possibly of the most widely utilized analytical instrument, which projects 3D nanoparticles into 2D pictures. Commonly, 2D-projection pictures are deficient to get a handle on how a nanoscale item's 3D design and capability interface. Electron tomography (ET) is a technique that separates 3D primary information from a grouping of shifted 2D projections. A strategy is consistently developing and has sub-nanometer goal.

2.7.1.1. Principle

The groundwork of ET is the discrete assortment of occurrence points in the scope of 80° for which a thing is pivoted inside an electron light emission magnifying lens. Each image that is caught is a projection of the incorporated amount of interest, for example, the sample's mass or synthetic sythesis along the electron venture, into the perception plane. These projections are exposed to the Radon change, which empowers the formation of a 3D grid portraying the sample in real space. This is finished by back extending each pertinent pixel's force from the arrangement of projection pictures onto an alternate voxel of the network that is being fabricated. As a general rule, quantitative or synthetic investigation may then be done utilizing the 3D voxel network that has been made.

2.7.1.2. Working

Any electron magnifying lens with a high slant test holder and obtaining programming that empowers robotized securing of the slant point series as well as control of both the defocusing and the sample float at each slant point can be utilized to play out the procurement of the slant series at the center of electron tomography. For the recorded pictures to precisely address the entire sample, the imaging field profundity of the magnifying instrument should constantly be more prominent than the sample thickness. While the spatial goal along the slant hub is kept up with as it was in the first 2D pictures, the most extreme slant point and the quantity of projections in all actuality do decide it in the two bearings opposite to the slant hub as shown in Figure 129 (Bharti, Chae, and Goyal 2020).

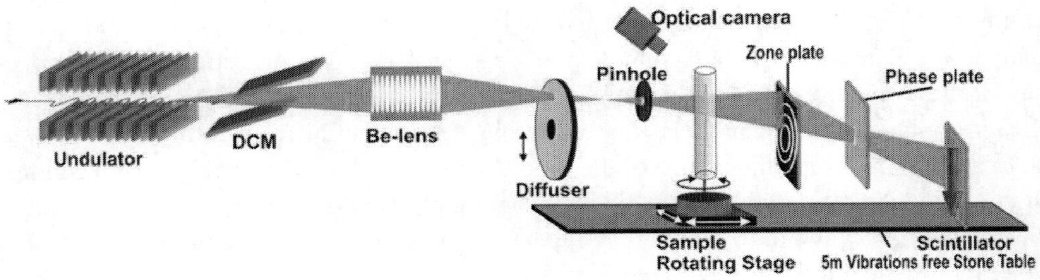

Figure 129. Schematic of monochromatic X-Ray nanotomography (Bharti, Chae, and Goyal 2020).

Three energy-filtered TEM pictures are typically acquired for each tilt angle when employing the analytical EFTEM tomography mode in order to reliably produce 3D chemical maps. This allows for correct extrapolation of the background to extract quantitative chemical signals. The recording sessions take longer as a result, but the evaluation of the backdrop and elemental projections is good.

The process of quantization is highly sample dependant. Other common analyses, such as establishing the precise morphology of materials at the nanoscale in terms of 3D shape, surface area, and 3D organisation state in particle agglomerations, can be carried out in addition.

2.7.1.3. 3D Nanoparticles Localization Using Electron Tomography

For electron tomography, restricting nanoparticles on and inside a help is critical. The general situations of constituent parts or their areas on a substrate essentially affect the elements of nanomaterials and nanoobjects. It has been noticed that keeping a substance inside a nanometric depression could energize the development of new metastable stages or change the substance's essential characteristics. For example, it has as of late been shown that restricting metal nanoparticles (NPs) inside a carbon nanotube channel improves the synergist productivity of the channel incredibly when contrasted with NPs that are not encased or forestalls extreme oxidation of the NPs. Just a 3D reproduction can straightforwardly uncover whether a molecule is "inside" or "outside" the channel, and these recreations even make it conceivable to exactly quantize the 3D spatial dispersion of the NPs to understand their mooring instrument.

Others techniques used to realize the 3D structure of a nanoparticle: AFM, SEM.

Conclusion

This chapter gives a succinct outline of the methods used to describe nanoparticles as far as their size, size dissemination, shape, surface region, surface charge, collection state, structure, synthetic piece, surface science, and focus. The objective of this work has been to assess and dissect current estimation and examination progress as intently as achievable as per standard working techniques. Not all new and unprecedented methods have been talked about.

All credible research that have been published have come to the conclusion that sample preparation techniques are one of the most significant causes of variability in the characterisation data of nanoparticles. The various specifications of measuring techniques make this worse. Others can only measure particles formed on a substrate, while still others require the particle to be in the form of a powder, an aerosol, or embedded in another substance. Some techniques can only measure nanoparticles in liquid suspension. Silica nanoparticles are a good illustration of how preparation processes that are either method- or material-dependent must ensure that the measurement outcomes are reflective of the material. The situation is perilous in this case, because sample preparation techniques are inconsistently described in the literature, and conclusive samples are mostly only accessible as a consequence of recent, sizable study.

It is important to take care while preparing samples and following prescribed methods since for many nanoparticulate materials, sample stability and homogeneity might be crucial for achieving accurate and repeatable findings. To guarantee that the sample is representative of the entire material, the sampling technique is also crucial. When various particles are distributed unevenly in space, as is true of powders and colloidal suspensions, this becomes crucial. Some analytical techniques may measure groups of nanoparticles, while others can focus on a single nanoparticle. The most effective strategy does not involve a single method, but rather a combination of carefully chosen and suitable approaches that should be used to guarantee that the necessary level of detail and statistically significant data on the desired qualities are collected.

It should be remembered that for all measuring techniques, the detected signal—whatever it may be—must be transformed into the desired quantitative attribute of the nanoparticles, which should ideally be a precise, verified outcome. This stage is important because it involves uncertainties that are frequently disregarded, underestimated, or even unknown. To accurately assess the relevant measurement uncertainties, further attention must be used. Reliable data analysis requires estimation of the uncertainty of findings acquired using various methodologies as well as comparability and compatibility of the outcomes. Such traceability chains have recently been constructed for the measurement of nanoparticle concentration, and they must be understood and recorded in order for the results to be traceable to reliable, ideally certified, reference materials or basic physical constants.

There are a few continuous progressions comparable to latest things in the estimation of relevant nanoparticle qualities. The most broadly utilized methods proceed to progress and advance, with more noteworthy edges for location, responsiveness, scientific degree, or spatial goal. A new shift toward benchtop instrumentation is another. There are a few creative instrumental systems and scientific procedures that commitment expanded insightful power, decreased costs, or the estimation of different qualities.

References

"2.3: BET Surface Area Analysis of Nanoparticles - Chemistry LibreTexts." n.d. Accessed October 30, 2022. https://chem.libretexts.org/Bookshelves/Analytical_Chemistry/Physical_Methods_in_Chemistry_and_Nano_Science_(Barron)/02%3A_Physical_and_Thermal_Analysis/2.03%3A_BET_Surface_Area_Analysis_of_Nanoparticles.

"14.11: Electronic Spectra: Ultraviolet and Visible Spectroscopy - Chemistry LibreTexts." n.d. Accessed October 31, 2022. https://chem.libretexts.org/Bookshelves/Organic_Chemistry/Map%3A_Organic_Chemistry_(Vollhardt_and_Schore)/14%3A_Delocalized_Pi_Systems%3A_Investigation_by_Ultraviolet_and_Visible_Spectroscopy/14.11%3A_%09Electronic_Spectra%3A__Ultraviolet_and__Visible__Spectroscopy.

"Atomic Force Microscopy - Nanoscience Instruments." n.d. Accessed October 30, 2022. https://www.nanoscience.com/techniques/atomic-force-microscopy/.

"Atomic Force Microscopy: General Principles and Applications." n.d. Accessed October 30, 2022. https://www.azooptics.com/Article.aspx?ArticleID=2083.

Bharti, Amardeep, Keun Hwa Chae, and Navdeep Goyal. 2020. "Real-Time Synthesis and Detection of Plasmonic Metal (Au, Ag) Nanoparticles under Monochromatic X-Ray Nano-Tomography." *Scientific Reports 2020 10:1* 10 (1): 1–11. https://doi.org/10.1038/s41598-020-77853-x.

Colpan, Can O., Yagmur Nalbant, and Mustafa Ercelik. 2018. "4.28 Fundamentals of Fuel Cell Technologies." *Comprehensive Energy Systems* 4–5 (January): 1107–30. https://doi.org/10.1016/B978-0-12-809597-3.00446-6.

Hughes, MA. 2007. "Modified Chalcogenide Glasses for Optical Device Applications." https://eprints.soton.ac.uk/47761/.

"IR Spectroscopy - Principle and Instrumentation of Infrared Spectroscopy." n.d. Accessed October 30, 2022. https://byjus.com/chemistry/infrared-spectroscopy/.

Marturi, Naresh. n.d. "Vision and Visual Servoing for Nanomanipulation and Nanocharacterization in Scanning Electron Microscope." Accessed October 31, 2022. https://tel.archives-ouvertes.fr/tel-01025318.

Ojeda, Jesús J., and Maria Dittrich. 2012. "Fourier Transform Infrared Spectroscopy for Molecular Analysis of Microbial Cells." *Methods in Molecular Biology* 881: 187–211. https://doi.org/10.1007/978-1-61779-827-6_8/COVER.

Pate, K., and P. Safier. 2016. "Chemical Metrology Methods for CMP Quality." *Advances in Chemical Mechanical Planarization (CMP)*, January, 299–325. https://doi.org/10.1016/B978-0-08-100165-3.00012-7.

"Photoluminescence Spectroscopy and Its Applications - Physical Methods in Chemistry and Nano Science - OpenStax CNX." n.d. Accessed October 31, 2022. https://cnx.org/contents/uieDnVBC@25.2:gbsDEZju@2/Photoluminescence-Spectroscopy-and-its-Applications.

Rocha, Fellipy S., Anderson J. Gomes, Claure N. Lunardi, Serge Kaliaguine, and Gregory S. Patience. 2018. "Experimental Methods in Chemical Engineering: Ultraviolet Visible Spectroscopy—UV-Vis." *The Canadian Journal of Chemical Engineering* 96 (12): 2512–17. https://doi.org/10.1002/CJCE.23344.

"Scanning Electron Microscope | Definition, Images, Uses, Advantages, & Facts | Britannica." n.d. Accessed October 29, 2022. https://www.britannica.com/technology/scanning-electron-microscope.

Seyama, H., M. Soma, and B. K. G. Theng. 2013. "X-Ray Photoelectron Spectroscopy." *Developments in Clay Science* 5 (C): 161–76. https://doi.org/10.1016/B978-0-08-098259-5.00007-X.

"Surface Characterization: BET Surface/Specific Surface | 3P Instruments." n.d. Accessed October 30, 2022. https://www.3p-instruments.com/measurement-methods/bet-surface-area/.

"The Raman Spectrophotometer." n.d. Accessed October 31, 2022. https://www.sas.upenn.edu/~crulli/TheRamanSpectrophotometer.html.

"XPS Analysis - X-Ray Photoelectron Spectroscopy, or ESCA." n.d. Accessed October 31, 2022. https://www.eag.com/techniques/xps-analysis/.

"Zeta Potential :: Anton Paar Wiki." n.d. Accessed October 31, 2022. https://wiki.anton-paar.com/en/zeta-potential/.

Zhang, Junsong, Shijie Hao, Daqiang Jiang, Yong Huan, Lishan Cui, Yinong Liu, Hong Yang, and Yang Ren. 2017. "*In Situ* Synchrotron High-Energy X-Ray Diffraction Study of Microscopic Deformation Behavior of a Hard-Soft Dual Phase Composite Containing Phase Transforming Matrix." *Acta Materialia* 130 (May): 297–309. https://doi.org/10.1016/J.ACTAMAT.2017.03.052.

Zhou, X. W., Y. D. Qi, X. D. Liu, J. J. Wei, and W. D. Wu. 2015. "Investigation on the Microstructure of As-Deformed NiCr Microwires Using TEM." *RSC Advances* 5 (110): 90852–57. https://doi.org/10.1039/C5RA19786D.

Chapter 7

Simulation and Analysis Tools for Nanosensors

Learning Outcomes

- This chapter brings the understanding of the fundamental principles underlying the operation of nanosensors.
- The learner will gain knowledge of various simulation techniques such as finite element analysis (FEA), molecular dynamics (MD), and Monte Carlo simulations.
- The advantages and limitations of different simulation techniques for modelling nanosensors is deeply understood.
- Knowledge of analysis tools, statistical methods, signal processing techniques, and machine learning algorithms for interpreting nanosensor data is acquired.
- A deep insight of the challenges associated with noisy and high-dimensional nanosensor data and learn strategies for data analysis and pattern recognition is obtained.
- The insights into the design optimization of nanosensors using simulation and analysis tools is gained.
- Recognize the challenges and future directions in the field of simulation and analysis tools for nanosensors.

Abstract

The design process for modern nanosensor integrated systems includes a number of processes that lead up to the complete system simulation using electrical and physical models before it is manufactured. The precise behavioural models that, before being physically implemented, accurately forecast the behaviour of the whole nanosensor system under various operating situations. The modelling and analysis of the nanosensor improve design optimisation, and the device's behaviour under ideal and unfavourable electrical and environmental circumstances will enable cost savings during the device's actual manufacture. The Internet of Things (IoT), a paradigm where devices may communicate with each other and transmit information acquired from themselves or from the environment to enhance the operation of the whole system, has evolved in conjunction with the emergence of new low-cost, optimised gadgets. The relevance, essential strategies, and numerous methods for simulating and assessing the nanosensor device are thoroughly explained in this chapter.

1. Introduction

By making it possible to detect and analyse nanoscale events with amazing accuracy and sensitivity, nanosensors have revolutionised the area of sensing. These small gadgets are used in a variety of fields, including as industrial operations, environmental monitoring, and healthcare. The use of simulation and analytic tools may be used to get a thorough knowledge of the behaviour and performance characteristics of nanosensors, which is necessary for their design and optimisation. This chapter examines the value of simulation and analysis in the context of nanosensors, emphasising how crucial they are to the development of the technology. By enabling researchers and engineers to forecast and comprehend the behaviour of nanosensors prior to physical production, simulation tools help them save time and money. They also make it easier to investigate the basic ideas that underlie the functioning of nanosensors by illuminating the intricate dynamics at the nanoscale. The use of analysis tools, on the other hand, enables researchers to enhance sensor designs, increase accuracy, and make wise judgements based on the data acquired by helping to extract relevant information from the output signals of nanosensors.

The first section of the chapter explores the foundations of nanosensor simulation. It gives an overview of several simulation techniques, such as finite element simulations, quantum mechanical simulations, and molecular dynamics simulations. The use of computational models, such as surface plasmon resonance models, diffusion models, and charge transport models, to simulate the interactions between nanosensors and target analytes is being investigated. Researchers are better equipped to understand and anticipate the behaviour of nanosensors when they comprehend these underlying concepts. The chapter then talks about simulation software designed especially for nanosensors. It offers a variety of open-source and paid software tools intended to replicate nanosensor activity. While open-source software gives freedom and customization choices, commercial software delivers complete functionality and user-friendly interfaces. Researchers and engineers may choose the best software choice for their unique simulation requirements by comparing the alternatives that are currently available.

In addition to modelling, analytical approaches are essential for comprehending and improving the performance of nanosensors. The chapter discusses numerous analysis methods for nanosensors, such as statistical analysis, noise reduction, filtering approaches, calibration, and error analysis. It also discusses signal processing and data analytics. With the use of these methodologies, researchers may characterise sensor performance, extract useful information from the output signals, lower noise and uncertainty levels, and enhance the precision and dependability of nanosensor measurements.

The interdependence of the simulation and analysis tools for nanosensors is emphasised throughout the whole chapter. Researchers and engineers may enhance nanosensor performance, verify experimental findings, and direct the creation of next-generation sensing devices by integrating simulation and analytic approaches. Researchers can speed up developments in nanosensor technology and open the door for novel applications in a variety of sectors by using the potential of these instruments.

A thorough review of simulation and analysis techniques for nanosensors is given in this chapter, in conclusion. Researchers and engineers will gain useful insights and tools to improve the design, performance, and application of nanosensors in a variety of domains by comprehending their significance, investigating the fundamentals of nanosensor simulation, looking at the simulation software that is available, and examining analysis techniques.

2. Significance of Simulation and Analysis of Nanosensors

The development of nanosensors has led to the development of sophisticated tools with unmatched sensitivity and selectivity for the analysis and detection of nanoscale events. In several industries, such as medical, environmental monitoring, and industrial operations, their special qualities make them indispensable. However, in order to build and optimise nanosensors, it is necessary to have a thorough knowledge of their behaviour and performance characteristics. This may be done by using simulation and analytic tools.

Because they can provide scientists and engineers foresight into how these devices will behave and operate, simulation and analysis are important tools in the field of nanosensors. With the use of simulation tools, nanosensor designs may be virtually explored, making it possible to optimise sensor settings, comprehend basic working principles, and forecast performance prior to physical production. To mimic the behaviour of nanosensors at the atomic and molecular sizes, researchers use computer models and simulation techniques including molecular dynamics simulations, quantum mechanical simulations, and finite element simulations. Researchers may improve their designs by using these technologies to acquire significant insights into the complex dynamics that control nanosensor performance.

On the other hand, analysis tools are essential for obtaining useful information from nanosensor output signals and interpreting the data that has been collected. They provide researchers the ability to verify experimental findings, improve sensor designs, and base choices on the data collected. Analytical methods include those for data analytics and signal processing, statistical analysis, noise reduction and filtering procedures, calibration, and error analysis. These methods support characterization of sensor performance, noise and uncertainty reduction, and increased precision and dependability of nanosensor readings.

In the field of nanosensors, the importance of simulation and analytic tools has been well acknowledged and documented in academic literature. Research studies have used simulation techniques to examine the behaviour of different nanosensors, including metal oxide nanowires, carbon nanotubes (CNTs), and sensors based on graphene, among others. For instance, gas molecule adsorption and desorption on CNT-based sensors have been studied using molecular dynamics simulations, revealing details about their sensing processes (Yeow and Wang 2009). To study the strain-induced performance improvement of graphene-based sensors, finite element simulations have been used (Chakraborthy et al. 2022). Wavelet analysis and machine learning methods have also been used to examine the signals produced by nanosensors, assisting in the identification and categorization of target analytes (Ye, Liu, and Li 2021; Domènech-Gil and Puglisi 2022).

The creation and improvement of nanosensors heavily rely on simulation and analytic technologies. Researchers may realistically investigate and forecast the behaviour of nanosensor designs by using simulation tools, and they can extract useful data from sensor output signals by using analytic tools. The importance of these tools resides in their capacity to confirm experimental findings, increase measurement precision, and give insightful information regarding the behaviour of nanosensors. These techniques have been widely used in scientific study, which shows how helpful they are for comprehending and improving nanosensor performance.

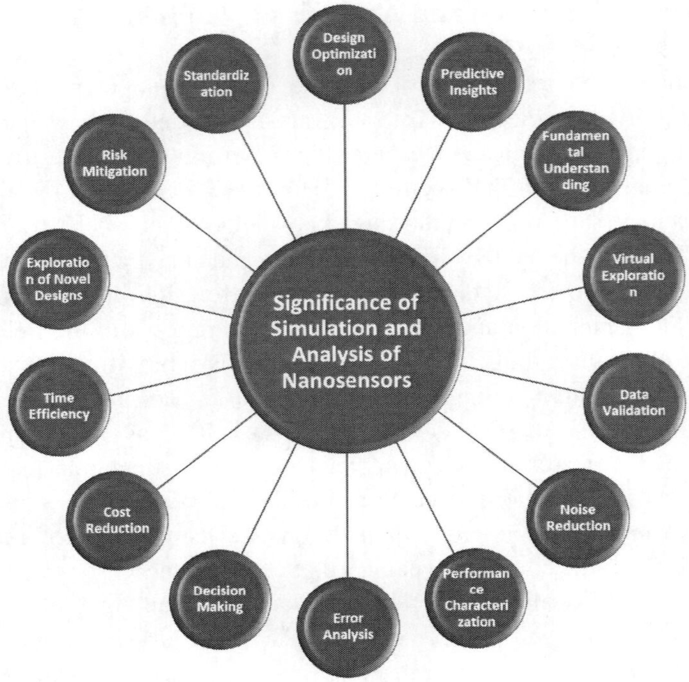

Figure 130. Significance of simulation and analysis of nanosensors.

The significance of simulation and analysis of nanosensors is given in Figure 130 and a few of them are explained below.

- *Design Optimization*: Simulation and analysis tools allow for the optimization of nanosensor designs, improving their performance and efficiency.
- *Predictive Insights*: Simulation tools provide predictive insights into nanosensor behaviour, aiding in understanding and predicting sensor responses.
- *Fundamental Understanding*: Simulation and analysis facilitate the exploration of fundamental principles governing nanosensor operation, enhancing understanding of their underlying dynamics.
- *Virtual Exploration*: Simulation tools enable virtual exploration of nanosensor designs, saving time and resources by assessing performance before physical fabrication.
- *Data Validation*: Analysis techniques validate experimental results, ensuring the reliability and accuracy of nanosensor measurements.
- *Noise Reduction:* Analysis tools assist in noise reduction and filtering, enhancing the quality of nanosensor data.
- *Performance Characterization*: Statistical analysis techniques aid in characterizing nanosensor performance and assessing measurement uncertainties.
- *Error Analysis:* Analysis tools enable calibration and error analysis, leading to improved accuracy and reliability of nanosensor measurements.
- *Decision Making*: Simulation and analysis tools support informed decision making in nanosensor design and optimization, maximizing performance and efficiency.

- *Cost Reduction:* Simulation tools help optimize nanosensor designs, reducing the need for costly and time-consuming experimental iterations.
- *Time Efficiency:* Simulation tools provide rapid evaluation of nanosensor designs, accelerating the development timeline and reducing time to market.
- *Exploration of Novel Designs*: Simulation tools enable the exploration of unconventional nanosensor designs, expanding the possibilities for innovative sensing approaches.
- *Risk Mitigation*: Simulation and analysis tools allow for the identification and mitigation of potential risks associated with nanosensor designs before fabrication.
- *Standardization:* Simulation and analysis contribute to the development of standardized methodologies and protocols for nanosensor characterization and performance evaluation.

3. Fundamentals of Nanosensor Simulation

The cornerstone of this chapter is a discussion of the foundations of nanosensor simulation since they provide the foundation for precisely modelling and forecasting nanosensor activity. Understanding the basic principles regulating nanosensor functioning and capturing the complex dynamics at the nanoscale need the use of simulation techniques. Researchers may investigate the interactions between nanosensors and target analytes, revealing the intricate workings of sensor response mechanisms, by using computational models and simulation tools.

This section explores the many simulation techniques, including molecular dynamics simulations, quantum mechanical simulations, and finite element simulations, that are used to nanosensor modelling. By monitoring the locations and velocities of atoms over time, molecular dynamics simulations make it possible to examine nanoscale events and gain knowledge about the atomic-scale behaviour of materials. To correctly simulate the electrical structure and characteristics of nanosensors, quantum mechanical simulations use the ideas of quantum mechanics. On the other hand, finite element simulations employ mathematical techniques to solve partial differential equations and mimic the behaviour of nanosensors under various circumstances.

In this part, computational methods for nanosensor simulation are also studied. These models include charge transport models, diffusion models, and surface plasmon resonance models, among others. A foundation for comprehending the optical characteristics of nanosensors is provided by surface plasmon resonance models, which mimic how light interacts with tiny structures. Diffusion models, which help in the understanding of analyte diffusion kinetics, explain the movement of analytes inside and around nanosensors. Charge transport models clarify how charge carriers migrate through nanosensors, allowing for the investigation of electrical conductivity and characteristics.

The use of simulation tools for nanosensor research is well documented in scientific literature. For instance, molecular dynamics simulations have been used to examine the effectiveness of nanosensors made from a variety of materials, such as metal oxide nanoparticles, carbon nanotubes, and graphene (PROYKOVA 2006; Kohn and Sham 1965). The electrical structure and characteristics of nanosensors have been studied using quantum mechanical models, which has aided in our comprehension of charge transfer processes and energy levels (Xu Zhang et al. 2017). Additionally, nanosensors' mechanical and electrical behaviour has been examined using finite element models, which have revealed details on their sensitivity, response time, and mechanical stability (Li et al. 2020).

To effectively simulate and anticipate nanosensor activity, a solid foundation in nanosensor simulation is essential. Researchers may investigate the intricate dynamics at the nanoscale, elucidate the underlying concepts regulating nanosensor functioning, and improve sensor designs by using simulation techniques and computer models. These simulation tools are widely used in scientific research, allowing improvements in nanosensor technology and improving their performance in numerous applications.

3.1. Simulation Approaches for Nanosensors

Nanosensors are very sensitive instruments capable of detecting and studying nanoscale occurrences, and simulation methods are crucial for understanding and forecasting their behaviour. Researchers may improve nanosensor designs and forecast their performance accurately by using computational models and simulation tools to acquire significant insights into the complex dynamics at the nanoscale. In-depth analysis of the technical details and practical scientific uses of the different simulation techniques utilised for nanosensor modelling is provided in this section.

The various methodology used in the simulation techniques used in nanosensor research capture various facets of nanosensor behaviour. A popular method for studying the behaviour of nanosensors at the atomic level is molecular dynamics (MD) simulations, which keep track of the locations and velocities of atoms throughout time. The mechanical characteristics, structural alterations, and interactions of nanosensors with analytes or their surroundings are all revealed by MD simulations (Hiratsuka et al. 2019; Tian 2008). On the other hand, quantum mechanical (QM) simulations use the concepts of quantum mechanics to precisely describe the electrical structure and characteristics of nanosensors. Studies of charge transfer processes, energy levels, and electrical characteristics of nanosensors benefit greatly from QM simulations (Zienert, Schuster, and Gessner 2013; Parvaiz et al. 2021).

The finite element method (FEM), a numerical method for solving partial differential equations, is a crucial modelling tool for nanosensors. The behaviour of nanosensors under various circumstances, such as mechanical stress, heat transport, or electrical potential, is modelled using FEM models. Researchers may investigate the mechanical and electrical characteristics of nanosensors, forecast their sensitivity, and assess their effectiveness using FEM models (Vazinishayan et al. 2018; Xiaoting Zhang et al. 2021).

Other simulation techniques, in addition to MD, QM, and FEM simulations, are pertinent to nanosensor research. Lattice-based models, multiscale simulations, and Monte Carlo simulations are a few examples. It is possible to examine phenomena like diffusion and adsorption on nanosensor surfaces using Monte Carlo simulations, which use random sampling to describe stochastic processes (Shariati, Roohi, and Ebrahimi 2023; Jiang, Klauda, and Sandler 2003). By discretizing space into lattice sites, lattice-based models streamline the atomic structure of nanosensors and make it easier to simulate large-scale systems (He and Li 2000; Xu et al. 2019). Through the integration of several simulation methodologies and models at various length and time scales, multiscale simulations enable the examination of nanosensor behaviour at various degrees of complexity (Anand et al. 2022; Gates et al. 2005a).

Simulation techniques are widely used in scientific research to enhance nanosensor technology. For instance, the behaviour of several nanosensors, including carbon nanotubes, metal nanoparticles, and nanostructured surfaces, has been examined using MD models (A. Santos et al. 2011; Lado-Touriño and Páez-Pavón 2021). Nanosensors' electrical characteristics, such as their conductance, bandgap, and charge transport pathways, have been studied using QM simulations (Maiti 2007; 2008). By analysing mechanical stress, strain distribution, and resonance behaviour of nanosensors using FEM simulations, their design and optimisation have improved (Luo et al. 2019).

Nanosensor research relies heavily on simulation techniques since they provide understanding of the behaviour and functionality of these devices. Researchers may investigate the behaviour of nanosensors at the atomic, electronic, mechanical, and multiscale levels using molecular dynamics, quantum mechanics, finite elements, and other simulation approaches. By using these simulation techniques, researchers may improve nanosensor designs, forecast their performance, and contribute to the creation of dependable and highly sensitive nanosensors.

3.2. Computational Models for Nanosensors

Understanding, optimising, and designing these very sensitive devices is made easier with the use of computational models, which are essential for modelling the behaviour of nanosensors. Researchers may mimic the interactions of nanosensors and their target analytes using mathematical and computational approaches, revealing the intricate dynamics at the nanoscale. In-depth analysis of the technical details and practical uses of the different computational models utilised in nanosensor simulation is provided in this part.

There are many different methods used in computational models for nanosensors that may capture various facets of sensor function. In order to mimic how light interacts with nanoscale objects, surface plasmon resonance (SPR) models are often utilised. These models provide a framework for comprehending the optical characteristics of nanosensors. SPR models take into account the interaction between incoming light and surface plasmons, which makes it possible to calculate resonance conditions and analyse changes in refractive index brought on by analyte binding (Homola 2008).

To mimic the movement and diffusion of analytes inside of and around nanosensors, diffusion models are used. These models provide insights into the kinetics of analyte diffusion and binding by taking into consideration variables including diffusion coefficients, analyte concentrations, and boundary conditions (Bodrenko, Milenkovic, and Ceccarelli 2022). Diffusion models may be used with other computer models to mimic the tying and untying processes between analytes and nanosensors, such as reaction kinetics models (Onaş et al. 2022).

To represent the movement of charge carriers in nanosensors, charge transport models are used. These models enable the analysis of electrical characteristics and conductivity by taking into account the electronic structure, energy levels, and transport processes inside nanoscale materials (Thomas et al. 2019). Quantum mechanical computations may be used with charge transport models to precisely depict the electrical behaviour of nanosensors (Fu et al. 2021).

Hybrid models and multiscale models are additional computational models that are useful for nanosensor simulation. In order to represent various length and time scales, hybrid models incorporate several modelling techniques, such as continuum models and molecular dynamics simulations. These models allow for more precise and effective modelling of nanosensor activity (Tien and Weng 2002). Multiscale models bridge the gap between atomistic simulations and macroscopic behaviour by integrating several computational tools and models to mimic nanosensor function at various degrees of complexity (Gates et al. 2005b).

Computational models are widely used in scientific research to simulate nanosensors and improve their performance while advancing our knowledge of sensor behaviour. To mimic how plasmonic nanosensors react to analyte binding, for instance, SPR models have been used, advancing the development of label-free sensing methods (Saleh and Aziz 2018). In order to better understand the dynamics of the sensor response, the kinetics of analyte diffusion and binding on different nanosensor surfaces have been studied using diffusion models (Antosiewicz and Käll 2016). To better understand charge transfer processes, conductance and electrical characteristics of nanosensors have been studied using charge transport models. Researchers can improve nanosensor designs, forecast their performance, and contribute to the creation of extremely sensitive and trustworthy nanosensors by using these computational models.

3.3. Challenges in Nanosensor Simulation

Although simulation is a useful tool for comprehending and improving the behaviour of nanosensors, there are still some difficulties with it. To get precise and trustworthy findings from modelling of nanosensors, researchers must overcome special technological and computational challenges. This chapter examines the difficulties in simulating nanosensor function, highlighting the constraints and difficulties researchers have when effectively modelling and forecasting nanosensor behaviour.

The proper depiction of nanoscale events is one of the main difficulties in nanosensor simulation. At the atomic and molecular level, where quantum mechanical effects are important, nanosensors function. Complex quantum mechanical models that precisely depict the electronic structure, charge transfer processes, and energy levels of nanosensors are needed to simulate these phenomena. It may be difficult to model big systems or run long-term simulations due to the high processing cost involved with quantum mechanical simulations.

The inclusion of accurate environmental conditions and analyte-sensor interactions is a substantial additional barrier. Nanosensors often function in challenging conditions like liquid media or gaseous atmospheres where variables like solvent effects, ionic strength, and temperature fluctuations may affect sensor performance. The creation of reliable and effective models that can capture the dynamic interactions between the sensor, analyte, and surrounding environment is necessary for effectively simulating various environmental variables.

Nanosensor devices provide extra difficulties due to their size and computational complexity. Nanosensors may display complex geometries, surface morphologies, and heterogeneities that need for accurate simulation modelling. Complex computational methods, such parallel computing or multiscale modelling, may be needed to simulate large-scale nanosensor arrays or intricate sensor designs in order to overcome computational constraints and provide correct findings.

Accurately incorporating noise and experimental errors is a difficulty in nanosensor modelling. Noise from a variety of causes, including temperature fluctuations, electrical noise, and measurement mistakes, may interfere with experimental observations of nanosensor activity. It may be difficult to capture and include various noise sources in simulation models, necessitating careful consideration of statistical and stochastic methods.

Furthermore, since nanosensor systems are so intricate, it might be difficult to validate simulation findings against experimental data. For simulation models to be accurate and reliable, experimental validation is required. However, the validation procedure may be challenging due to the complex makeup of nanosensors and the scarcity of extensive experimental datasets. To meaningfully validate their simulation models, researchers must carefully plan experiments and collect accurate data.

To get precise and trustworthy findings, researchers must overcome a number of difficulties presented by nanosensor simulation. To meet these problems, improved quantum mechanical models must be created, realistic environmental variables must be included, computational complexity and scale must be addressed, noise and uncertainties must be captured, and simulation findings must be validated against experimental data. Researchers can increase the precision and predictive capacity of nanosensor simulations by addressing these issues, which will improve the comprehension, optimisation, and design of nanosensor devices.

4. Simulation Software for Nanosensors

Nanosensor design, optimisation, and analysis all heavily rely on simulation software. These software tools provide scientists and engineers the computational power and user-friendly interfaces they need to precisely model the behaviour of nanosensors. This section provides a thorough review of the commercial and open-source software solutions available as it investigates the landscape of modelling software that is especially designed for nanosensors. Prior to actual production, simulation software enables researchers to realistically investigate nanosensor designs, forecast their behaviour, and improve their performance. By giving insights into the intricate dynamics at the nanoscale and simplifying the investigation of different design parameters, it conserves time and resources. Additionally, simulation software directs the creation of novel sensing devices and assists in understanding the basic working principles of nanosensors. Below, a number of commercial and open-source programmes are thoroughly explored.

4.1. Commercial Software for Nanosensor Simulation

For academics and engineers, nanosensor simulation software is available with a variety of capabilities and an intuitive user interface. These computer applications provide in-depth modelling and analysis of nanosensors. The pros and cons of commercial nanosensor modelling software are covered in this section.

Molecular dynamics (MD) modelling must be included in commercial nanosensor simulation software. MD simulations track atom positions and motion to study the atomic level behaviour of nanosensors. Using force fields, commercial MD simulation engines accurately depict atomic interactions. Using their extensive nanosensor material libraries, researchers may mimic metal oxide nanowires, graphene-based sensors, and carbon nanotubes. The graphical user interfaces (GUIs) of these software packages make it simple for non-programmers to set up, run, and analyse MD simulations.

MD and QM simulations are features of commercial nanosensor simulation software. The electrical structure and properties of nanosensors are faithfully simulated by QM simulations. These simulations calculate the electronic energies, charge densities, and transport properties of nanosensors using density functional theory (DFT) and time-dependent DFT. QM simulations need effective computing techniques and algorithms. A broad variety of basis sets, functionals, and exchange-correlation potentials are available for researchers to select from when simulating nanosensors. To evaluate the atomistic behaviour and electrical properties of nanosensors, commercial software combines QM and MD simulations.

FEA is often included in commercial nanosensor simulation tools. To simulate the mechanical, thermal, and electrical behaviour of nanosensors, FEA solves partial differential equations. Using advanced solvers and meshing tools, these software model complex geometries, heterogeneities, and boundary conditions. FEA simulations may look at the electrical response, resonance behaviour, and mechanical stress distribution in nanosensors. Researchers may assess simulation results and enhance nanosensor devices with the use of post-processing and visualisation tools in commercial software.

Commercial simulation software for nanosensors aids in the design, tuning, and analysis of nanosensor activity. Researchers may use methods from molecular dynamics, quantum mechanics, and finite element analysis to assess nanosensor activity at various length and time scales. Commercial software's user-friendly interfaces and extensive libraries of preset materials and models make it accessible to academics and engineers without prior computational skills. Using commercial software, researchers may create novel sensing devices, enhance performance, and hasten the creation of nanosensors.

4.1.1. COMSOL Multiphysics

For modelling and analysis of nanosensors, COMSOL Multiphysics is a potent commercial software programme that is often utilised. This programme offers a thorough framework for simulating and resolving coupled physics issues, making it appropriate for researching the intricate behaviour of nanosensors. The capabilities and scientific uses of COMSOL Multiphysics are highlighted in this part, which also examines the software's features and characteristics.

The multiphysics technique used by COMSOL Multiphysics enables researchers to model the interactions between many physical processes at once. This capacity is especially important for nanosensor simulations since the behaviour of the sensor may be affected by a variety of physical processes, including fluid dynamics, heat transfer, and electrostatics. The programme offers a large variety of physics modules that are already preset and may be used together to precisely mimic the behaviour of nanosensors. These modules provide researchers the ability to replicate many facets of nanosensor functioning by offering choices for simulating electromagnetic fields, diffusion processes, chemical reactions, and more.

The versatility of COMSOL Multiphysics in handling intricate geometries and boundary conditions is one of its main benefits. Because of the software's sophisticated meshing capabilities, which can handle complex nanosensor geometries, researchers are now able to precisely model sensor systems at the nanoscale. Additionally, by including predetermined material characteristics and boundary conditions, COMSOL Multiphysics enables the introduction of realistic environmental conditions. Researchers are now able to examine the behaviour of nanosensors in situations that are more like real-world ones, such as when fluids or intricate surface interactions are present.

A user-friendly graphical user interface (GUI) is another feature of COMSOL Multiphysics that makes it easier to set up and run simulations. Through an easy-to-use interface, users may create and alter simulation parameters, specify material attributes, and establish boundary conditions. Researchers can quickly assess and understand simulation findings using the software's robust post-processing and visualisation features. These technologies make it possible to extract useful information, including the distribution of analyte concentrations, electric field profiles, or heat transfer patterns, to better understand the behaviour of nanosensors.

COMSOL Multiphysics is widely used in scientific research for simulation and analysis of nanosensors. Applications for the programme include surface plasmon resonance sensors (Almawgani et al. 2023), graphene-based sensors (as illustrated in Figure 131) ("Simulating a 3D Multilayered Graphene Biosensor Design | COMSOL Blog" n.d.), and nanowire sensors ("Design and Simulation of 3D ZnO Nanowire Based Gas Sensors for Conductivity Studies" n.d.). Researchers have investigated the optical response, sensitivity, and detection limitations of nanosensors using COMSOL Multiphysics, yielding important insights for sensor optimisation and design advancements.

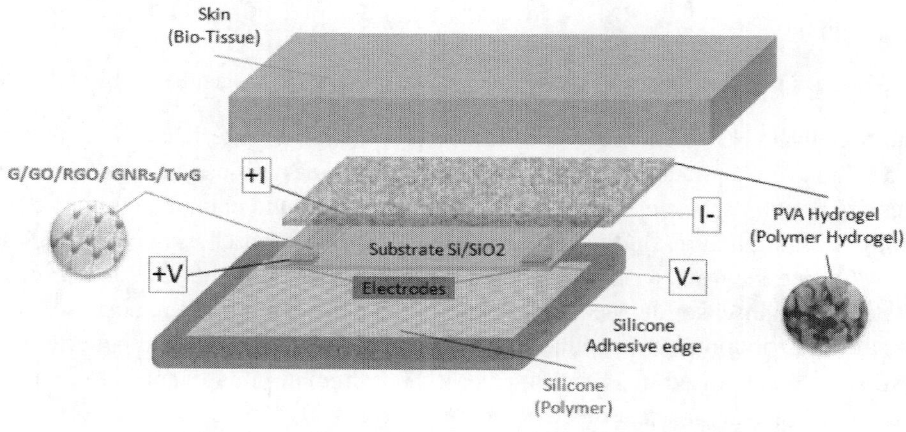

Figure 131. A multilayered Graphene biosensor device using COMSOL Multiphysics ("Simulating a 3D Multilayered Graphene Biosensor Design | COMSOL Blog" n.d.).

4.1.2. ANSYS

Advanced simulation capabilities are provided by the widely used commercial software programme ANSYS for the study and design of nanosensors. This programme offers a complete environment for physics modelling and problem-solving, making it appropriate for researching the behaviour of nanosensors. This section examines ANSYS's attributes while emphasising its capabilities and practical uses in the study of nanosensors. Researchers can precisely model and simulate the mechanical, thermal, and electrical behaviour of nanosensors using ANSYS' finite element analysis (FEA) methodologies. The programme enables researchers to mimic several facets of nanosensor functioning by offering a large variety of preconfigured physics modules, such as structural mechanics, heat transport, and electromagnetics. To represent the multiphysics aspect of nanosensors, where several physical phenomena interact and affect sensor performance, these modules may be connected together.

An important feature of ANSYS is its strong meshing skills, which can handle intricate boundary conditions and nanosensor geometries. With the use of adaptive meshing techniques, the programme ensures correct depiction of sensor components at the nanoscale by automatically enhancing the mesh in locations of interest. Additionally, ANSYS offers sophisticated material models and constitutive equations that enable researchers to take into consideration the unique mechanical and thermal characteristics of nanosensor materials. The modelling of nanosensor behaviour under various loading situations, temperature gradients, and environmental parameters is made possible by this capacity.

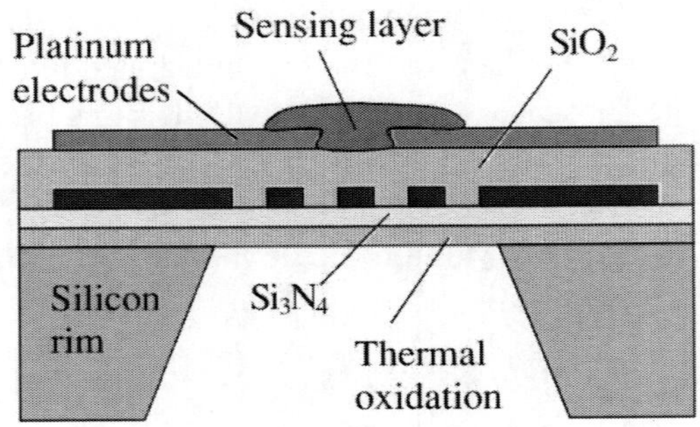

Figure 132. Cross-sectional sketch of stacked membrane in ANSYS (Puigcorbé et al. 2003).

Additionally, ANSYS provides an intuitive graphical user interface (GUI) that makes it easier to set up, run, and analyse simulations. Through an easy-to-use interface, users may set and alter simulation parameters, describe material attributes, and declare boundary conditions. In order to assess simulation findings and get useful insights, researchers may use the software's robust post-processing and visualisation features. These instruments enable the design and optimisation of nanosensor devices by facilitating the understanding of nanosensor behaviour, such as stress distributions, temperature profiles, and electromagnetic field patterns.

ANSYS is widely used in scientific research for simulation and analysis of nanosensors. Numerous nanosensor applications, including piezoresistive sensors (Thawornsathit et al. 2022), biosensors, and nanowire-based sensors (as illustrated in Figure 132) (Puigcorbé et al. 2003), have used the software. The mechanical behaviour, thermal responsiveness, and electrical performance of nanosensors have been studied by researchers using ANSYS, yielding useful information for sensor optimisation and design advancements. For the modelling and analysis of nanosensors, ANSYS is a potent software programme. It is a useful tool for researching the behaviour of nanosensors because to its finite element analysis capabilities, reliable meshing techniques, and user-friendly interface. Researchers may model a variety of physical events using ANSYS, which can also handle complicated geometries and include accurate material characteristics and boundary conditions.

4.1.3. Silvaco TCAD

Silvaco TCAD software is a potent tool used in the semiconductor industry for simulating and modelling different semiconductor devices and processes. TCAD stands for Technology Computer-Aided Design. TCAD software enables scientists and engineers to examine semiconductor devices' electrical, thermal, and optical behaviour, improve device performance, and forecast device characteristics prior to the actual manufacturing process. A full range of simulation tools for many areas of semiconductor device design and manufacture are included in the Silvaco TCAD programme. Device physics, process simulation, and circuit simulation are all covered by its components. Researchers may model and simulate the electrical behaviour of numerous semiconductor devices, such as transistors, diodes, and sensors, using the device physics modules. Users may simulate the manufacturing process phases, including as deposition, etching, and diffusion, in order to better understand how they affect device performance. The circuit simulation modules make it possible to analyse and improve electrical systems and integrated circuits.

To solve the mathematical models describing device physics and process events, Silvaco TCAD software uses a variety of numerical approaches and algorithms. Among the often used numerical techniques for resolving partial differential equations and modelling the behaviour of semiconductor devices and processes are the Finite Element Method (FEM), the Finite Difference Method (FDM), and the Monte Carlo Method. These numerical techniques allow precise and effective simulations, offering insightful information on the functionality and behaviour of the device.

To precisely depict the features and attributes of various semiconductor materials and devices, Silvaco TCAD software also offers comprehensive libraries of material models, device models, and process models. These models range from recombination models to band structure models to transport models. Accurate and realistic simulations are made possible by the software's ability to customise and adapt these models to particular device designs and process technologies.

For developing, optimising, and assessing semiconductor devices and processes, Silvaco TCAD software has been extensively used in academia, research, and the semiconductor industry. It has helped progress a number of industries, including sensor technologies, photonics, power devices, and microelectronics. The programme is a favourite among academics and engineers that work in semiconductor device design and manufacture because of its adaptability, precision, and user-friendly interface.

4.1.4. CST Studio Suite

In the subject of computational electromagnetics, the CST Studio Suite is a complete software package that is often used for electromagnetic simulation and analysis. The programme, created by Computer Simulation Technology (CST), offers a number of potent tools for designing and improving electromagnetic devices and systems. A range of simulation modules are available in CST Studio Suite that are designed to simulate and study electromagnetic phenomena in complicated structures. These modules include time-domain and frequency-domain solutions such the transmission line matrix (TLM) method, finite integration technique, and finite element method (FEM). With the use of the programme, scientists and engineers may model a variety of electromagnetic effects, such as electromagnetic compatibility (EMC), signal integrity, antenna design, microwave devices, and multiphysics simulations.

Maxwell's equations, which regulate electromagnetic fields, are solved by CST Studio Suite using sophisticated numerical methods and algorithms. These methods consist of eigenmode analysis, moments method, and finite element analysis. The programme gives users the ability to examine electromagnetic fields, assess performance indicators, and improve device designs by precisely solving Maxwell's equations. A vast array of libraries with material parameters, component models, and measurement information are provided by the programme. With the aid of these libraries, users may precisely simulate a wide range of substances, parts, and gadgets, including conductors, dielectrics, magnetics, and nonlinear components. The accuracy and realism of simulations in CST Studio Suite are improved by having the ability to specify complicated material attributes and component models. For the design and analysis of electromagnetic devices and systems, CST Studio Suite has been extensively used in academia, research, and business. It has been a popular option for academics and professionals working in fields including wireless communication, microwave engineering, antenna design, and electromagnetic compatibility because to its flexibility, user-friendly interface, and potent modelling capabilities.

4.1.5. IntelliSuite

For device simulation, modelling, and design in the realm of micro and nanotechnology, there is a complete software package called IntelliSuite. The IntelliSuite analysis and optimisation suite for microelectromechanical systems (MEMS), nanoelectromechanical systems (NEMS), and integrated photonics devices was created by IntelliSense. The extensive range of simulation modules available from IntelliSuite are created to simulate and investigate different physical processes in micro- and nanoscale devices. Finite element analysis (FEA), computational fluid dynamics (CFD), coupled electromechanical simulations, and optical simulations are some of the modules that fall under this category. These simulations provide scientists and engineers a way to analyse how devices behave under various operating settings, improve device performance, and provide design guidance.

In order to solve the mathematical models describing the physical events happening in micro- and nanoscale devices, IntelliSuite makes use of cutting-edge numerical techniques and algorithms. These numerical approaches include the boundary element method (BEM), the finite element method (FEM), and other numerical strategies appropriate for tackling challenging multi-physics issues. These techniques provide precise and effective simulations of device behaviours and offer perceptions into device functionality. To correctly depict the properties and traits of diverse materials and devices, the programme also offers comprehensive libraries of material models, device models, and process models. These models enable users to very realistically recreate device behaviour since they integrate mechanical, electrical, and optical aspects. With the help of IntelliSuite, it is possible to customise and adapt these models to particular device designs and process technologies, enabling precise and realistic simulations.

Lumerical, CoventorWare, SolidWorks Simulation, MSC Nastran, Sentarus TCAD, QuantumWise ATK, Remcom XFdtd, and ConventorMP are more examples of comparable commercial software for nanosensor simulations.

4.2. Open-Source Software for Nanosensor Simulation

Using open-source software, researchers may mimic and test the operation of nanosensors. These software solutions come with a variety of customizable algorithms and features. The advantages and applications of open-source nanosensor simulation software are covered in this section. Software for simulating nanosensors that is open-source is adaptable and expandable. By having access to the source code, these software packages allow researchers to modify methods and models. This adaptability may be used by researchers to develop novel simulation techniques, models, and algorithms to address challenges in nanosensor research. Open-source software allows researchers to contribute their updates and improvements, which benefits a broader user base and promotes interdisciplinary research.

Many different modelling approaches and methodologies are often available in open-source nanosensor simulation software. These include calculations based on quantum mechanics, finite element analysis, and molecular dynamics simulations, among others. Atomic interactions, charge transport, diffusion dynamics, and sensor response in nanosensors may all be replicated using these skills. Using well-documented APIs and scripting languages, users may implement high-throughput calculations, automate simulations, and build bespoke processes. Software that is open source is available. These free software options may be used by researchers. Open-source software allows for the verification of its source code and algorithms, which encourages scientific repeatability and openness. Cooperation, innovation, and the growth of the nanosensor research community are all facilitated by accessibility.

Open-source software for simulating nanosensors benefits from a vibrant user base. Thanks to a large number of researchers and developers, the programme changes and becomes better. This community-driven methodology assures regular updates, issue resolutions, and state-of-the-art simulation techniques. Users may share knowledge, seek for assistance, and exchange experiences on forums, email lists, and collaborative platforms. Flexibility, extensibility, and a variety of modelling techniques are all features of open-source nanosensor simulation software. They aid scholars in creating simulation algorithms specifically for their research. Software for simulating nanosensors that is open-source encourages collaboration, consistency, and innovation. Using open-source software allows researchers to enhance nanosensor modelling tools, sensor design, and behaviour.

4.2.1. NanoTCAD ViDES

A potent software programme called NanoTCAD ViDES (Virtual Device Simulation) was created with the goal of modelling and studying nanoscale devices using the principles of quantum mechanics. NanoTCAD ViDES, created by Synopsys, offers engineers and scientists a complete platform for analysing the electrical, thermal, and optical behaviour of nanoscale devices. A selection of simulation modules designed exclusively for nanoscale device modelling are available via NanoTCAD ViDES. These modules combine cutting-edge quantum mechanical modelling methods including non-equilibrium Green's function (NEGF) approaches and density functional theory (DFT). The programme enables modelling and analysis of several nanoscale devices, including graphene-based structures, nanowires, nanotubes, and quantum dots. It allows researchers to investigate innovative device ideas, optimise device performance, and examine device behaviour under various operating situations.

The quantum mechanical equations regulating the behaviour of nanoscale devices are solved using cutting-edge numerical approaches by NanoTCAD ViDES. The Poisson equation, the Schrödinger equation, and transport equations are examples of self-consistent field methods. The programme allows researchers to study the electrical structure, carrier transport, and quantum effects in nanoscale devices by precisely solving these equations. In order to faithfully depict the characteristics of nanoscale materials, the programme also offers a complete collection of material models, such as band structure models and electron-phonon interaction models. Users may simulate the behaviour of many material systems, such as semiconductors, metals, and insulators, using these models. Researchers may investigate how material characteristics affect the performance and design of devices using NanoTCAD ViDES. It has been extensively used for the design and study of nanoscale devices in academia, research, and industry. It has become a useful tool for scientists and engineers working in fields like nanoelectronics, quantum computing, and developing nanoscale technologies because of its sophisticated simulation capabilities, quantum mechanical modelling approaches, and vast material models.

4.1.2. QuantumATK

Developed by Synopsys, QuantumATK is a potent software suite for atomistic and quantum mechanical simulations in materials science and nanotechnology. It gives scientists, engineers, and other professionals cutting-edge tools for researching the electrical composition, characteristics, and behaviour of materials and nanostructures. With the use of tight-binding techniques, density functional theory (DFT), and classical molecular dynamics, QuantumATK can simulate a variety of systems, including bulk materials, surfaces, interfaces, and nanoscale devices. A variety of simulation modules are available from QuantumATK that are developed to investigate different facets of materials and nanostructures. These modules consist of Molecular Dynamics (MD), Atomistic ToolKit (ATK), and Quantum Mechanics (QM). The electronic structure and characteristics of materials are examined by the QM module using DFT computations. For modelling and investigating nanostructures, surfaces, and interfaces, ATK offers a flexible framework. Researchers may run conventional molecular dynamics simulations using the MD module to examine material dynamics and thermodynamics.

To resolve the underlying physics equations, QuantumATK uses a variety of numerical methods and algorithms. QuantumATK employs effective algorithms for density functional theory (DFT) computations, including the plane-wave basis set, pseudopotentials, and self-consistent field techniques. Accurate estimates of the electrical structure, band gaps, charge densities, and other characteristics of materials and nanostructures are made possible by these approaches. Additionally, it offers a large collection of material databases that include experimental data as well as material attributes and preconfigured crystal structures. These databases provide users access to a variety of materials and attributes, making it easier to build up simulations and allowing accurate and effective material research. In the fields of nanoelectronics, catalysis, energy storage, and material discovery, quantumATK has found extensive usage in academia, research, and business. Researchers and engineers working in materials science and nanotechnology value it for its ability to simulate materials at the atomic level, analyse electronic structure, and anticipate material characteristics.

4.1.3. LAMMPS

A popular piece of software for simulating molecular dynamics is called LAMMPS (Large-scale Atomic/Molecular Massively Parallel Simulator). Researchers at Sandia National Laboratories created LAMMPS, a flexible and scalable framework for modelling materials' atomic- and molecular-scale behaviour. Using a variety of interatomic potentials and modelling techniques, it enables researchers to explore a broad range of systems, including solids, liquids, gases, and biomolecules. It provides a wide variety of simulation capabilities, such as molecular dynamics (MD), Monte Carlo simulations, and hybrid MD/MC techniques. With the use of these techniques, scientists can mimic how atoms and molecules behave and interact in various settings. A wide range of materials, including metals, semiconductors, polymers, and biomolecules, may be simulated using LAMMPS because it offers a complete collection of interatomic potentials that explain the interactions between atoms.

In order to produce high-performance simulations on a variety of computing architectures, from single processors to massively parallel supercomputers, LAMMPS makes use of effective algorithms and parallel computing approaches. The programme uses message-passing interfaces and domain decomposition techniques to divide the computing workload across many processors, allowing simulations of massive systems containing millions to billions of atoms. A user-friendly input script language in this programme enables users to define simulation settings, system setup, and analytic tasks in order to offer wide capabilities. As a result of the software's support for a broad variety of analytic tools and output formats, researchers may derive valuable knowledge from the simulation regarding the system's characteristics, dynamics, and thermodynamic behaviour. Researchers from a variety of disciplines, such as materials science, chemistry, biophysics, and nanotechnology, have enthusiastically embraced it. It is a useful tool for modelling and comprehending the behaviour of materials at the atomic and molecular size because of its adaptability, scalability, and large user community.

4.1.4. NanoHUB

An extensive selection of software tools and other resources for study and instruction in nanotechnology are available on the internet platform NanoHUB. NanoHUB, a collaborative platform created by the Network for Computational Nanotechnology (NCN), provides researchers, teachers, and students with access to cutting-edge simulation and modelling tools, instructional resources, and community-driven content. It has a large selection of software tools and simulations that deal with many facets of nanotechnology, such as calculations of electrical structure, device modelling, quantum transport simulations, and simulations of nanoscale materials. Leading researchers and universities in the area have created these tools, which enable users to explore and evaluate complicated nanosystems using cutting-edge computational methods.

High-performance computing resources are used by NanoHUB to allow effective and scalable simulations. For computationally intensive simulations, it makes use of parallel computing methods and a cloud-based infrastructure to provide quick and precise results. NanoHUB enables users to do simulations on complicated nanosystems with improved performance and faster turnaround time by dividing the computational burden over numerous processors and clusters. This programme also acts as a platform for collaboration amongst academics, students, and researchers in the field of nanotechnology. It promotes a thriving community where users may utilise resources that are shared, work together on projects, and participate in debates. This cooperative setting fosters knowledge sharing and facilitates the creation and adoption of fresh nanotechnology simulation tools and approaches. It has developed into a useful tool for academics, students, and researchers in the area of nanotechnology. By giving users access to cutting-edge simulation tools, instructional resources, and collaborative platforms, it has significantly contributed to the advancement of research and education. The platform has proven crucial in advancing multidisciplinary cooperation, advancing nanotechnology research, and fostering the next generation of scientists and engineers in the area.

4.1.5. Quantum ESPRESSO

One of the most popular software programmes for doing quantum mechanical simulations of materials at the atomic scale is called Quantum ESPRESSO (Quantum opEn-Source Package for Research in Electronic Structure, Simulation, and Optimisation). Quantum ESPRESSO, created by a global team of researchers, offers a complete set of tools for ab-initio calculations of electronic structure, molecular dynamics simulations, and materials optimisation. It provides a variety of simulation modules to explore the electrical structure and characteristics of materials using plane-wave basis sets and density functional theory (DFT). These modules consist of pw.x for calculations of electronic structure, cp.x for calculations of Car-Parrinello molecular dynamics, and ph.x for calculations of phonons. With the use of the programme, scientists may examine a variety of materials, including nanostructures, surfaces, interfaces, and crystals, and get knowledge on their electrical, thermal, and mechanical characteristics.

In the context of DFT, Quantum ESPRESSO uses effective numerical methods and algorithms to resolve the Kohn-Sham equations. Pseudopotentials, plane-wave basis sets, self-consistent field methods, and iterative diagonalization procedures are some of these approaches. Accurate computations of electronic structure, total energies, forces, and numerous material characteristics are made possible by the programme, giving users important new insights into how materials behave at the atomic level. The programme offers a wide range of exchange-correlation functionals and pseudopotentials that precisely characterise the electronic characteristics and interactions in materials. Users may represent a variety of elements and compounds with varying degrees of precision and computing efficiency thanks to these pseudopotentials and functionals. The programme also enables the development of fresh pseudopotentials and functionals for particular research requirements. It has been extensively used in research, education, and business for a variety of purposes, including materials science, solid-state physics, chemistry, and nanotechnology. For scientists and engineers attempting to comprehend and create materials at the quantum level, it has proven to be an invaluable tool because to its open-source nature, potent simulation capabilities, and large user community.

OpenMD, GROMACS, OpenFOAM, NWChem, MEEP, Gmsh, OOMMF, OpenModelica, and NEMO5 are more comparable open-source programmes for nanosensor simulations.

5. Analysis Techniques for Nanosensors

For the purpose of identifying and evaluating diverse compounds at the nanoscale, nanosensors have become an effective instrument. Numerous industries, including healthcare, environmental monitoring, and security, might be completely transformed by these little gadgets. The analytical methods used in nanosensor research are essential for deciphering sensor outputs, locating target analytes, and enhancing sensor functionality. These approaches include a wide variety of scientific and mathematical procedures that provide researchers the ability to correctly analyse, evaluate, and interpret the gathered data. These methods allow for the discovery of hidden patterns, the reduction of noise, and an improvement in the overall accuracy of nanosensor data.

In order to aggregate and integrate data from several nanosensors, data fusion and multi-sensor analysis approaches are used. The integration of sensor outputs is facilitated by data fusion methods, including as Kalman filtering, Bayesian inference, and fuzzy logic, which increase accuracy, dependability, and resilience. By using the advantages of individual sensors and making up for their weaknesses, multi-sensor analysis provides thorough analyte detection and analysis. The performance and dependability of nanosensors may be improved by researchers by using a variety of analytical approaches. These methods help make the detection and analysis of analytes at the nanoscale more sensitive, selective, and accurate. The entire potential of nanosensor technology may be realised by researchers thanks to the integration of signal processing, machine learning algorithms, electrochemical impedance spectroscopy, and data fusion techniques.

5.1. Signal Processing and Data Analytics

In order for researchers to properly interpret and analyse the signals gathered by nanosensors, useful information must be extracted from the data collected. This is where signal processing and data analytics come into play. Due to their high sensitivity and miniaturised design, nanosensors provide complicated data that calls for sophisticated approaches in order to improve signal quality, identify pertinent characteristics, and derive insightful information. Analysing nanosensor data requires the use of basic signal processing methods. With the use of these approaches, researchers may take complicated and noisy data and extract useful information by preprocessing, filtering, and analysing the signals using a sequence of mathematical operations. Signal processing techniques that are often used include Fourier analysis, wavelet analysis, statistical algorithms like principal component analysis (PCA) and independent component analysis (ICA), and time-frequency analysis. These methods help with feature extraction, noise reduction, and pattern identification, which increases the precision and dependability of nanosensor data processing.

By allowing researchers to draw conclusions and patterns from the processed data, data analytics techniques are a useful supplement to signal processing methods in nanosensor analysis. In nanosensor data analytics, machine learning methods including Support Vector Machines (SVM), Random Forests, Artificial Neural Networks (ANN), and K-Nearest Neighbours (KNN) are often utilised. With the help of these algorithms, target analytes may be classified, identified, and quantified with accuracy because they can learn from the processed data and generate predictions based on patterns and correlations. To further group related data points and find underlying patterns in the nanosensor data, clustering methods like k-means and hierarchical clustering may be used.

In nanosensor data analytics, feature selection and dimensionality reduction approaches are essential to get rid of extraneous or unnecessary characteristics, boost computational effectiveness, and boost model performance. For feature selection and dimensionality reduction, methods including Recursive Feature Elimination (RFE), Principal Component Analysis (PCA), and t-distributed Stochastic Neighbour Embedding (t-SNE) are often used. These methods aid in lowering data complexity, increasing interpretability, and boosting the effectiveness of machine learning models.

Additionally, data fusion techniques integrate data from many nanosensors to increase the analysis's overall accuracy and dependability. Utilising the complimentary strengths of several sensors, fusion methods like Kalman filtering, Bayesian inference, and fuzzy logic are used to combine data from various sensors. Researchers can improve the sensitivity, selectivity, and resilience of nanosensors by data fusion, enabling more precise target analyte detection and analysis. Noise reduction and feature extraction in nanosensor data are made possible by signal processing methods such as Fourier analysis, wavelet analysis, and statistical algorithms. Pattern identification, classification, and analyte quantification are made easier by data analytics methods including feature selection, dimensionality reduction, and machine learning algorithms. To improve accuracy and dependability, data fusion approaches integrate data from many nanosensors. By using these methods, scientists can maximise the capabilities of nanosensors and develop applications across a variety of industries, from healthcare to environmental monitoring.

5.2. Statistical Analysis of Nanosensor Data

To get relevant insights from nanosensor data, statistical analysis is essential. Nanosensors provide complicated data due to their high sensitivity and miniaturised design, which calls for exact statistical approaches for interpretation and analysis. Researchers may establish solid judgements using statistical analysis tools, analyse the variability in the data, and identify relationships between variables. Hypothesis testing is one of the core statistical analytic methods used in nanosensor development. Researchers may analyse the significance of observed differences via hypothesis testing and draw statistical conclusions regarding the correlations between variables. T-tests, analysis of variance (ANOVA), and chi-square tests are often used statistical tests. These tests provide information on the changes brought on by various causes or experimental settings and assist in determining if observed variances in sensor responses are statistically significant.

A further statistical method that is often used with data from nanosensors is regression analysis. Researchers may better understand links between independent variables (like sensor response, analyte concentration, and environmental conditions) and dependent variables (like desired analyte concentration or response) by using regression models. Regression methods that are often employed include linear regression, logistic regression, and non-linear regression. Regression analysis makes it possible to estimate model inputs, forecast results, and pinpoint the variables affecting nanosensor responses.

In the study of nanosensor data, Principal Component study (PCA) is a popular multivariate statistical approach. By splitting the original variables into a smaller group of uncorrelated variables known as principle components, PCA decreases the dimensionality of the data. Researchers may visualise trends, spot outliers, and simplify data by using PCA to discover the primary components that account for the majority of the variation in the data. When working with huge datasets and trying to identify the connections between various variables, PCA is quite helpful.

To delve further into the data from nanosensors, other approaches may be used, such as multivariate analysis, clustering algorithms, and time series analysis. These techniques improve the precision and dependability of nanosensor data analysis by assisting in the discovery of hidden patterns, data classification, and behaviour prediction. The interpretation and analysis of nanosensor data greatly benefits from the use of statistical analytic techniques, such as hypothesis testing, regression analysis, principal component analysis (PCA), and advanced statistical approaches. Researchers may improve a variety of application areas by using these methodologies to analyse the complex data produced by nanosensors in order to find correlations, pinpoint driving forces, and draw firm conclusions.

5.3. Noise Reduction and Filtering Techniques

Techniques for noise reduction and filtering are crucial for the processing of nanosensor data because they improve signal quality, decrease undesired noise, and increase measurement accuracy. Due to their extreme sensitivity, nanosensors are vulnerable to a variety of noise sources that might reduce the quality of the signals they collect. Digital filtering is one of the main noise reduction methods used in the processing of nanosensor data. To deliberately weaken certain frequency components in the signal, digital filters including low-pass filters, high-pass filters, bandpass filters, and adaptive filters are used. These filters assist in removing noise sources that are present outside of the required frequency range, allowing scientists to concentrate on the important signal constituents. The characteristics of the noise and the intended signal determine the filter design and its settings.

Signal averaging is another popular method for reducing noise in nanosensor data. To lessen the influence of random noise, signal averaging entails gathering many measurements of the same signal and average them. The intended signal may be improved by averaging away the noise elements that are not associated between samples. When the signal of interest is recurrent or can be replicated by many measurements, this approach is very helpful.

Wavelet denoising is a potent method used in the interpretation of data from nanosensors. It makes use of the wavelet transforms' multi-resolution analytical capabilities to isolate noise from the signal in both the temporal and frequency domains. Signal quality may be improved by identifying and successfully suppressing the noise components by breaking the signal down into several wavelet sizes. To get the best noise reduction performance, the wavelet basis and denoising threshold must be chosen carefully.

In addition to these strategies, real-time nanosensor data are adaptively estimated and noise is suppressed using adaptive filtering techniques like the Kalman filter and the Wiener filter. These filters effectively reduce noise and provide accurate signal estimate by dynamically adjusting their filter parameters based on a mathematical model of the noise and the signal. The quality and accuracy of nanosensor data must be improved using noise reduction and filtering methods, such as digital filtering, signal averaging, wavelet denoising, and adaptive filtering. These methods aid in reducing noise sources, improve the signal-to-noise ratio, and let researchers concentrate on the important signal elements. These methods allow for the meaningful extraction of data from nanosensors and the advancement of several applications.

5.4. Calibration and Error Analysis

In order to establish precise connections between sensor responses and analyte concentrations and to quantify and reduce measurement uncertainties, calibration and error analysis are essential elements of nanosensor research. Calibration is the process of comparing the output of the nanosensor to established reference values through a series of measurements. Error analysis entails evaluating and quantifying measurement mistakes to guarantee accurate and consistent findings. In order to establish a quantifiable link between the sensor response and the analyte concentration, calibration is an essential step in nanosensor research. In order to create a calibration curve or model, a collection of data for various known analyte concentrations must be gathered. Future sensor data may then be utilised to calculate the analyte concentration using this curve or model. Depending on how complicated the connection is between the sensor response and analyte concentration, other calibration approaches may be used, including linear regression, polynomial regression, and multivariate regression.

The goal of error analysis is to assess and reduce the risks involved with nanosensor readings. It entails evaluating numerous sources of errors, such as instrument noise, ambient influences, drift, and cross-sensitivity, that may have an impact on the measurement's accuracy and precision. To assess and characterise the measurement errors, statistical techniques including uncertainty analysis, statistical regression analysis, and error propagation are used. Researchers may increase the accuracy and reproducibility of nanosensor readings by studying the causes of inaccuracy and their contributions.

Error analysis includes not just calculating measurement mistakes but also evaluating the accuracy, precision, linearity, and sensitivity of the sensor. Researchers may assess and contrast various nanosensor designs and optimisation techniques using these performance indicators, which provide insights into the sensor's capabilities and limits. Furthermore, uncertainty analysis, a crucial component of error analysis, offers a numerical indicator of the level of assurance related to nanosensor readings. The calculation of the measurement uncertainty within a certain confidence interval is provided by uncertainty analysis, which takes into consideration a variety of sources of uncertainties, including random and systematic mistakes. In order to establish precise correlations between sensor responses and analyte concentrations while evaluating and reducing measurement uncertainties, calibration and error analysis are crucial elements of nanosensor research. These methods improve measurement accuracy, provide useful information about how well nanosensors operate, and support the creation of improved sensing systems.

Conclusion

We have thoroughly examined the function and significance of simulation and analysis tools in the creation and improvement of nanosensors in this chapter. We have developed a thorough grasp of the importance of these tools in the area of nanosensors by diving into the underlying concepts, simulation methodologies, modelling approaches, analytical tools, design optimisation methods, and real-world case studies. In order to simulate the behaviour and functionality of nanosensors, simulation methods including finite element analysis (FEA), molecular dynamics (MD), and Monte Carlo simulations have shown to be quite useful. They make it possible for researchers to examine different design factors, simulate complicated processes, and enhance sensor performance. Each simulation approach has its own benefits and drawbacks, and the choice of strategy relies on the particular needs of the nanosensor under investigation. The underlying processes and interactions between analytes and nanosensors may be better understood by simulating nanosensor performance. Researchers may simulate and examine important performance metrics, such as sensitivity, selectivity, reaction time, and detection limits, by precisely capturing aspects like surface functionalization and charge transport. This modelling strategy helps with the design and optimisation of nanosensors for better efficiency and performance.

The enormous volumes of data produced by nanosensors must be interpreted, and analysis tools are essential for this. Data analysis, pattern identification, and data-driven decision making are made easier by statistical methodologies, signal processing techniques, and machine learning algorithms. With the use of these technologies, researchers may get insightful knowledge, spot patterns and trends, and improve the precision and dependability of nanosensor data. However, it is crucial to take into account the difficulties presented by noisy and high-dimensional data and create reliable analytic techniques to deal with these problems. To get the most performance out of nanosensors, their design must be optimised. Researchers may carry out parameter sweep evaluations, sensitivity assessments, and use optimisation algorithms to fine-tune and optimise sensor designs by using simulation and analytic tools. Researchers may systematically hone sensor settings and enhance their sensitivity, selectivity, and overall performance thanks to the repetitive nature of design and simulation cycles. Real-world case studies have shown how simulation and analytic tools may be used effectively in a variety of industries. These case studies have shown how these technologies are successful in solving particular problems, refining sensor designs, and offering insightful data on sensor behaviour and performance. Researchers may better grasp the constraints and potential related to the creation and use of nanosensors by looking at real-world examples.

Tools for modelling and analysis are crucial in the area of nanosensors, to sum up. Researchers can simulate, model, and optimise sensor performance, comprehend complicated data, and improve sensor designs thanks to their integration. However, there are still issues that need to be resolved, such as improving data analysis algorithms, creating more precise simulation models, and combining different simulation methods. A broad variety of applications will benefit from the development of highly effective and dependable nanosensors as a result of ongoing research and innovation in this field. The area of nanosensors is developing quickly, and simulation and analysis tool development must keep up with new developments and trends. Researchers may progress nanosensor technology by using these methods, which will result in better sensing capabilities, greater accuracy, and broader application domains. The development of nanosensors must include the integration of modelling and analytic tools in order for scientists to fully exploit the capabilities of these extraordinary devices.

References

Almawgani, Abdulkarem H.M., Partha Sarkar, Amrindra Pal, Gaurav Srivastava, Arun Uniyal, Adam R.H. Alhawari, and Arjuna Muduli. 2023. "Titanium Disilicide, Black Phosphorus–Based Surface Plasmon Resonance Sensor for Dengue Detection." *Plasmonics*, April, 1–10. https://doi.org/10.1007/S11468-023-01856-3/METRICS.

Anand, G., Swarnava Ghosh, Liwei Zhang, Angesh Anupam, Colin L. Freeman, Christoph Ortner, Markus Eisenbach, and James R. Kermode. 2022. "Exploiting Machine Learning in Multiscale Modelling of Materials." *Journal of The Institution of Engineers (India): Series D*, November, 1–11. https://doi.org/10.1007/S40033-022-00424-Z/FIGURES/3.

Antosiewicz, Tomasz J., and Mikael Käll. 2016. "A Multiscale Approach to Modeling Plasmonic Nanorod Biosensors." *Journal of Physical Chemistry C* 120 (37): 20692–701. https://doi.org/10.1021/ACS.JPCC.6B01897/ASSET/IMAGES/LARGE/JP-2016-018979_0008.JPEG.

Bodrenko, Igor V., Stefan Milenkovic, and Matteo Ceccarelli. 2022. "Diffusion of Molecules through Nanopores under Confinement: Time-Scale Bridging and Crowding Effects via Markov State Model." *Biomolecular Concepts* 13 (1): 207–19. https://doi.org/10.1515/BMC-2022-0019/ASSET/GRAPHIC/J_BMC-2022-0019_FIG_004.JPG.

Chakraborthy, Aniket, Suresh Nuthalapati, Anindya Nag, Nasrin Afsarimanesh, Md Eshrat E. Alahi, and Mehmet Ercan Altinsoy. 2022. "A Critical Review of the Use of Graphene-Based Gas Sensors."

Chemosensors 2022, Vol. 10, Page 355 10 (9): 355. https://doi.org/10.3390/CHEMOSENSORS10090355.

"Design and Simulation of 3D ZnO Nanowire Based Gas Sensors for Conductivity Studies." n.d. Accessed July 14, 2023. https://www.comsol.com/paper/design-and-simulation-of-3d-zno-nanowire-based-gas-sensors-for-conductivity-stud-11484.

Domènech-Gil, Guillem, and Donatella Puglisi. 2022. "A Virtual Electronic Nose for the Efficient Classification and Quantification of Volatile Organic Compounds." *Sensors 2022, Vol. 22, Page 7340* 22 (19): 7340. https://doi.org/10.3390/S22197340.

Fu, Tianren, Kathleen Frommer, Colin Nuckolls, and Latha Venkataraman. 2021. "Single-Molecule Junction Formation in Break-Junction Measurements." *Journal of Physical Chemistry Letters* 12 (44): 10802–7. https://doi.org/10.1021/ACS.JPCLETT.1C03160.

Gates, T. S., G. M. Odegard, S. J.V. Frankland, and T. C. Clancy. 2005a. "Computational Materials: Multi-Scale Modeling and Simulation of Nanostructured Materials." *Composites Science and Technology* 65 (15–16): 2416–34. https://doi.org/10.1016/J.COMPSCITECH.2005.06.009.

———. 2005b. "Computational Materials: Multi-Scale Modeling and Simulation of Nanostructured Materials." *Composites Science and Technology* 65 (15–16): 2416–34. https://doi.org/10.1016/J.COMPSCITECH.2005.06.009.

He, Xiaoyi, and Ning Li. 2000. "Lattice Boltzmann Simulation of Electrochemical Systems." *Computer Physics Communications* 129 (1–3): 158–66. https://doi.org/10.1016/S0010-4655(00)00103-X.

Helder, A. Santos, Luis M. Bimbo, Vesa-Pekka Lehto, Anu J. Airaksinen, Jarno Salonen, and Jouni Hirvonen. 2011. "Multifunctional Porous Silicon for Therapeutic Drug Delivery and Imaging." *Current Drug Discovery Technologies* 8 (3): 228–49. https://doi.org/10.2174/157016311796799053.

Hiratsuka, Masaki, Motoki Emoto, Akihisa Konno, and Shinichiro Ito. 2019. "Molecular Dynamics Simulation of the Influence of Nanoscale Structure on Water Wetting and Condensation." *Micromachines 2019, Vol. 10, Page 587* 10 (9): 587. https://doi.org/10.3390/MI10090587.

Homola, Jiří. 2008. "Surface Plasmon Resonance Sensors for Detection of Chemical and Biological Species." *Chemical Reviews* 108 (2): 462–93. https://doi.org/10.1021/CR068107D/ASSET/CR068107D.FP.PNG_V03.

Jiang, Jianwen W., Jeffery B. Klauda, and Stanley I. Sandler. 2003. "Monte Carlo Simulation of O2 and N2 Adsorption in Nanoporous Carbon (C168 Schwarzite)." *Langmuir* 19 (8): 3512–18. https://doi.org/10.1021/LA020319S.

Kohn, W., and L. J. Sham. 1965. "Self-Consistent Equations Including Exchange and Correlation Effects." *Physical Review* 140 (4A). https://doi.org/10.1103/PHYSREV.140.A1133.

Lado-Touriño, Isabel, and Alicia Páez-Pavón. 2021. "Interaction between Graphene-Based Materials and Small Ag, Cu, and Cuo Clusters: A Molecular Dynamics Study." *Nanomaterials* 11 (6): 1378. https://doi.org/10.3390/NANO11061378/S1.

Li, Chao Yu, Sai Duan, Jun Yi, Chen Wang, Petar M. Radjenovic, Zhong Qun Tian, and Jian Feng Li. 2020. "Real-Time Detection of Single-Molecule Reaction by Plasmon-Enhanced Spectroscopy." *Science Advances* 6 (24): 6012–22. https://doi.org/10.1126/SCIADV.ABA6012/SUPPL_FILE/ABA6012_SM.PDF.

Luo, Shaohong, Yarjan Abdul Samad, Vincent Chan, and Kin Liao. 2019. "Cellular Graphene: Fabrication, Mechanical Properties, and Strain-Sensing Applications." *Matter* 1 (5): 1148–1202. https://doi.org/10.1016/J.MATT.2019.10.001.

Maiti, Santanu K. 2007. "Quantum Transport through Organic Molecules." *Chemical Physics* 331 (2–3): 254–60. https://doi.org/10.1016/J.CHEMPHYS.2006.10.013.

———. 2008. "Quantum Transport through Single Phenalenyl Molecule: Effect of Interface Structure." *Physica E: Low-Dimensional Systems and Nanostructures* 40 (8): 2730–35. https://doi.org/10.1016/J.PHYSE.2007.12.028.

Onaş, Andra Mihaela, Constanţa Dascălu, Matei D. Raicopol, and Luisa Pilan. 2022. "Critical Design Factors for Electrochemical Aptasensors Based on Target-Induced Conformational Changes: The Case of Small-Molecule Targets." *Biosensors 2022, Vol. 12, Page 816* 12 (10): 816. https://doi.org/10.3390/BIOS12100816.

Parvaiz, M. Shunaid, Khurshed A. Shah, H. Alrobei, G. N. Dar, Farooq A. Khanday, S. Muzaffar Ali Andrabi, and Rabia Hamid. 2021. "Modeling and Simulation of Carbon Nanotube Amino-Acid Sensor: A First-Principles Study." *Computational and Theoretical Chemistry* 1204 (October): 113402. https://doi.org/10.1016/J.COMPTC.2021.113402.

Proykova, Ana. 2006. "Molecular Dynamics Simulation of Gas Adsorption and Absorption in Nanotubes." *Carbon Nanotubes*, May, 187–207. https://doi.org/10.1007/1-4020-4574-3_31.

Puigcorbé, J., D. Vogel, B. Michel, A. Vilà, I. Gràcia, C. Cané, and J. R. Morante. 2003. "Thermal and Mechanical Analysis of Micromachined Gas Sensors." *Journal of Micromechanics and Microengineering* 13 (5): 548–56. https://doi.org/10.1088/0960-1317/13/5/304.

Saleh, Norsyahidah Md, and A. Abdul Aziz. 2018. "Simulation of Surface Plasmon Resonance on Different Size of a Single Gold Nanoparticle." *Journal of Physics: Conference Series* 1083 (1): 012041. https://doi.org/10.1088/1742-6596/1083/1/012041.

Shariati, Vahid, Ehsan Roohi, and Amin Ebrahimi. 2023. "Numerical Study of Gas Flow in Super Nanoporous Materials Using the Direct Simulation Monte-Carlo Method." *Micromachines 2023, Vol. 14, Page 139* 14 (1): 139. https://doi.org/10.3390/MI14010139.

"Simulating a 3D Multilayered Graphene Biosensor Design | COMSOL Blog." n.d. Accessed July 14, 2023. https://www.comsol.com/blogs/simulating-a-3d-multilayered-graphene-biosensor-design/.

Thawornsathit, Phongsakorn, Ekachai Juntasaro, Hwanjit Rattanasonti, Putapon Pengpad, Karoon Saejok, Chana Leepattarapongpan, Ekalak Chaowicharat, and Wutthinan Jeamsaksiri. 2022. "Enhancing Performance of a MEMS-Based Piezoresistive Pressure Sensor by Groove: Investigation of Groove Design Using Finite Element Method." *Micromachines 2022, Vol. 13, Page 2247* 13 (12): 2247. https://doi.org/10.3390/MI13122247.

Thomas, James O., Bart Limburg, Jakub K. Sowa, Kyle Willick, Jonathan Baugh, G. Andrew D. Briggs, Erik M. Gauger, Harry L. Anderson, and Jan A. Mol. 2019. "Understanding Resonant Charge Transport through Weakly Coupled Single-Molecule Junctions." *Nature Communications* 10 (1). https://doi.org/10.1038/S41467-019-12625-4.

Tian, Pu. 2008. "Molecular Dynamics Simulations of Nanoparticles." *Annual Reports Section "C" (Physical Chemistry)* 104 (0): 142–64. https://doi.org/10.1039/B703897F.

Tien, Chang Lin, and Jian Gang Weng. 2002. "Molecular Dynamics Simulation of Nanoscale Interfacial Phenomena in Fluids." *Advances in Applied Mechanics* 38 (C): 95–146. https://doi.org/10.1016/S0065-2156(02)80103-X.

Vazinishayan, Ali, Dasaradha Rao Lambada, Shuming Yang, Guofeng Zhang, Biyao Cheng, Yonas Tesfaye Woldu, Shareen Shafique, Yiming Wang, and Ndahimana Anastase. 2018. "Effects of Mechanical Strain on Optical Properties of ZnO Nanowire." *AIP Advances* 8 (2): 025306. https://www.academia.edu/76238319/Effects_of_mechanical_strain_on_optical_properties_of_ZnO_nanowire.

Xu, Han, Yu Chen, Jun Hyuk Kim, Zheng Dang, and Meilin Liu. 2019. "Lattice Boltzmann Modelling of the Coupling between Charge Transport and Electrochemical Reactions in a Solid Oxide Fuel Cell with a Patterned Anode." *International Journal of Hydrogen Energy* 44 (57): 30293–305. https://doi.org/10.1016/J.IJHYDENE.2019.09.086.

Ye, Zhenyi, Yuan Liu, and Qiliang Li. 2021. "Recent Progress in Smart Electronic Nose Technologies Enabled with Machine Learning Methods." *Sensors 2021, Vol. 21, Page 7620* 21 (22): 7620. https://doi.org/10.3390/S21227620.

Yeow, John T.W., and Yun Wang. 2009. "A Review of Carbon Nanotubes-Based Gas Sensors." *Journal of Sensors* 2009. https://doi.org/10.1155/2009/493904.

Zhang, Xiaoting, Jose Villafuerte, Vincent Consonni, Jean Fabien Capsal, Pierre Jean Cottinet, Lionel Petit, and Minh Quyen Le. 2021. "Characterizing and Optimizing Piezoelectric Response of ZnO Nanowire/PMMA Composite-Based Sensor." *Nanomaterials 2021, Vol. 11, Page 1712* 11 (7): 1712. https://doi.org/10.3390/NANO11071712.

Zhang, Xu, Shuyan Liu, Han Liu, Jinwen Zhang, and Xiaoning Yang. 2017. "Molecular Dynamics Simulation of the Mechanical Properties of Multilayer Graphene Oxide Nanosheets." *RSC Advances* 7 (87): 55005–11. https://doi.org/10.1039/C7RA10273A.

Zienert, Andreas, Jörg Schuster, and Thomas Gessner. 2013. "Comparison of Quantum Mechanical Methods for the Simulation of Electronic Transport through Carbon Nanotubes." *Microelectronic Engineering* 106 (June): 100–105. https://doi.org/10.1016/J.MEE.2012.12.018.

Chapter 8

Nanosensors in Healthcare Monitoring: Biosensors

Learning Outcomes

- This chapter explains how nanosensors may be used to monitor healthcare, including their great sensitivity, selectivity, and capacity to detect minute quantities of analytes or biomarkers.
- The student becomes more used to the basics of biosensors, such as the concepts of biosensing, transducers for signal conversion, and signal amplification methods.
- A thorough understanding of the many uses for nanosensors in healthcare monitoring is attained, including point-of-care diagnostics, ongoing patient health monitoring, wearable biosensors, early illness diagnosis, personalised medications, and drug delivery systems.
- There is a greater understanding of the difficulties in creating, enhancing sensitivity and specificity, miniaturising, and commercialising nanosensors for medical monitoring.
- The learner gains insight into prospective new nanomaterials, improved sensitivity and specificity, issues associated with miniaturisation, and successful commercialisation as they relate to future developments in nanosensor research.
- An understanding of how biosensors and nanosensors have the power to revolutionise personalised medicine, illness diagnosis, and the monitoring of medical treatments.

Abstract

Nanosensors have become effective instruments for monitoring healthcare, with the promise to provide real-time, precise, and non-invasive readings. Nanosensors, which are a subset of biosensors, are essential in this situation. They use highly sensitive and selective biological recognition elements and transducers to identify and measure the target analytes. The potential of nanosensors in healthcare monitoring is explored in this chapter, with a focus on its benefits including sensitivity, selectivity, and the capacity to detect minuscule levels of analytes or biomarkers. The basic ideas of biosensing are covered, including target identification, signal transduction processes, and signal amplification strategies used in biosensors. In-depth research is done on the use of nanosensors in healthcare monitoring. Their function in point-of-care diagnostics, ongoing patient health monitoring, wearable biosensors, early illness diagnosis, personalised medications, and drug delivery systems is examined in this chapter. It focuses on how nanosensors might improve patient care by improving illness diagnosis and therapy monitoring. It also discusses the difficulties in developing, miniaturising, sensitivity, specificity, and commercialising nanosensors for healthcare monitoring. It emphasises the need of overcoming these difficulties to guarantee their effective implementation in healthcare settings. The chapter concludes with insights into potential directions for nanosensor research, including novel nanomaterial creation, improvement of sensitivity and specificity, difficulties with miniaturisation, and successful commercialisation. These next developments have the power to promote personalised treatment and further revolutionise healthcare monitoring. For academics, practitioners, and stakeholders interested in using nanosensors to improve patient care and healthcare outcomes and comprehending the function of biosensors in healthcare monitoring, the chapter serves as a thorough reference.

1. Introduction

The use of nanosensors, especially biosensors, has led to notable breakthroughs in the monitoring of healthcare in recent years. The detection and analysis of biological markers has been revolutionised by these nanoscale devices, which also make very sensitive, non-invasive, real-time monitoring of numerous health indices possible. This chapter examines the crucial function of nanosensors in healthcare monitoring, with an emphasis on biosensors.

A specialised kind of nanosensor called a biosensor combines a transducer and a biological recognition component to transform a biological signal into a quantifiable signal. They provide a direct link between biological and electrical systems, enabling the identification and measurement of certain biomolecules or biological activities. Due to their great specificity, sensitivity, and adaptability, biosensors have drawn significant interest in healthcare monitoring.

The development of nanotechnology has greatly improved the capacities of biosensors in monitoring medical treatment. To magnify the detection signals and enhance overall performance, biosensor systems use nanomaterials like nanoparticles, nanowires, or nanotubes. The distinctive qualities of nanomaterials, such as their huge surface area, adjustable optical characteristics, or increased electron transport, help biosensors be more sensitive and selective.

Numerous uses for nanobiosensors in healthcare monitoring have been discovered, advancing personalised medicine and enhancing patient care. With the use of nanobiosensors, disease-specific biomarkers may be quickly and sensitively detected, enabling early and precise diagnosis of a variety of ailments, including cancer, infectious diseases, and cardiovascular problems. Rapid on-site testing of biological samples is made possible by the incorporation of nanobiosensors into lightweight, portable devices that are simple to use. This shortens turnaround times and enables prompt decision-making in clinical settings or situations with limited resources.

Wearable biosensors that can continually monitor vital signs, blood sugar levels, hormone concentrations, or other health-related factors have been made possible thanks to nanotechnology. These gadgets enable users to actively control their health and well-being since they provide real-time monitoring. Nanobiosensors are essential for assessing the effectiveness and security of therapeutic treatments. Healthcare professionals may optimise treatment regimens and guarantee individualised patient care by tracking medication concentrations, detecting metabolite levels, or monitoring tissue reactions.

Despite the enormous promise of nanobiosensors in healthcare monitoring, a number of issues need to be resolved before they can be widely used. These difficulties include those related to biocompatibility, long-term stability, affordability, regulatory issues, and moral ramifications. To further improve the dependability, precision, and accessibility of nanobiosensors, future research should concentrate on solving these challenges.

In conclusion, this chapter examines how nanobiosensors are revolutionising healthcare monitoring. These nanosensors provide unmatched sensitivity, selectivity, and mobility by fusing nanotechnology with biosensing principles. Nanobiosensors have a wide range of uses, from illness diagnosis to treatment monitoring to wearable technology, all of which are advancing personalised healthcare and enhancing patient outcomes. The necessity for ongoing study and innovation in this fascinating sector is emphasised as the chapter goes on to address the difficulties and prospects for the future.

2. Potential of Nanosensors in Healthcare Monitoring

Due to their special powers at the nanoscale, nanosensors and biosensors have a tremendous potential to revolutionise healthcare monitoring. To detect and analyse numerous physiological and biochemical data, these gadgets make use of biological sensing and nanotechnology principles. Their great sensitivity, specificity, and capacity to communicate with biological systems, together with their small size, provide them technological advantages.

Devices made at the nanoscale called nanosensors are used to monitor and detect certain physical or chemical characteristics. They may be created using a variety of substances, including carbon nanotubes, nanowires, and nanoparticles, which have extraordinary capabilities at the nanoscale. These nanomaterials often have superior surface-to-volume ratios, allowing for better interactions with target analytes. Nanosensors may bind to target biomolecules or analytes with specificity by functionalizing their surfaces with certain receptors or molecules.

On the other hand, biosensors are sensor systems that combine transducing elements (such as electrodes or optical components) with biological elements (such as enzymes, antibodies, or nucleic acids). They have the ability to transform a biological signal into an electrical or optical signal that can be measured. A vast variety of analytes, including glucose, proteins, DNA, and even certain diseases, may be detected and quantified using biosensors.

Nanosensors are perfect for continuous and real-time readings due to their tiny size, which enables minimally invasive or non-invasive monitoring. They provide the opportunity for long-term monitoring of numerous physiological indicators and may be incorporated into wearable devices or implanted directly into the body. For instance, nanosensors built into a wearable patch may continually monitor the wearer's blood pressure, respiration rate, and other vital indicators without making them uncomfortable.

Nanosensors and biosensors are very sensitive due to a number of factors. The likelihood of analyte interactions is increased by the high surface-to-volume ratio seen in many nanomaterials. Additionally, some receptors or molecules may be added to nanosensors to increase their selectivity and affinity for the target analytes. This makes it possible to identify analytes down to individual molecules and at very low concentrations. Such sensitivity is essential for finding disease-related biomarkers, which enables early detection and treatment.

For precise monitoring and diagnosis, the specificity of nanosensors and biosensors is crucial. They may selectively bind to target analytes or biomarkers by using certain receptors or biomolecules, which lowers the possibility of false-positive or false-negative findings. When finding chemicals that are unique to a disease, such as cancer biomarkers or viral antigens, this specificity is very important. Accurate detection and quantification of these compounds may help in the early identification of illnesses and enhance the effectiveness of treatments.

Additionally, nanosensors and biosensors are compatible with wireless technology and cutting-edge data processing methods. Data transfer in real time and remote monitoring are made possible by this integration. In order to find trends or abnormalities, the gathered data may be analysed using algorithms and machine learning methods, enabling early diagnosis of health issues and prompt action. The ability to monitor patients remotely allows medical practitioners to provide better access to treatment and enable proactive actions.

Due to its capacity to function at the nanoscale, high sensitivity, specificity, and integration with cutting-edge technology, nanosensors and biosensors have the potential to revolutionise healthcare monitoring. These characteristics allow accurate, real-time, and remote monitoring, enabling illness early identification, individualised care, and better patient outcomes.

3. Fundamentals of Biosensors

In order to detect and quantify certain analytes or biomolecules, sensor devices known as "biosensors" merge biological components with transducing elements. They depend on the core ideas of signal transduction, molecular recognition, and data processing.

A bioreceptor, a transducer, and a signal processor are the essential parts of a biosensor. The bioreceptor is a biological component that interacts with the target analyte in a targeted manner. Depending on the application, this might be a complete cell, nucleic acid, enzyme, or antibody. A biological reaction is started when the bioreceptor identifies and binds to the analyte.

The biological reaction is transformed into a quantifiable signal by the transducer. It could be a thermal, optical, piezoelectric, or electrochemical device. An electrochemical reaction is sparked by the binding of the analyte to the bioreceptor in an electrochemical biosensor, for instance, and produces a detectable current or voltage signal.

The transducer signal is amplified, filtered, and analysed by the signal processor to provide useful information. Techniques for digital or analogue signal processing may be used in this. To facilitate further analysis and interpretation, the data may subsequently be displayed, saved, or sent.

3.1. Principles of Biosensing

The overall working principle of a biosensor involves the following steps:

- Recognition: The bioreceptor recognises the target analyte in the sample and binds to it specifically. Usually, particular interactions, such the binding of an antigen to an antibody or an enzyme to a substrate, provide the basis for this binding event.
- Transduction: When an analyte binds to a bioreceptor, a biological or physicochemical reaction is set off. This reaction is transformed into a quantifiable signal by the transducer, which may be electrical, optical, or mechanical in nature.
- Signal conversion is the third step, and it transforms the transduced signal into a form that is simple to measure and analyse. To increase the signal's detectability and accuracy, this often entails amplification, filtering, or modulation of the signal.

- Data Processing: Using the proper techniques or methodologies, the signal that has been processed is analysed and interpreted. To derive useful information about the analyte concentration or other pertinent parameters, this may entail calibration, data normalisation, or statistical analysis.
- Output and Interpretation: The biosensor's ultimate output is often represented as a numerical number, a graphic, or a binary answer (such as the presence or absence of a certain analyte). By comparing the data to established diagnostic criteria, thresholds, or reference levels, the results may be further analysed.

The schematic of the principle of biosensing is shown in Figure 133.

Biosensors may be created for a variety of purposes, such as environmental monitoring, food safety, clinical diagnostics, and bioprocess monitoring. They are useful instruments in many domains because of their adaptability, sensitivity, and specificity, which allows for quick and accurate detection of target analytes with little sample preparation.

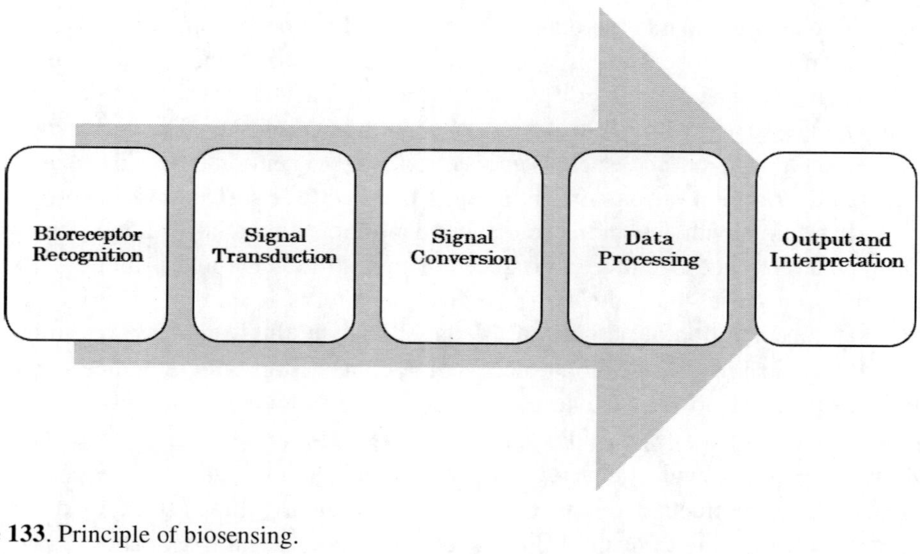

Figure 133. Principle of biosensing.

3.2. Transducers and Signal Amplification in Biosensors

The two key processes of every biosensor that help it achieve its goals are transduction and signal amplification. According to their use, transducers generally make use of changes in some physical quantity that correlate to the biosensor's reaction to a certain kind of analyte upon detection. The transducers are divided into many varieties according to their capacity to detect certain kinds of physical changes, which are described in more detail below.

3.2.1. Calorimetric Transducers

A calorimetric transducer is a particular kind of transducer used in biosensors to detect and measure analytes or biomolecules using heat-based measurements. The method is based on the idea that certain biological processes produce or absorb heat, and that this change in temperature may be connected to the concentration or activity of the target analyte. A biochemical reaction is started in a calorimetric biosensor when the bioreceptor selectively interacts with the target analyte. The temperature of the sensor system changes as a consequence of this reaction, which either releases or absorbs heat. A sensitive transducer, such as a thermistor or a microfabricated calorimetric chip, measures this temperature change.

The target analyte's concentration or activity may be ascertained by processing and analysing the electrical signal that the transducer converted from the temperature change. This may be accomplished either by employing calibration curves or by comparing the signal produced from the sample with reference signals acquired from known concentrations. Calorimetric biosensors are advantageous because they do not need extra labelling or amplification stages, and they are simple, sensitive, and allow for label-free detection. They have uses in a variety of industries, including as clinical diagnostics, environmental monitoring, and food safety, where it is desirable to quickly and quantitatively analyse analytes.

3.2.2. Potentiometric Transducers

An electrical potential (voltage) shift brought on by a particular biochemical interaction between a bioreceptor and a target analyte is measured by a potentiometric transducer, a kind of transducer used in biosensors. It works on the premise that when an analyte binds to a bioreceptor, an electrical potential difference is created. This electrical potential difference is measured and associated with the analyte's concentration or activity. The bioreceptor is often immobilised on a solid surface, such as an electrode, in a potentiometric biosensor. A change in the charge distribution or ion concentration at the electrode surface results from the target analyte's interaction with the bioreceptor. A quantifiable potential difference between the electrode and a reference electrode is produced by this shift in charge distribution.

A transducer senses the potential difference and turns it into an electrical signal. To ascertain the concentration or activity of the target analyte, this signal is next processed and analysed. It is typical to utilise calibration curves or comparisons with reference signals to link the observed potential to the analyte concentration. Potentiometric biosensors come with a number of benefits, such as ease of use, label-free detection, and excellent specificity. They are capable of detecting a wide range of analytes, including ions, gases, enzymes, and other biomolecules. Potentiometric biosensors are used in many different industries, such as environmental monitoring, industrial process control, and clinical diagnostics, where precise and targeted analyte detection is necessary.

3.2.3. Amperometric Transducers

A particular form of transducer used in biosensors called an amperometric transducer detects the electric current produced by a biochemical interaction between a bioreceptor and a target analyte. It works on the premise that when an analyte is reduced or oxidised, an electrical current is produced. This electrical current may be measured and connected to an analyte's concentration or activity. The bioreceptor is generally immobilised on an electrode surface in an amperometric biosensor. The target analyte engages in an electrochemical reaction with the bioreceptor that includes the transfer of electrons. This causes the creation of a current that is directly inversely proportional to the analyte concentration.

A transducer, which changes the electrochemical signal into an electrical signal, is used to measure the current. To ascertain the concentration or activity of the target analyte, this signal is next processed and analysed. Common methods for connecting the observed current to the analyte concentration include calibration curves and comparisons with reference signals.

Amperometric biosensors include benefits including high sensitivity, quick reaction times, and a broad spectrum of analytes that may be detected. Glucose, enzymes, neurotransmitters, and other chemicals, as well as environmental contaminants, may all be detected using them. Applications for amplimetric biosensors include clinical diagnostics, food safety, environmental monitoring, and bioprocess control, all of which need accurate and quick analyte analysis.

3.2.4. Optical Transducers

An optical transducer is a particular form of transducer used in biosensors that makes use of measurements based on light to find and examine certain analytes or biomolecules. It works on the premise that when a target analyte interacts with a bioreceptor, it causes detectable changes in light intensity, wavelength, or polarisation that may be attributed to the analyte's concentration or activity. The bioreceptor is often immobilised on a surface or integrated into a sensing component in an optical biosensor. The optical characteristics of the system are altered when the target analyte binds to the bioreceptor. Several optical methods, including absorbance, fluorescence, surface plasmon resonance (SPR), or interferometry, may be used to identify and measure these changes.

To ascertain the concentration or activity of the target analyte, the detected optical signal is generally processed and analysed. This may be accomplished either by employing calibration curves or by comparing the signal produced from the sample with reference signals acquired from known concentrations. High sensitivity, real-time monitoring, and the possibility for multiplexed detection are some benefits of optical biosensors. They may be used to find a variety of analytes, such as infections, poisons, biomolecules, or environmental contaminants. Clinical diagnostics, biotechnology, environmental monitoring, and food safety are just a few of the areas where optical biosensors are used to accurately and quickly analyse analytes.

3.3. Selectivity and Sensitivity in Biosensing

Selectivity is a key factor in determining a biosensor's effectiveness and dependability. The term "selectivity" describes a biosensor's capacity to recognise and isolate the target analyte of interest from other interfering chemicals in a sample. A biosensor's readings will be more precise and dependable the greater their selectivity. In the absence of interfering molecules, a highly selective biosensor makes sure that it reacts solely to the target analyte and does not generate false positive signals. As a consequence, there is less likelihood that the data would be interpreted incorrectly, and measurement findings are more confident. Sensitivity and selectivity go hand in hand. The interference brought on by other compounds may be completely eliminated or drastically reduced using a biosensor that is focused on the target analyte. As a result, the target analyte may be detected and quantified at lower concentrations, increasing the biosensor's overall sensitivity. The biosensor readings' accuracy and precision are aided by selectivity. The biosensor may give more accurate and exact measurements of the analyte concentration or activity, producing trustworthy and repeatable findings by reducing or eliminating interference from non-target chemicals. The biosensor's selectivity allows it to selectively recognise and separate the target analyte from molecules with similar structures or functions. This is especially crucial for complicated samples since there might be a large number of possible interferents. The biosensor may offer precise information about the target analyte by binding to it specifically while being unaffected by other chemicals in the sample. The possibility of false negative findings, when the target analyte is missed owing to interference or non-specific binding, is reduced by a biosensor with high selectivity. This is essential in situations like illness diagnosis or environmental monitoring when the presence of the target analyte is critical.

Sensitivity describes a biosensor's capacity to identify and quantify minute changes in the target analyte, such as low concentrations. A biosensor's sensitivity determines how precisely it can quantify the analyte and provide accurate results. Even at extremely low concentrations, a highly sensitive biosensor can identify and measure the target analytes. This is crucial for applications like early illness detection or environmental monitoring when the analyte is present in low quantity. Greater sensitivity enables earlier detection and action, which enhances results. The accuracy of the readings made by the biosensor is directly related to sensitivity. Small variations in analyte concentration may be picked up by a more sensitive biosensor, leading to more precise and accurate results. For tracking dynamic changes in analyte levels over time or in reaction to outside stimuli, this is crucial. Sensitivity aids in amplifying the target analyte's intended signal while reducing background noise or other types of interference. High sensitivity biosensors have a greater signal-to-noise ratio because they can distinguish between minor analyte-induced signal changes and noise fluctuations. As a result, the measurements made by the biosensor are more trustworthy and accurate. Biosensors can work across a large dynamic range and a variety of analyte concentrations because to their sensitivity. This eliminates the requirement for sample dilution or concentration and enables the detection and measurement of both low and high analyte concentrations. The adaptability and usability of the biosensor in varied contexts are improved by a broad dynamic range. Smaller sample quantities or lower analyte concentrations are needed for reliable readings with highly sensitive biosensors. When working with small sample sizes, uncommon analytes, or priceless samples, this may be very helpful. Reduced sample needs aid in minimally invasive or non-invasive monitoring as well as cost and time effectiveness.

4. Application of Nanosensors in Healthcare Monitoring

Among other uses, nanosensors may be used to detect infections, track vital signs, and diagnose illnesses. Nanosensors can identify certain biomarkers that signify the presence of diseases or abnormal situations in the body for diagnosing illnesses. Nanosensors can detect these indicators and provide early illness diagnosis, enabling prompt intervention and therapy. Nanosensors can track the development of a disease's treatment, giving real-time information on a therapy's efficacy. Nanosensors can monitor vital indicators including heart rate, blood pressure, and oxygen levels in healthcare monitoring. These sensors may be incorporated into wearable technology, enabling ongoing patient health status monitoring. Additionally, nanosensors can keep track of drug levels in the body to ensure proper dosage and avoid any possible overdoses or underdoses. Pathogens, such as bacteria or viruses, may be found and identified using the nanosensors in biological samples. The area of controlling and monitoring infectious diseases benefits greatly from having this skill. Nanosensors may help in early outbreak containment and prevention by rapidly and precisely detecting microorganisms. Overall, the use of nanosensors in healthcare monitoring shows considerable potential for enhancing patient outcomes, therapeutic effectiveness, and diagnostic accuracy. Various applications of nanosensors in healthcare monitoring is given in Figure 134. Further developments and applications in healthcare are anticipated to result from ongoing research and development in this area.

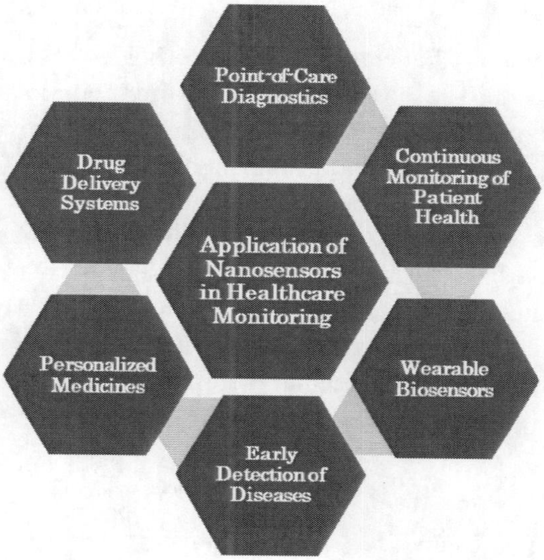

Figure 134. Application of nanosensors in healthcare monitoring.

4.1. Point-of-Care Diagnostics

The development and use of nanotechnology-based biosensors for quick and precise illness detection and monitoring at the patient's bedside or in non-laboratory settings is referred to as point-of-care (POC) diagnostics in the healthcare industry. Because of their tiny size, high sensitivity, and capacity to identify and track certain biological markers, nanosensors, which are sensor devices with nanoscale dimensions, have significant benefits over other types of healthcare monitoring technologies (Noah and Ndangili 2019).

Recent scientific studies have highlighted the usefulness of nanosensors in POC diagnosis. Due to their distinctive features, nanomaterials including gold and silver nanoparticles, carbon nanomaterials, and hybrid nanocomposites are increasingly employed in POC testing (Wang, Li, and Guo 2020). The use of nanosensors in healthcare has shown significant promise for enhancing diagnostics and patient outcomes. These nanoparticles may act as carriers for biorecognition components or as labels for signal creation in different POC diagnostics. Numerous uses of nanosensors in POC diagnostics, such as illness diagnosis, vital sign monitoring, and pathogen detection, have been emphasised in recent scientific papers (Wang, Li, and Guo 2020). Early illness identification, treatment progress monitoring, and accurate medication dosage are all possible using nanosensors (Alemu and Alemu 2022). They are perfect for point-of-care settings because of their compact size and great sensitivity since prompt decision-making and management of patient care depend on swift and accurate diagnoses (Noah and Ndangili 2019). The use of nanosensors in the medical field has significant potential for enhancing patient outcomes, therapeutic effectiveness, and diagnostic accuracy.

4.2. Continuous Monitoring of Patient Health

Nanosensor-based continuous patient health monitoring is a growing field of healthcare research and development. The possibility for real-time, non-invasive, and individualised monitoring of many physiological indicators and biomarkers is provided by nanosensors. The real-time, non-invasive detection and measurement of numerous physiological characteristics or biomarkers to offer continuous updates on the patient's health is referred to as continuous monitoring of a patient's health using nanosensors. In this area, nanosensors, which are tiny devices or materials that can detect and translate certain signals, offer significant promise. They provide more regular and accurate monitoring, which enables early identification of health concerns and prompt action. The schematic of various examples of continuous monitoring of patients health using nanosensors is given in Figure 135.

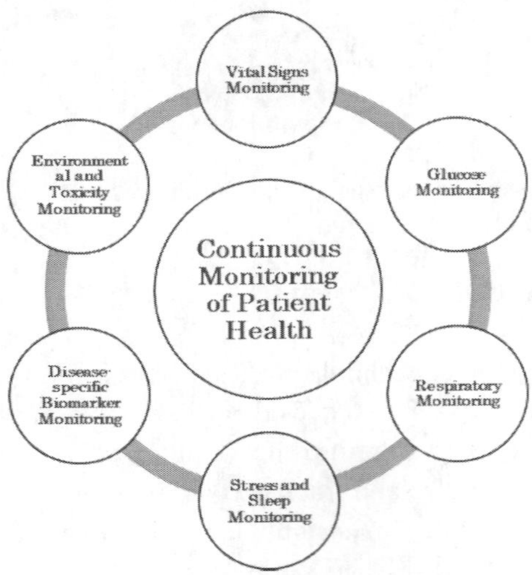

Figure 135. Examples of continuous monitoring of patients health using nanosensors.

The developments and prospective uses of nanosensors in continuous health monitoring have been emphasised in recent research studies. For instance, a research showed the creation of implanted nanosensors that can track glucose levels in diabetic patients with excellent sensitivity and accuracy (Ruckh and Clark 2014). The nanosensor patch measures sweat glucose concentrations and provides real-time feedback on blood sugar levels using a biocompatible graphene-based substance. Another research (Shiwani et al. 2023) showed how nanosensors may be used to continuously monitor cardiac biomarkers, leading to earlier diagnosis of heart problems and better patient outcomes. The wearable device's nanosensors can identify certain biomarkers linked to cardiac problems and wirelessly convey the information to medical specialists. This makes it possible to control heart health proactively and identify prospective cardiac problems before they happen. Additionally, a study revealed the creation of nanosensors for ongoing respiratory parameter monitoring, providing insightful information on respiratory health and disorders (Nicolò et al. 2020). The nanosensors can identify changes in respiration rate, oxygen saturation, and other important characteristics when they are integrated into a wearable device or included in clothes. This makes it possible to remotely monitor patients with chronic respiratory disorders' respiratory health.

These latest scientific studies show how nanosensors have the potential to revolutionise healthcare by providing personalised and proactive patient monitoring. By delivering real-time data on numerous physiological markers, they open up the potential of personalised and proactive therapy. Improved healthcare results may result from continuous monitoring utilising nanosensors, which can give useful information for the early identification of health conditions, individualised treatment programmes, and remote patient monitoring. It is crucial to remember that more study and development are still required to enhance the functionality, dependability, and incorporation of nanosensors in therapeutic settings.

4.3. Wearable Biosensors

In the field of healthcare, wearable biosensors are little gadgets that may be worn on the body to track and gather physiological data. These biosensors are made to provide continuous, real-time data regarding a range of health factors, including heart rate, blood pressure, oxygen levels, level of activity, sleep patterns, and more. Wearable biosensors have a lot of promise for use in healthcare. The illustrations of various wearable biosensors is given in Figure 136. By allowing early illness identification, tracking chronic ailments, delivering individualised healthcare, and enabling remote patient monitoring, they have the potential to revolutionise healthcare. Better health outcomes may be achieved by using wearable biosensors to assist patients and healthcare providers in monitoring and managing medical problems. To find biomarkers in the body, these biosensors employ a variety of technologies, including electrochemical and optical sensors. They are often linked to cloud platforms or mobile applications that provide consumers and healthcare professionals access to and analysis of the gathered data. The information may be utilised for early diagnosis, individualised treatment strategies, and keeping track of how well therapies are working.

These gadgets, which are often built into wearable clothes or accessories, provide real-time information that may help with illness management, wellness monitoring, and individualised treatment. For those with diabetes, for instance, continuous glucose monitoring is essential. A handy and continuous monitoring option is provided by wearable biosensors. Recently, a wearable biosensor patch that used enzyme-based glucose sensors was created for non-invasive, continuous glucose monitoring. With the use of the biosensor patch, glucose readings were precise and dependable, leading to better diabetes control (Hanna et al. 2020).

Heart rate, blood pressure, and electrocardiogram (ECG) are just a few of the cardiovascular indicators that wearable biosensors can track. For continuous cardiovascular monitoring, a wearable biosensor with integrated photoplethysmography and ECG sensors has been created. Real-time monitoring and early diagnosis of anomalies were made possible by the biosensor's precise recording of cardiac impulses (Mukhopadhyay, Suryadevara, and Nag 2022). These biosensors can monitor respiratory variables including breathing rate and lung capacity. For instance, a wearable biosensor that used strain sensors to track breathing patterns was created in 2021. The biosensor correctly recorded breathing data, giving insightful information on the state of the respiratory system (Andreozzi et al. 2021). Fitness tracking and activity tracking are common uses for wearable biosensors. Such biosensor applications include a wearable biosensor that tracks physical activity using accelerometers and gyroscopes. The biosensor reliably captured movement patterns, giving information on training regimens and general levels of fitness (R. Liu et al. 2022). Similar to this, wearable biosensors may monitor stress and sleep quality-related metrics. For example, a wearable biosensor that tracks stress levels makes use of sensors that measure skin temperature and galvanic skin reaction. The biosensor helped regulate stress and encourage improved sleep hygiene by providing insightful data on stress patterns (Yoon, Sim, and Cho 2016).

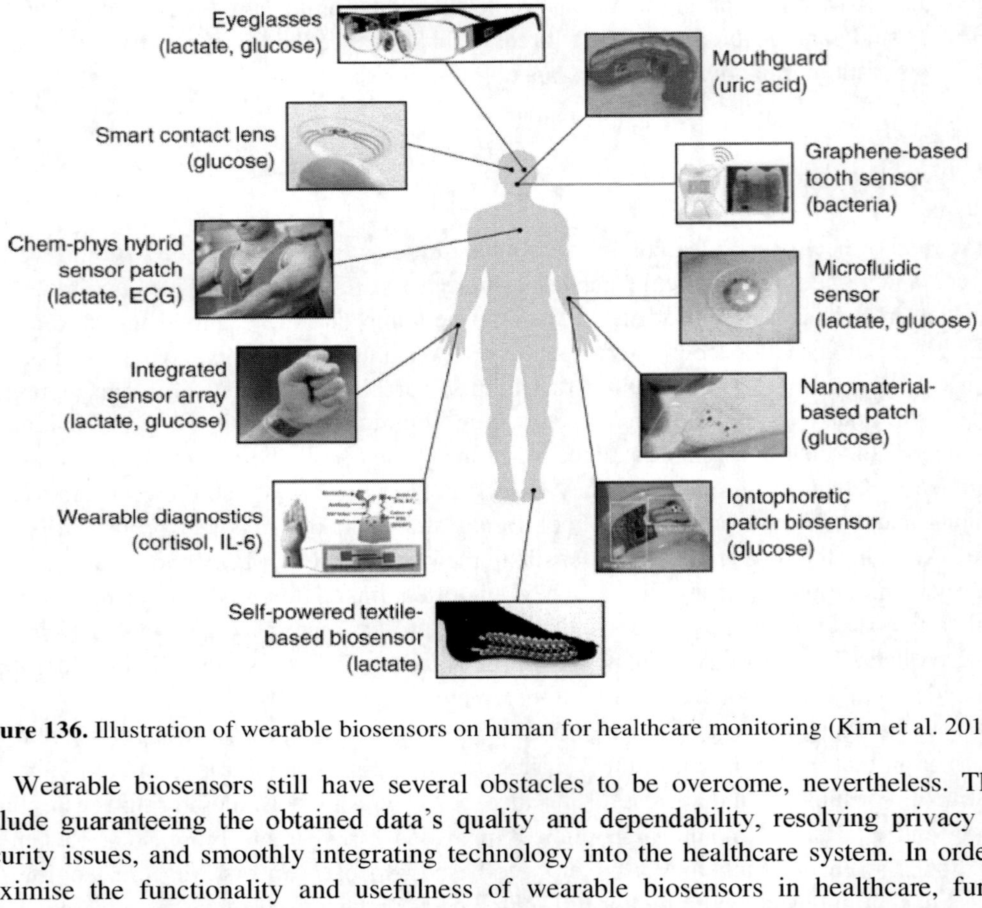

Figure 136. Illustration of wearable biosensors on human for healthcare monitoring (Kim et al. 2019).

Wearable biosensors still have several obstacles to be overcome, nevertheless. These include guaranteeing the obtained data's quality and dependability, resolving privacy and security issues, and smoothly integrating technology into the healthcare system. In order to maximise the functionality and usefulness of wearable biosensors in healthcare, further research and development is required. With their ability to provide continuous monitoring and tailored treatment, wearable biosensors have the potential to revolutionise healthcare. These gadgets are anticipated to become more important in illness management, preventative healthcare, and enhancing general wellbeing as technology develops and more study is done.

4.4. Early Detection of Diseases

Due to the special characteristics of nanosensors, such as their high sensitivity, selectivity, and capacity for quick findings, early illness detection utilising nanosensors in healthcare has emerged as a potential technique ("Nanotechnology and Early Cancer Detection and Diagnosis - NCI" n.d.). These sensors use biosensing and nanomaterials to identify illnesses early, allowing for prompt treatment and better patient outcomes and various examples are given in Figure 137.

Nanosensors are important in the area of cancer detection. By identifying particular biomarkers linked to the illness, nanosensors have shown promise in the early identification of cancer. These nanosensors may be created to specifically attach to chemicals that are linked to cancer, such tumor-associated antigens or genetic markers. Nanosensors may identify the early stages of cancer before symptoms appear by detecting the presence of these biomarkers. Prostate-specific antigen (PSA) may be detected with great sensitivity and specificity using a nanosensor-based platform for the early detection of prostate cancer (B. Liu et al. 2022). Magnetic resonance imaging (MRI), positron emission tomography (PET), and computed tomography (CT) conventional scanning technologies have been used to target and improve the identification of tumours in vivo utilising nanotechnology-based imaging contrast agents ("Nanotechnology and Early Cancer Detection and Diagnosis - NCI" n.d.). These nanoscale imaging systems allow for earlier treatment start by increasing the accuracy of diagnosis and enabling the early identification of malignant growths.

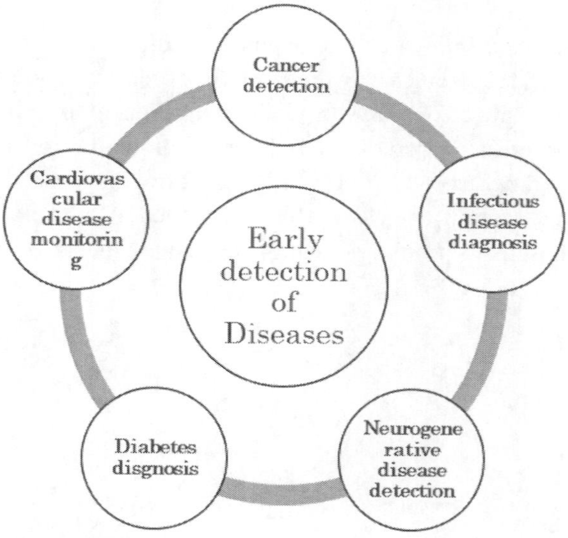

Figure 137. Examples of nanosensors used in early detection of diseases.

By identifying certain biomarkers linked to infections, nanosensors have been investigated for the early diagnosis of infectious disorders. These nanosensors enable for the quick and accurate detection of infectious pathogens since they may focus on nucleic acids or pathogen-specific antigens. High sensitivity in identifying sepsis-related biomarkers was shown using a nanosensor-based test for the early detection of sepsis, enabling early intervention and better patient outcomes (Alba-Patiño et al. 2022). Nanosensors make it possible to detect infectious illnesses in a sensitive, targeted, and time-efficient manner by fusing the benefits of nanomaterials with biosensing technologies (Deng et al. 2021). Controlling the spread of infectious illnesses requires prompt and precise identification of affected persons, and nanosensors provide a viable method for accomplishing early detection. Nanosensors may also be used in healthcare surveillance. For instance, nanosensors have revolutionised the detection of glucose levels in blood glucose monitoring, providing better accuracy and convenience (Kaushik, Soni, and Skotti 2022).

Nanosensors have the potential to be used in the early identification of neurodegenerative illnesses like Parkinson's and Alzheimer's. The buildup of certain protein aggregates, such as amyloid-beta in Alzheimer's disease, is a hallmark of these disorders. These protein aggregates may be specifically detected by nanosensors, allowing for disease early-stage identification. High sensitivity in detecting amyloid-beta peptides was shown using a nanosensor-based technique for the early diagnosis of Alzheimer's disease, simplifying the illness's early-stage identification (Vrahatis et al. 2023). These instances show how nanosensors may be used to identify particular biomarkers or compounds linked to a certain state in order to diagnose illnesses early. Nanosensors have the potential to revolutionise illness diagnosis and enhance patient outcomes via early intervention and individualised treatment plans by offering sensitive and selective detection.

4.5. Personalized Medicines

An approach to patient care known as personalised medicine aims to customise therapies to each patient's particular requirements in light of their genetic, environmental, and lifestyle characteristics (Wong et al. 2020). By providing real-time monitoring of biomarkers, drug targets, and disease progression, nanosensors have the potential to play a significant role in personalised medicine (Abazari et al. 2020). This will allow for focused and precise medication delivery. The following section discusses a few instances of personalised medicine utilising nanosensors. The nanosensors finds application in many aspects of personalized medicines which is illustrated in Figure 138.

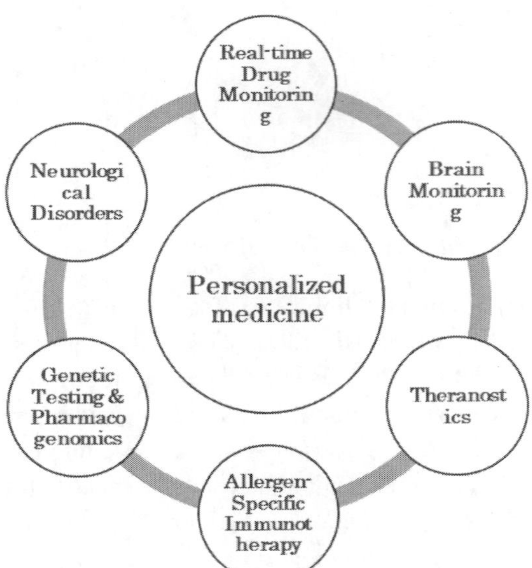

Figure 138. Examples of nanosensors used in personalized medicines.

Theranostics is the use of nanosensors to integrate therapeutic and diagnostic capabilities into a single platform. Imaging and treatment are combined into one process in the discipline of theranostics. It uses a variety of approaches to deliver a thorough diagnosis, molecular imaging within the body, and individualised and targeted treatment sstrategies, going beyond conventional imaging and therapeutic procedures. Recently, scientists have been investigating how to create new platforms and techniques for application in clinical settings by using cutting-edge materials and nanotechnologies. An imaging and targeted medication delivery theranostic nanosensor technology for cancer therapy (Choi et al. 2012).

By tracking nutritional status and enabling customised food advice, nanosensors may support personalised treatment. Nanosensors are able to identify certain vitamins, metabolites, or minerals in the body and provide real-time information on nutritional status. By taking into account things like vitamin deficits or particular dietary limitations, this information may be utilised to customise dietary programmes to suit individual requirements. Wearable nanosensors for vital nutrient monitoring in people with special dietary needs, such vegans or gluten-free dieters, allowed for personalised nutritional monitoring and provided suggestions for dietary changes to maintain optimum health (Sempionatto et al. 2021).

For the diagnosis and monitoring of neurological illnesses, nanosensors have tremendous promise in personalised treatment. Early identification and individualised treatment plans are often needed for these illnesses. Specific biomarkers linked to brain health may be found using nanosensors, which can reveal disease progression and particular therapeutic responses. Specific biomarkers for neurodegenerative illnesses including Alzheimer's, Autism Spectrum Disorder, Multiple Sclerosis, and Parkinson's may be found using nanosensors for personalised diagnosis and therapies in such conditions. The nanosensors allow early illness identification and give crucial information for individualised therapy approaches by tracking the levels of these biomarkers. This makes it possible for customised treatment strategies, including early interventions or focused medication administration, to be used, which enhances patient outcomes and quality of life (Ngowi et al. 2021).

Through pharmacogenomics and genetic testing, nanosensors have a substantial impact on personalised medicine. Genetic differences in people may affect drug response and metabolism, thus choosing and administering medications individually is crucial. Specific genetic variants or mutations linked to drug metabolism enzymes or drug targets may be found using nanosensors. Clinicians may carry out genetic testing using nanosensors to find genetic variants that affect medication responsiveness. To improve treatment results, this information enables personalised medicine selection and dose modifications. Specific genetic differences associated to drug metabolism enzymes may be found using a nanosensor for genetic testing and pharmacogenomics, giving useful knowledge to inform individual treatment choices (Chawla et al. 2021).

By progressively exposing patients to higher concentrations of the allergen, allergen-specific immunotherapy seeks to desensitise patients to certain allergens. Through the identification of individual allergens and the observation of immune responses, nanosensors may be a key component of personalised immunotherapy. A nanosensor can identify particular IgE antibodies linked to allergies to enable personalised allergen-specific treatment. The nanosensors may assist in identifying the precise allergens causing allergy responses in different individuals by analysing the IgE antibody profile (Johnson, Duschl, and Himly 2020). With the use of this knowledge, medical professionals may create individualised immunotherapy plans that specifically target the relevant allergens, resulting in better treatment results and enhanced patient wellbeing.

These examples highlight the role that nanosensors play in personalised medicine in several fields, including theranostics, neurological condition diagnosis and monitoring, genetic testing and pharmacogenomics, and allergen-specific immunotherapy. Nanosensors have the potential to transform healthcare by optimising treatment results and enhancing patient care by offering targeted and individualised treatments.

4.6. Drug Delivery Systems

The precise distribution of therapeutic agents combined with real-time monitoring of drug release and effectiveness is known as drug delivery utilising nanosensors and the examples are given in Figure 139. Drug delivery systems with nanosensor integration provide targeted and controlled drug release while also reporting drug concentrations and therapeutic response. The therapeutic effectiveness, side effects, and patient compliance of this tailored drug delivery system are all improved (Patra et al. 2018).

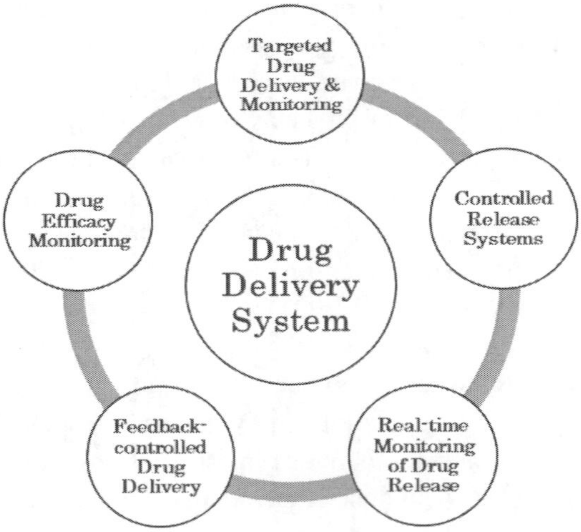

Figure 139. Examples of nanosensors used in Drug Delivery Systems.

Through the incorporation of certain targeting ligands on the surface of nanoparticles, nanosensors play a significant part in the delivery of targeted drugs. These ligands enable precise delivery of therapeutic medicines by selectively binding to receptors expressed on target cells or organs. Real-time monitoring of drug release and treatment response is also made possible by nanosensors. In order to administer anticancer medications selectively to tumour cells, a nanosensor for targeted drug delivery and monitoring employed antibodies as targeting ligands. The same nanosensors might measure medication release and evaluate therapy effectiveness in real-time thanks to imaging capabilities (Wei et al. 2022).

Controlled release systems may use nanosensors to provide accurate and on-demand medication delivery. These systems react to a variety of stimuli, including pH changes, temperature changes, or the presence of certain biomarkers, causing the release of therapeutic drugs. In order to facilitate the regulated release of medications precisely in the acidic environment of tumours, a nanosensor for controlled drug release was created to react to pH variations in the tumour microenvironment. While minimising off-target effects, this focused and regulated medication release increased therapeutic effectiveness (Alsawaftah et al. 2022).

Nanosensors provide real-time monitoring of medication release, providing insightful data on the kinetics and effectiveness of drug delivery. It is possible to design them with fluorescent or optical probes that react to changes in drug concentration or the immediate surroundings. To monitor the release of anticancer medications from nanoparticles in real time, nanosensors with fluorescent probes were incorporated (Qiu et al. 2014). The nanosensors enabled the optimisation of medication delivery systems and dose regimens by visualising drug release in real-time.

With the use of feedback control systems and nanosensors, real-time therapeutic monitoring may be used to provide adaptive medication administration. To get the best therapeutic results, these systems continually monitor medication concentrations and modify drug dose or release rate. Drug concentrations were continually monitored by feedback control mechanisms built into a feedback-controlled drug delivery system that used nanosensors (Cicha et al. 2022). With the use of this personalised and adaptable drug delivery strategy, specific therapy modifications might be made to achieve desired therapeutic results.

Due to their ability to provide real-time input on medication concentrations and patient reaction, nanosensors provide personalised treatment monitoring. They enable individualised monitoring of therapy effectiveness by detecting certain biomarkers linked to therapeutic response or illness development. In rheumatoid arthritis patients, a nanosensor for personalised treatment monitoring identified biomarkers linked to medication response (Conti et al. 2020). The nanosensors permitted personalised therapeutic monitoring and directed customised therapy modifications to improve patient outcomes by continually monitoring medication concentrations and illness biomarkers.

The aforementioned justifications show the potential of nanosensors in drug delivery, including targeted distribution, controlled release systems, real-time drug release monitoring, feedback-controlled drug delivery, and customised therapeutic monitoring. Drug delivery systems with nanosensor integration provide accurate drug distribution, on-demand release, and real-time medication concentration and effectiveness monitoring. This offers chances for improved treatment results, diminished side effects, and personalised medical strategies.

5. Challenges Associated with Using Nanosensors in Healthcare Systems

The use of nanosensors in healthcare monitoring is fraught with difficulties but also holds promise for the future. Transforming nanosensors from basic research or prototype projects into commercial goods is one of the main hurdles as shown in Figure 140. Technical difficulties with this approach include scalability, dependability, and manufacturability. To assure the viability and usefulness of nanosensors in healthcare applications, significant research and development efforts are needed to address these issues. The integration of nanosensors into current healthcare infrastructure and systems presents another difficulty. To guarantee smooth integration and efficient use of nanosensors in healthcare monitoring, compatibility with existing monitoring technologies and data management systems is essential. Nevertheless, despite these difficulties, the use of nanosensors in healthcare monitoring seems to have a bright future.

Nanosensors are little instruments that can detect and gauge very minute objects. When dealing with nanosensors, scientists and engineers do encounter several issues or difficulties. One difficulty is referred to as "drift." As a result, even if the object being monitored doesn't change over time, the measurements made by the nanosensor could. This might make it more difficult to believe that the sensor is accurate. Finding a reliable calibration method for nanosensors is another difficulty. Making ensuring the sensor is providing correct readings is known as calibration. Because nanosensors are so tiny and sensitive, this may be challenging. Before employing the nanosensor, scientists will sometimes wish to focus or isolate the substances they are seeking to measure. Creating techniques to achieve this is another difficulty since it might be challenging to perform with such little equipment. Another significant problem is integrating nanosensors with other sensor system components. To create a comprehensive sensor package, nanosensors must collaborate with other elements, but ensuring sure everything fits and functions properly together may be challenging.

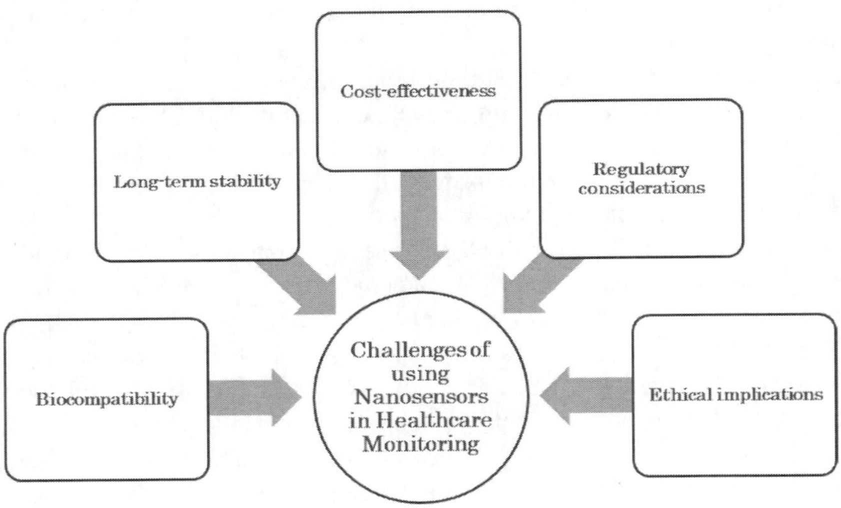

Figure 140. Illustration of challenges faced in using the nanosensors in healthcare monitoring.

Because they are a relatively new technology, nanosensors have a limited use in biological systems. This indicates that not all of their potential applications in living things have been properly investigated by science. Last but not least, certain nanosensors may affect the cells they are monitoring. They may alter how the cells function, making it difficult to determine if the changes are the result of the sensor or something that is occurring naturally in the cells. This makes it difficult to distinguish between the sensor's effects and regular biological activities. Nanosensors have a lot of promise, but there are still a number of obstacles to clear before they can be extensively and consistently employed in a variety of applications, particularly in biological systems.

5.1. The Development of New Nanomaterials for Nanosensors

To enable the effective integration of novel nanomaterials for nanosensors into sensor devices, a number of obstacles must be addressed as mentioned in Figure 141. The creation of nanomaterials with exact characteristics and regulated architectures is one of the main problems. To create nanomaterials that are constant in quality, purity, and repeatability, it is imperative to design dependable and scalable production processes. Nanomaterials must have their size, shape, composition, and surface properties under control in order to function well in nanosensors.

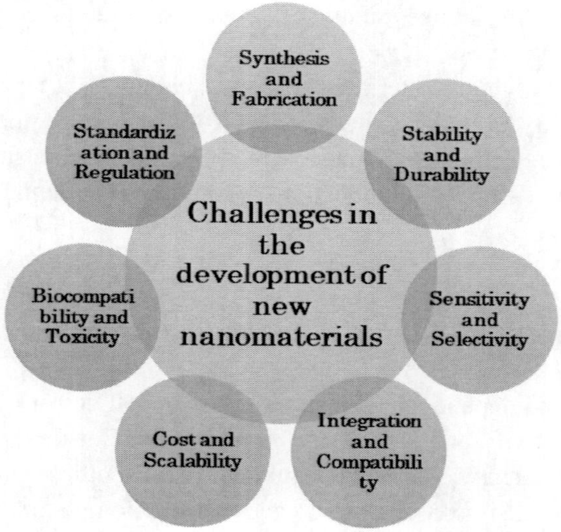

Figure 141. Illustration of the challenges in the development of new nanomaterials for the nanosensors.

For lengthy usage and exposure to a variety of environmental factors, nanosensors need to be stable and robust. Over time, nanomaterials may degrade or undergo property changes that have an impact on the performance and dependability of the sensor. The problem of ensuring long-term stability and endurance of nanomaterials calls for careful material selection and protective coatings. The target analyte or biomarker must be very sensitive to and selectible by the nanomaterials utilised in nanosensors. Designing nanomaterials with the required qualities, such as a large surface area, improved charge transport, or particular interactions with the target analyte, is necessary to achieve high sensitivity. Selectivity is the capacity to differentiate between the target analyte and interference-causing compounds. For precise and trustworthy sensing, high sensitivity and selectivity nanomaterials are essential.

Nanomaterials must generally be integrated with other parts of nanosensors, such as transducers, electronics, or signal processing components. Due to variations in material characteristics, interfaces, and production techniques, it may be difficult to ensure compatibility and smooth integration between nanomaterials and other components. The overall effectiveness, stability, and usefulness of the nanosensor device may be hampered by integration problems. Significant obstacles include the scalability and cost-effectiveness of nanomaterial synthesis and production techniques. Some nanomaterials could be difficult or costly to manufacture, which prevents their mainstream acceptance. The commercialization and widespread usage of nanomaterials in nanosensor applications depend on the development of scalable, affordable syntheses.

For biomedical or biological applications, nanosensors must adhere to rigorous biocompatibility standards. Some nanomaterials could be poisonous to living things or have negative impacts on them, which would prevent their usage in biological systems. To guarantee the safe and effective use of nanomaterials in biological applications such as healthcare monitoring, it is essential to comprehend and reduce their potential toxicity. It is crucial to establish standardised methods, testing processes, and laws for nanosensors based on nanomaterials. Standardisation guarantees constant performance and makes it easier to compare various nanosensor systems. In especially for medical or healthcare applications, regulatory frameworks must address safety, effectiveness, and ethical issues related to the use of nanomaterials in nanosensors.

Collaborations involving material scientists, chemists, engineers, and biomedical researchers are necessary to address these issues. By overcoming these obstacles, improved nanomaterials may be created that can improve the functionality, sensitivity, and dependability of nanosensors for a variety of applications, from environmental sensing to healthcare monitoring and beyond.

5.2. The Improvement of the Sensitivity and Specificity of Nanosensors

To guarantee precise and dependable detection of target analytes or biomarkers, it is essential that nanosensors' selectivity and specificity be improved. But developing nanosensors with great selectivity and specificity comes with a number of difficulties as mentioned in Figure 142. Dealing with background signals or interference-causing compounds in the sample is one of the key obstacles to enhancing selectivity and specificity. Multiple components in intricate biological or environmental samples may interact with the nanosensor, producing false-positive or false-negative readings. The creation of selective identification components or signal processing methods is required to distinguish the target analyte from interference-causing chemicals.

When a nanosensor reacts to analytes other than the target analyte as a result of comparable properties or structures, this is known as cross-reactivity. False detections and decreased specificity may arise from this. Selecting and designing recognition components with high affinity and specificity towards the target analyte while minimising interactions with other compounds is necessary to overcome cross-reactivity. Even when a certain analyte is present, nanosensors might react differently. The accuracy and dependability of measurements may be impacted by elements like batch-to-batch fluctuations in the synthesis of nanomaterials, surface alterations, or ambient variables. Achieving reliable and consistent outcomes requires reducing response variability.

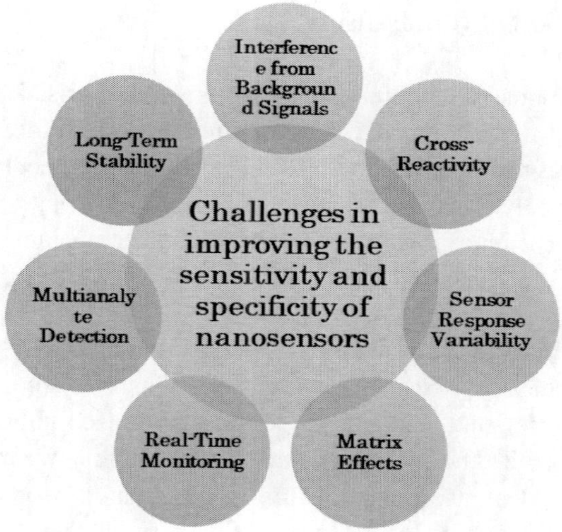

Figure 142. Illustration of challenges in improving the sensitivity and specificity of nanosensors.

The term "matrix effects" describes how the sample matrix affects how the nanosensor reacts. Complex combinations of chemicals, ions, or particulates are often present in biological or environmental samples, which might interact with the nanosensor and impair its efficiency. To minimise the influence of matrix components on the nanosensor response, rigorous sample preparation procedures must be used, such as sample cleaning, preconcentration, or separation methods. Real-time analyte monitoring is preferred in several applications. It may be difficult to provide real-time monitoring with great selectivity and specificity, however. Concentrations of analytes in a sample may fluctuate quickly, thus it's important to have quick and sensitive nanosensor systems that can collect and transduce the analyte signals with precision.

Multiple analytes or biomarkers must be detected simultaneously in many applications. It is difficult to achieve high selectivity and specificity in multianalyte detection because the nanosensor must be able to distinguish between a variety of targets while reducing cross-reactivity. The problem of multiplexing nanosensors and adding selective recognition components for several analytes calls for careful design and optimisation. For nanosensors to be useful, their selectivity and specificity need to be maintained throughout time. The selectivity and specificity of nanosensors may be impacted by changes in their characteristics, such as deterioration, fouling, or loss of activity. For nanosensors to remain stable over time and continue to function properly, it is crucial to develop stable nanomaterials, surface coatings, or regeneration techniques.

Innovative material design, cutting-edge surface engineering methods, signal processing algorithms, and extensive characterization and validation of nanosensors are all necessary to meet these difficulties. To overcome these obstacles and progress the area of nanosensors with greater selectivity and specificity, close cooperation between researchers from several disciplines, including chemistry, materials science, biology, and engineering, are required.

5.3. The Miniaturization of Nanosensors

Due to the limitations imposed by their smaller size, the miniaturisation of nanosensors, which makes them smaller and more compact, faces a number of difficulties as given in Figure 143. The nanosensor's signal-to-noise ratio (SNR) may drop as it becomes smaller. The signal produced by the target analyte may weaken as the sensor size shrinks, whilst noise sources like thermal noise or electrical interference may become more noticeable. It becomes difficult to maintain a high SNR, necessitating the creation of sensitive detecting systems and noise reduction strategies.

It may be difficult for little nanosensors to achieve high sensitivity and low detection limits. The fewer target molecules that may interact with the sensor as a consequence of the smaller size may lead to a lower signal. Innovative designs, optimised nanomaterials, and sensitive transduction techniques with the ability to capture and amplify weak signals are required to increase sensitivity and decrease detection limits. The nanosensor must be combined with additional elements, such as transducers, microelectronics, power sources, or communication interfaces, in order to be miniaturised. Space limitations, heat dissipation, power consumption, and component compatibility are among issues that this integration raises. For miniaturised nanosensors to operate and be reliable, effective packaging methods and improved integration strategies are necessary.

Figure 143. Illustration of the challenges in the miniaturization of the nanosensors.

It may be difficult to produce miniaturised nanosensors with excellent consistency and yield. The manufacturing process gets more difficult and sensor variability rises as sensor dimensions become smaller. A huge number of nanosensors must be produced using sophisticated manufacturing methods, exacting quality control procedures, and rigorous process optimisation in order to achieve consistent and dependable performance. Nanosensors that have been scaled down often have insufficient power sources. Small-scale sensors may not be suitable for conventional power sources due to their size or practicality. Therefore, guaranteeing a sufficient and long-lasting power source is difficult. To allow the long-term operation of miniature nanosensors, techniques like energy harvesting, low-power electronics, or integration with miniature power sources (such as batteries or micro-scale energy devices) must be investigated.

To assure the performance and dependability of miniature nanosensors, precise calibration and validation processes are necessary. It may be difficult to calibrate tiny sensors, nevertheless, because of their low volume and restricted access. For precise measurements and dependable sensor functioning, repeatable calibration techniques, including reference standards or reference materials, must be developed. Nanosensors that have been reduced in size could be more sensitive to changes in the environment, such those in the humidity, temperature, or electromagnetic fields. These environmental interactions may skew measurements, degrade sensor stability, and add noise. To lessen the impact of environmental interactions on miniature nanosensors, shielding methods, environmental control systems, and cutting-edge signal processing algorithms are required.

Multidisciplinary teams made up of materials scientists, engineers, physicists, and chemists, among others, are needed to overcome these obstacles. To successfully miniaturise nanosensors and enable their widespread use in a variety of applications, including healthcare, environmental monitoring, and smart devices, advanced nanofabrication techniques, creative sensor designs, effective integration methods, and reliable calibration and validation protocols are imperative.

5.4. The Commercialization of Nanosensors for Healthcare Monitoring

To enable the effective acceptance and broad usage of nanosensors for medical monitoring, a number of hurdles must be overcome in the commercialization of these devices that are mentioned in Figure 144. To guarantee their safety, effectiveness, and compliance with applicable standards, nanosensors destined for healthcare applications must go through stringent regulatory clearance procedures. It may be difficult and time-consuming to get regulatory clearances from organisations like the Food and Drug Administration (FDA) in the US or other regulatory authorities in other nations. Successful commercialization depends on meeting the high standards for dependability, performance, and safety.

The economic feasibility of nanosensors depends on their affordability and scalability. The hurdles of commercialising nanosensors include creating efficient production procedures, finding economical materials, and reaching economies of scale. It is crucial to produce nanosensors in large numbers while keeping consistent quality and performance in order to lower manufacturing costs and increase their marketability. Nanosensors used in medical monitoring should be stable and reliable over the long term. It is difficult to ensure that nanosensors retain their performance and accuracy over long periods of time without deterioration or drift. Building confidence in the functioning of the nanosensor and allowing its commercial success require addressing problems such material deterioration, sensor fouling, or signal fluctuations over time.

It is crucial to establish standardisation procedures and make sure that various nanosensor platforms and healthcare systems can communicate with one another. To facilitate the smooth integration of nanosensors with current healthcare infrastructure, electronic health records (EHRs), and data management systems, standardised formats, communication protocols, and data exchange methods are required. The utility of nanosensors for healthcare monitoring is increased overall and is facilitated by achieving interoperability by healthcare providers. Sensitive patient data is generated and sent via nanosensors, making data privacy and security crucial. The commercialization of nanosensors has several difficulties, including safeguarding patient privacy, assuring secure data transmission and storage, and resolving cybersecurity issues. To foster trust and preserve patient information, it is crucial to adhere to data protection laws, adopt strong encryption techniques, and set up secure data management systems.

The acceptance and adoption of nanosensors by users in healthcare settings are essential for their successful commercialization. It is important for both healthcare experts and end users to comprehend the advantages, usefulness, and worth of nanosensors for healthcare monitoring. User acceptability and wider implementation depend on removing obstacles like user training, incorporating nanosensors into current processes, and proving clinical efficacy and cost-effectiveness. For nanosensors to be successfully commercialised, ethical and societal issues must be addressed. Careful thought must be given to matters like informed consent, data ownership, patient autonomy, and possible discrepancies in access to nanosensor-based healthcare monitoring. The effective commercialization of nanosensors depends on ethical debates and the creation of rules for their ethical development, usage, and deployment in the healthcare industry.

Collaborations between academics, industry stakeholders, regulatory agencies, healthcare providers, and legislators are necessary to address these issues. To overcome these obstacles and advance the commercialization of nanosensors for healthcare monitoring, continued research and development efforts, regulatory backing, strategic alliances, and proactive interaction with end users and stakeholders are essential.

Figure 144. Illustration of the challenges in the commercialization of nanosensors.

6. Future Trends Associated with Using Nanosensors in Healthcare Systems

By making it possible to assess several biological parameters in real-time, with high accuracy, and without intrusive procedures, nanosensors have the potential to revolutionise healthcare monitoring. Future developments for nanosensors in healthcare monitoring are interesting and have the potential to have a big influence on patient care and overall health outcomes as technology develops. It is anticipated that future nanosensors will have greater sensitivity and specificity, enabling the detection of even minute quantities of biomarkers or analytes. This greater precision will make it possible to diagnose diseases early, track the effectiveness of treatments, and use more personalised medical techniques. Nanosensors will probably be able to identify many biomarkers or analytes in the same sample at the same time. A more complete knowledge of a patient's state would be possible because to this multiplexing capacity, which would provide a thorough examination of their health status. Continuous monitoring of physiological parameters, biomarkers, and vital signs will be possible thanks to the incorporation of nanosensors into wearable and implantable devices. With the ability to provide real-time data for preventative care, managing chronic diseases, and early intervention, these devices might be smoothly incorporated into daily life as given in Figure 145.

With the development of non-invasive monitoring methods made possible by nanosensors, patient pain and the necessity for invasive treatments may be decreased. For instance, biomarkers might be tracked using sweat or skin contact instead of blood samples thanks to nanosensors built into wearable patches or smart fabrics. Rapid, transportable, and point-of-care diagnostic tools may be made possible using nanosensors. These tools would make it possible for medical professionals to identify illnesses or ailments at the patient's bedside or in distant locations with little access to healthcare facilities. Applications for telehealth and remote patient monitoring may greatly benefit from the use of nanosensors. Nanosensors may allow healthcare practitioners to remotely monitor patients and give timely treatments, improving patient outcomes and lowering healthcare costs. These data can be sent in real-time on vital signs, medication adherence, or illness development.

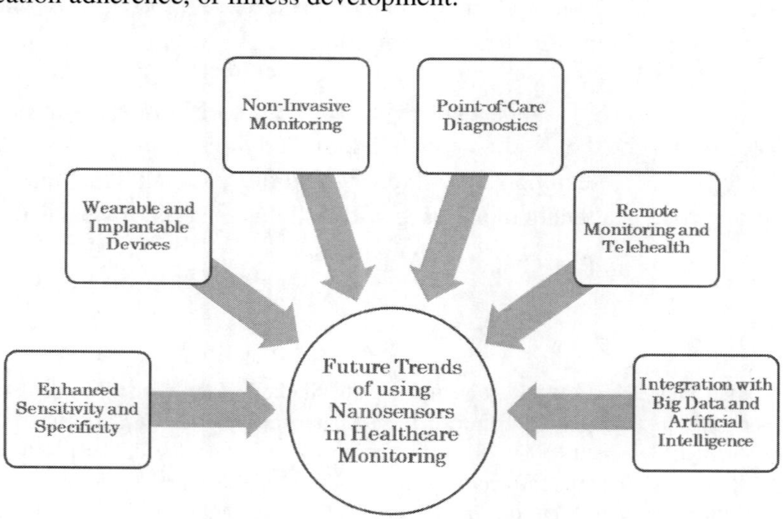

Figure 145. Illustration of future trends of using nanosensors in healthcare monitoring.

Healthcare monitoring has a lot of promise when big data analytics and artificial intelligence (AI) algorithms are combined with nanosensors. Nanosensors can provide actionable insights, predictive analytics, and personalised therapy recommendations based on specific patient data by using huge datasets and AI algorithms. The way illnesses are identified, treated, and controlled has the potential to change as a result of these emerging developments in nanosensors for healthcare monitoring. Nanosensors may help with better patient care, early illness identification, and more individualised healthcare methods by providing real-time, precise, and non-invasive monitoring.

Conclusion

This chapter examines the potential, foundations, uses, difficulties, and emerging trends of nanosensors in monitoring healthcare, with an emphasis on biosensors. It starts with an introduction that emphasises the value of biosensors in giving in-the-moment, precise readings that don't need intrusive procedures for better patient care. The chapter highlights the potential of nanosensors in monitoring healthcare, including their high sensitivity, selectivity, and capacity to detect traces of biomarkers or analytes. The basic tenets of biosensing are covered, including target analyte identification, signal conversion transducers, and signal amplification strategies. In-depth research is done on the use of nanosensors in healthcare monitoring. This covers their function in drug delivery systems, wearable biosensors, early illness detection, personalised medications, and point-of-care diagnostics. Each use case emphasises the advantages and potential influence of nanosensors in enhancing medical results.

The chapter discusses the difficulties in developing, miniaturising, commercialising, and making nanosensors sensitive, particular, and effective. It emphasises the need of overcoming these obstacles to guarantee the broad deployment of nanosensors in healthcare settings. In the chapter's conclusion, potential developments in nanosensor technology are discussed. This involves the creation of novel nanomaterials, sensitivity and specificity enhancements, difficulties with miniaturisation, and the commercialization of nanosensors. The capabilities and uses of nanosensors in healthcare monitoring might be further improved by these advancements.

Overall, the chapter offers a thorough review of the use of biosensors in nanosensors for healthcare monitoring. It shows their potential, underlying ideas, uses, difficulties, and directions for the future. Researchers, practitioners, and other stakeholders interested in using nanosensors to improve healthcare monitoring will find this material to be a useful resource.

References

Abazari, Mohammad Foad, Navid Nasiri, Fatemeh Nejati, Shohreh Zare Karizi, Mojdeh Amini Faskhodi, Ehsan Saburi, Nasrin Aghapur, Mohammad Reza Mahdavi, Abdolreza Ardeshirylajimi, Seyed Ehsan Enderami, Fatemeh Soleimanifar. 2020. "Comparison of Human-Induced Pluripotent Stem Cells and Mesenchymal Stem Cell Differentiation Potential to Insulin Producing Cells in 2D and 3D Culture Systems in Vitro." *Journal of Cellular Physiology* 235 (5): 4239–46. https://doi.org/10.1002/JCP.29298.

Alba-Patiño, Alejandra, Andreu Vaquer, Enrique Barón, Steven M. Russell, Marcio Borges, and Roberto de la Rica. 2022. "Micro- and Nanosensors for Detecting Blood Pathogens and Biomarkers at Different Points of Sepsis Care." *Mikrochimica Acta* 189 (2): 74. https://doi.org/10.1007/S00604-022-05171-2.

Alemu, Anshebo G., and Anshebo T. Alemu. 2022. "Recent Advances of Nanomaterial Sensor for Point-of Care Diagnostics Applications and Research." *Advanced Nanomaterials for Point of Care Diagnosis and Therapy*, January, 181–202. https://doi.org/10.1016/B978-0-323-85725-3.00009-X.

Alsawaftah, Nour M., Nahid S. Awad, William G. Pitt, and Ghaleb A. Husseini. 2022. "PH-Responsive Nanocarriers in Cancer Therapy." *Polymers* 14 (5). https://doi.org/10.3390/POLYM14050936.

Andreozzi, Emilio, Jessica Centracchio, Vincenzo Punzo, Daniele Esposito, Caitlin Polley, Gaetano D. Gargiulo, and Paolo Bifulco. 2021. "Respiration Monitoring via Forcecardiography Sensors." *Sensors 2021, Vol. 21, Page 3996* 21 (12): 3996. https://doi.org/10.3390/S21123996.

Chawla, Ruchi, Varsha Rani, Mohini Mishra, Krishan Kumar, Ruchi Chawla, Varsha Rani, Mohini Mishra, and Krishan Kumar. 2021. "Integrated Role of Nanotechnology and Pharmacogenetics in Diagnosis and Treatment of Diseases." *Pharmacogenetics*, May. https://doi.org/10.5772/INTECHOPEN.97643.

Choi, Ki Young, Gang Liu, Seulki Lee, and Xiaoyuan Chen. 2012. "Theranostic Nanoplatforms for Simultaneous Cancer Imaging and Therapy: Current Approaches and Future Perspectives." *Nanoscale* 4 (2): 330. https://doi.org/10.1039/C1NR11277E.

Cicha, Iwona, Ronny Priefer, Patrícia Severino, Eliana B. Souto, and Sona Jain. 2022. "Biosensor-Integrated Drug Delivery Systems as New Materials for Biomedical Applications." *Biomolecules 2022, Vol. 12, Page 1198* 12 (9): 1198. https://doi.org/10.3390/BIOM12091198.

Conti, Valeria, Graziamaria Corbi, Maria Costantino, Emanuela De Bellis, Valentina Manzo, Carmine Sellitto, Berenice Stefanelli, Francesca Colucci, and Amelia Filippelli. 2020. "Biomarkers to Personalize the Treatment of Rheumatoid Arthritis: Focus on Autoantibodies and Pharmacogenetics." *Biomolecules* 10 (12): 1–24. https://doi.org/10.3390/BIOM10121672.

Deng, Jinqi, Shuai Zhao, Yuan Liu, Chao Liu, and Jiashu Sun. 2021. "Nanosensors for Diagnosis of Infectious Diseases." *ACS Applied Bio Materials* 4 (5): 3863–79. https://doi.org/10.1021/ACSABM.0C01247.

Hanna, Jessica, Moussa Bteich, Youssef Tawk, Ali H. Ramadan, Batoul Dia, Fatima A. Asadallah, Aline Eid, Rouwaida Kanj, Joseph Costantine, and Assaad A. Eid. 2020. "Noninvasive, Wearable, and Tunable Electromagnetic Multisensing System for Continuous Glucose Monitoring, Mimicking Vasculature Anatomy." *Science Advances* 6 (24): 5320–30. https://doi.org/10.1126/SCIADV.ABA5320.

Johnson, Litty, Albert Duschl, and Martin Himly. 2020. "Nanotechnology-Based Vaccines for Allergen-Specific Immunotherapy: Potentials and Challenges of Conventional and Novel Adjuvants under Research." *Vaccines* 8 (2). https://doi.org/10.3390/VACCINES8020237.

Kaushik, Suresh, Vijay Soni, and Efstathia Skotti. 2022. "Nanosensors for Futuristic Smart and Intelligent Healthcare Systems." *Nanosensors for Futuristic Smart and Intelligent Healthcare Systems*, June. https://doi.org/10.1201/9781003093534/NANOSENSORS-FUTURISTIC-SMART-INTELLIGENT-HEALTHCARE-SYSTEMS-SURESH-KAUSHIK-VIJAY-SONI-EFSTATHIA-SKOTTI.

Kim, Jayoung, Alan S. Campbell, Berta Esteban Fernández de Ávila, and Joseph Wang. 2019. "Wearable Biosensors for Healthcare Monitoring." *Nature Biotechnology 2019 37:4* 37 (4): 389–406. https://doi.org/10.1038/s41587-019-0045-y.

Liu, Bojing, Yinxi Wang, Philippe Weitz, Johan Lindberg, Johan Hartman, Wanzhong Wang, Lars Egevad, Henrik Grönberg, Martin Eklund, and Mattias Rantalainen. 2022. "Using Deep Learning to Detect Patients at Risk for Prostate Cancer despite Benign Biopsies." *IScience* 25 (7). https://doi.org/10.1016/j.isci.2022.104663.

Liu, Rex, Albara Ah Ramli, Huanle Zhang, Erik Henricson, and Xin Liu. 2022. "An Overview of Human Activity Recognition Using Wearable Sensors: Healthcare and Artificial Intelligence." *Lecture Notes in Computer Science (Including Subseries Lecture Notes in Artificial Intelligence and Lecture Notes in Bioinformatics)* 12993 LNCS: 1–14. https://doi.org/10.1007/978-3-030-96068-1_1/COVER.

Mukhopadhyay, Subhas Chandra, Nagender Kumar Suryadevara, and Anindya Nag. 2022. "Wearable Sensors for Healthcare: Fabrication to Application." *Sensors 2022, Vol. 22, Page 5137* 22 (14): 5137. https://doi.org/10.3390/S22145137.

"Nanotechnology and Early Cancer Detection and Diagnosis - NCI." n.d. Accessed July 6, 2023. https://www.cancer.gov/nano/cancer-nanotechnology/detection-diagnosis.

Ngowi, Ebenezeri Erasto, Yi Zhen Wang, Lei Qian, Yasmeen Ahmed Saleheldin Hassan Helmy, Bright Anyomi, Tao Li, Meng Zheng, et al. 2021. "The Application of Nanotechnology for the Diagnosis and Treatment of Brain Diseases and Disorders." *Frontiers in Bioengineering and Biotechnology* 9 (March): 629832. https://doi.org/10.3389/FBIOE.2021.629832/BIBTEX.

Nicolò, Andrea, Carlo Massaroni, Emiliano Schena, and Massimo Sacchetti. 2020. "The Importance of Respiratory Rate Monitoring: From Healthcare to Sport and Exercise." *Sensors 2020, Vol. 20, Page 6396* 20 (21): 6396. https://doi.org/10.3390/S20216396.

Noah, Naumih M., and Peter M. Ndangili. 2019. "Current Trends of Nanobiosensors for Point-of-Care Diagnostics." *Journal of Analytical Methods in Chemistry* 2019. https://doi.org/10.1155/2019/2179718.

Patra, Jayanta Kumar, Gitishree Das, Leonardo Fernandes Fraceto, Estefania Vangelie Ramos Campos, Maria Del Pilar Rodriguez-Torres, Laura Susana Acosta-Torres, Luis Armando Diaz-Torres, et al. 2018. "Nano Based Drug Delivery Systems: Recent Developments and Future Prospects." *Journal of Nanobiotechnology 2018 16:1* 16 (1): 1–33. https://doi.org/10.1186/S12951-018-0392-8.

Qiu, Feng, Dali Wang, Qi Zhu, Lijuan Zhu, Gangsheng Tong, Yunfeng Lu, Deyue Yan, and Xinyuan Zhu. 2014. "Real-Time Monitoring of Anticancer Drug Release with Highly Fluorescent Star-Conjugated Copolymer as a Drug Carrier." *Biomacromolecules* 15 (4): 1355–64. https://doi.org/10.1021/BM401891C/SUPPL_FILE/BM401891C_SI_001.PDF.

Ruckh, Timothy T., and Heather A. Clark. 2014. "Implantable Nanosensors: Towards Continuous Physiologic Monitoring." *Analytical Chemistry* 86 (3): 1314. https://doi.org/10.1021/AC402688K.

Sempionatto, Juliane R., Victor Ruiz Valdepeñas Montiel, Eva Vargas, Hazhir Teymourian, and Joseph Wang. 2021. "Wearable and Mobile Sensors for Personalized Nutrition." *ACS Sensors* 6 (5): 1745–60. https://doi.org/10.1021/ACSSENSORS.1C00553.

Shiwani, Muhammad Ali, Timothy J. A. Chico, Fabio Ciravegna, and Lyudmila Mihaylova. 2023. "Continuous Monitoring of Health and Mobility Indicators in Patients with Cardiovascular Disease: A Review of Recent Technologies." *Sensors 2023, Vol. 23, Page 5752* 23 (12): 5752. https://doi.org/10.3390/S23125752.

Vrahatis, Aristidis G., Konstantina Skolariki, Marios G. Krokidis, Konstantinos Lazaros, Themis P. Exarchos, and Panagiotis Vlamos. 2023. "Revolutionizing the Early Detection of Alzheimer's Disease through Non-Invasive Biomarkers: The Role of Artificial Intelligence and Deep Learning." *Sensors 2023, Vol. 23, Page 4184* 23 (9): 4184. https://doi.org/10.3390/S23094184.

Wang, Xu, Feng Li, and Yirong Guo. 2020. "Recent Trends in Nanomaterial-Based Biosensors for Point-of-Care Testing." *Frontiers in Chemistry* 8 (October): 586702. https://doi.org/10.3389/FCHEM.2020.586702/BIBTEX.

Wei, Zongyi, Yuxin Zhou, Rongjie Wang, Jin Wang, and Zhenhua Chen. 2022. "Aptamers as Smart Ligands for Targeted Drug Delivery in Cancer Therapy." *Pharmaceutics* 14 (12): 2561. https://doi.org/10.3390/PHARMACEUTICS14122561.

Wong, Martin C. S., Junjie Huang, Paul S. F. Chan, Veeleah Lok, Colette Leung, Jingxuan Wang, Clement S. K. Cheung, Wing Nam Wong, Ngai Tseung Cheung, Chung Ping Ho, Chung Ping Ho, Eng Kiong Yeoh. 2020. "The Perceptions of and Factors Associated With the Adoption of the Electronic Health Record Sharing System Among Patients and Physicians: Cross-Sectional Survey." *JMIR Medical Informatics* 8 (5). https://doi.org/10.2196/17452.

Yoon, Sunghyun, Jai Kyoung Sim, and Young Ho Cho. 2016. "A Flexible and Wearable Human Stress Monitoring Patch." *Scientific Reports 2016 6:1* 6 (1): 1–11. https://doi.org/10.1038/srep23468.

Chapter 9

Nanosensors in Environmental Monitoring: Gas Sensors

Learning Outcomes

- The reader will have a deep insight into the significance and classification of nanosensors used to detect various gases.
- Various technologies deployed to synthesis the nanosensors and functionalization of those materials to perfectly fit in the desired gas sensing application can be identified.
- The reader will possess an ample knowledge of deploying the gas sensors in environmental monitoring systems.
- The reader will have a clear idea about the Internet of Thing (IoT) platform to which the gas sensors can be integrated to meet the real world usage in a more sophisticated and customized manner.
- The reader will be able to independently develop a gas sensor and design and develop an IoT platform to make it a complete sensing system.

Abstract

Gas sensors are effectively used in a variety of settings, from environmental management to routine monitoring of operations including public safety, motor efficiency, medical treatment, and many others. Sensing materials are required for this procedure. Air pollution and inhalation of different toxic gasses such as H_2S, CO, NH_3 and NO_2 poses immediate and long-term threats to human health, leading to premature deaths even. Therefore, to safeguard human life, an effective surveillance system is required to monitor air quality and control heating, ventilation and air conditioning. Also due to rising automobile and industrial emissions, there is a significant need for active monitoring of the quality of air in locations of living and work. Among greenhouse gases, carbon dioxide (CO_2) is of great concern as the toxological CO_2 inhalation in humans can cause headaches, increased respiratory and heart rate, dizziness, coma and even death depending on exposure concentration and duration. The development of low-cost and highly sensitive sensors to detect CO_2 concentration in the atmosphere is therefore very important. During the last several years, nanostructuring of sensing layers has been an active and competitive research field. Nanoscience and nanotechnology, in particular, make it possible to grow well-known gas-sensitive materials with better-controlled morphology, crystalline structure, and chemical composition. In order to create an air pollutant monitoring device, the sensor should be sensitive to air pollutants in the range of interest, the device should have single processing, data storage and communication network, be reliable and respond in real-time. In this chapter, commercially available gas sensors for detecting multiple characteristics were employed to construct a stand-alone real-time air quality monitoring system. The created system, which is based on the Internet of Things (IoT), integrates monitoring and emergency alerts, enabling users to check the air quality around their house, place of employment, or industrial facility from any location. This chapter provides a deep insight into the needs and classifications of gas sensors, various nanomaterials used to fabricate gas sensors, their advantages and disadvantages and the technology behind the integration of gas sensors with the IoT platform.

Keywords: metal oxide, carbon nanotube, nanocomposites, flexible semiconductors, environmental monitoring, IoT

1. Introduction

Concerns about environmental pollution, efficiency in different combustion processes, and the need to monitor potentially dangerous gases have lately prompted significant gas sensor research and development. Ever since the industrial revolution, atmospheric pollution has become particularly serious. Outdoor air pollution is projected to be a substantial risk factor for lung cancer (Turner et al. 2020) and to cause around 3.3 million unnecessary deaths per year, especially in Asia. Currently, manufacturing operations necessitate the use and synthesis of extremely hazardous compounds, notably gaseous toxic and combustible gases. Unavoidable gas flights put the industrial facility, its workers, and the locals in the area in danger. Daily reminders of this problem are provided by many asphyxiation, explosion, and human mortality situations that occur all over the globe. As a result, gas sensors are crucial in many technical processes where gas management and analysis are required since they assist to avoid the aforementioned problems. Urban populations are projected to rise quite quickly, from 3.8 billion in 2015 to 5.5 billion in 2050, and the health effects will worsen as a result of growing air pollution concentrations (Lelieveld et al. 2015).

Monitoring and controlling the atmosphere at home and at work has become more important as public awareness of the importance of living and working in a healthy environment has grown. People who operate in various situations lack sufficient knowledge about the components of the air they breathe to determine whether or not they are in danger. These conditions call for the need of wearable gas sensors with continuous monitoring. Portable and wearable gas sensors that can analyse air quality and notify users when they reach new terrain with insufficient information are required.

Gas sensors are being used more frequently in a variety of fields, including petrochemical, chemical, and environmental research, healthcare, including hospitals, food and beverage processing, microelectronics, including semiconductor manufacturing and telecommunications, agriculture, and manufacturing, including the ship, engine, and aviation industries. An alert from a gas detector beside the leak could also go off, allowing staff to leave the area. The Table 3 shows some examples of how gas sensors have been employed. Gas sensors can thereby avoid accidents caused by gas leakage, safeguarding both machinery and people. Gas detectors can detect fuel and dangerous gases. This apparatus, which is often seen in gas canisters, trunk gas lines, and oil rigs, may also be used to put out fires.

Rapid advancement is being marked in the development of sensors for CO_2, NO_2, NO, SO_2, ozone and fluorocarbons. Among greenhouse gases, carbon dioxide (CO_2) is of great concern concerned as it is the main ingredient in carbon-containing fuel combustion products and it is expected to have double the concentrations of carbon dioxide in the atmosphere by the mid-22 century (Hoang, Chu, and Tran 2017).

Table 3. Applications of gas sensors

Field of Application	Function	Example of Gas Detected
Environment	Monitoring of pollution, weather stations, and harmful gas levels in the atmosphere caused by industrial emissions	CO, CH_4, humidity, CO_2, O_3, NOx, VOCs, SOx, HCs, NH_3, H_2S
Security in the workplace	Control of interior air quality; monitoring of harmful gases in a workplace, such as a hazardous chemicals firm	Toxic gases, combustible gases, O_2
Protection of the home or an application for the home	Hazardous gas or smoke detection in houses as a consequence of fires or explosions, smart refrigerators or ovens, fire alarms, gas heating, leak detection, air quality control, air purifiers, and culinary control are all examples of home safety features	CO, CO_2, humidity, CH_4, VOCs
Safety in car	Controlling the vehicle's ventilation, detecting gasoline vapour, and performing alcohol breath tests	CO, LPG, CH_4, VOCs
Security of the public	Control of indoor air quality and detection of dangerous compounds for public safety	Toxic gases, combustible gases, flammable gases, explosives, O_2
Medical safety	Medical technology includes, for example, diagnostics breath analysis, sickness diagnosis, patient monitoring at the point of care, medication monitoring, artificial organs and prostheses, and drug research	O_2, NH_3, NOx, CO_2, H_2S, H_2, Cl_2, anesthesia gases
Agriculture	Diagnostics of plants and animals, testing of water and soil, inspection of meat and poultry, and tracking of waste and sewage	NH_3, amines, humidity, CO_2
Food quality control	Identification of certain compounds created when food begins to degrade and becomes unsafe for consumption	humidity, CO_2, etc
Utilities/automotives/ power plants	The gas concentration in the engine and gas boiler is adjusted to ensure that the combustion process is as efficient as feasible. Because combustion generates energy, power crops may profit from this approach as well	O_2, CO, HCs, NOx, SOx, CO2, H_2
Industry: Petrochemical, steel, oil and gas, water treatment, semiconductor	Process monitoring, quality assurance, workplace tracking, waste stream tracking, and leak detection alarms are some of the services offered	O_2, CO, CO_2, H_2, Cl_2CH_4, conventional pollutants, HCs, H_2S
Defense/military	Chemical, biological, and toxin warfare agent detection; treaty compliance	Agents, explosives, propellants
Aerospace	Monitoring of oxygen and combustible and hazardous gases in the atmosphere	H_2, O_2, CO_2, humidity
Traffic/ tunnels/carpark	Control and management of city traffic, monitoring of air quality in tunnels or subterranean parking	CO, O_3, NOx, SO_2, CH_4, LPG

Table 4. Effects of CO_2 exposure on human health (Satish et al. 2012)

CO_2 Concentration and Exposure Time	Effect on Health
0.035%	No discernible impact at the approximate atmospheric concentration
3.3-5.4% for 15 minutes	Increased depth of breathing
7.5% for 15 minutes	Breathing difficulty, rapid heartbeat, headache, dizziness, perspiration, restlessness, confusion, and distorted vision
3% over 15 hour	Decreased color sensitivity and night vision
10% for 1.5 hour	Eye flickering, increased muscle activity, twitching
More than 10%	Breathing problems, hearing problems, nausea, vomiting, a tangled feeling, perspiration, and after 15 minutes, a loss of awareness
30%	Convulsions and unconsciousness. Several fatalities linked to CO2 levels of more than 20%

Carbon dioxide in its vapour and fluid state is non-inflammable, odourless, colourless and not easily detectable. With an average volume of between 0.036 percent and 360 ppm, it is a minor yet significant component of the environment. It's also a common by product of animal and human metabolism. Gaseous CO_2 is found in greater quantities even at low levels due to its 1.5 times higher density than air. CO_2 sensing is required for the measurement or management of bio-related activities, as well as for the automated ventilation of air in living spaces and automobiles. Guidelines and regulatory controls have been implemented over time to maintain an acceptable level of exposure in household and industrial uses. Only 15% of the CO_2 quantity in the atmosphere is hazardous to life owing to its negative effects. Toxicological CO_2 inhalation symptoms in humans can range from headaches to higher respiratory and cardiac rates, disorientation, unconsciousness, and death, as seen in Table 4. As a result, it is critical to develop exceptionally precise, low-cost sensors capable of measuring atmospheric CO_2 levels.

Trace detection requires simple and dependable gas sensors in a variety of applications, including environmental monitoring, industrial, transportation, power, agricultural, and medical diagnostics. Metal oxides and carbon-based materials, among others, have witnessed a boom in popularity as gas sensing nanomaterials in the last 10 years. It stands to reason that increasing the contact surfaces between analytes and sensing materials will increase sensitivity greatly, given that the most prevalent gas sensing process includes the adsorption and desorption of gas molecules on sensing materials.

Electronics and nanotechnology advancements enable the use of sensors with ever-decreasing size and power needs. However, decreasing the size of the active region may make it less responsive. As a result, research into nanostructuring of active layers has been vigorous and competitive in recent years (J. Zhang et al. 2016; Fratoddi et al. 2015). Thanks to nanoscience and nanotechnology, well-known gas-sensitive materials may be created with better regulated morphology, crystalline size, and chemical composition. These materials (such as nanowires, nanotubes, and nanorods) offer a number of innovative and odd properties in addition to having a substantially greater surface-to-volume ratio and increased sensitivity.

2. Classification of Gas Sensors

There are currently a vast range of gas detection devices available, each based on a distinct material and functioning on a different principle. A chemical sensor is composed of two parts: (i) a receptor, or responsive layer, which turns chemical information into energy, and (ii) a physical transducer, which is a device that can convert the energy conveying the chemical information from the specimen into a usable analytical signal.

Many ideas exist to explain how the receptor component of a chemical sensor works: (i) there is no chemical reaction in a physical. Examples include those based on absorption, refractive index, conductivity, temperature, or mass change measurement, (ii) chemical, in which an analytical signal is generated by a chemical reaction involving the analyte, and (iii) biochemical, in which the analytical signal is generated by a biochemical method. Two popular examples are immunosensors and microbial potentiometric sensors. They are a subgroup of chemicals. Biosensors are the technical term for these devices. Chemical sensors are classified into different categories based on the transducer component, such as optical sensors, electrochemical sensors, calorimetric sensors, capacitance sensors, mass sensors, solid state ionic sensors, quartz microbalance sensors, and so on.

The analyte's contact with the receptor component causes changes in optical phenomena that optical devices convert. This category may be further divided by the optical qualities utilised in chemical detectors, such as absorbance, reflectance, luminescence, florescence, refractive index, optothermal effect, and light scattering. The electrochemical reaction between an analyte and an electrode is converted into a useful signal by electrochemical devices. Such impacts may interact with the zero-current condition either spontaneously or electrically. Its subcategories include potentiometric solid state electrolyte gas sensors, chemically sensitised field effect transistors, voltammetric sensors, and potentiometric sensors, among others. Electrical sensors, such as metal oxide semiconductors, organic semiconductor gas sensors, electrolytic conductivity sensors, etc., rely on measurements in which the signal is the result of changes in the electrical characteristics produced by the analyte's interaction with the material. The mass shift is converted by mass sensitive devices into a change in the property of the support material at a specifically altered surface. Piezoelectric technology or a surface acoustic wave device may be used to detect the mass shift that results from the accumulation of analytes. Magnetic sensors operate by altering the paramagnetic properties of the gas under investigation. Several types of oxygen monitors depict this. The heat effects of a specific chemical reaction or adsorption involving the analyte are the basis for thermometric equipment. Heat's impacts may be evaluated in a variety of ways in this area. For example, heat created by a combustion process may be monitored using catalytic sensors, and an enzyme reaction can be analysed using a thermistor.

Metal oxide-based electrical gas sensors have gained the most attention among the many types of gas sensors due to their low cost and high sensitivity. The basic design of conductometric or resistive electric sensors is based on the notion that certain chemical species in contact with the sensor can influence its electrical conductivity (or electrical resistance). The key aspect of sensor operation is the shift in resistance caused by processes (adsorption, chemical reactions, diffusion, catalysis, inflammation) occurring on the surface or in the majority of the detecting layer. Chemical species interact with the thin layer, allowing them to change conductivity. Adsorption or reactivity of a gas on the semi-conducting metal-oxide surface affects the density of conductive electrons in the sensor element (Barsan and Weimar 2001). The gas sensor's gas sensitivity is really caused by electron mobility between the semiconductor and the gas. Figure 146 is a schematic of a commercially available semiconductor gas sensor.

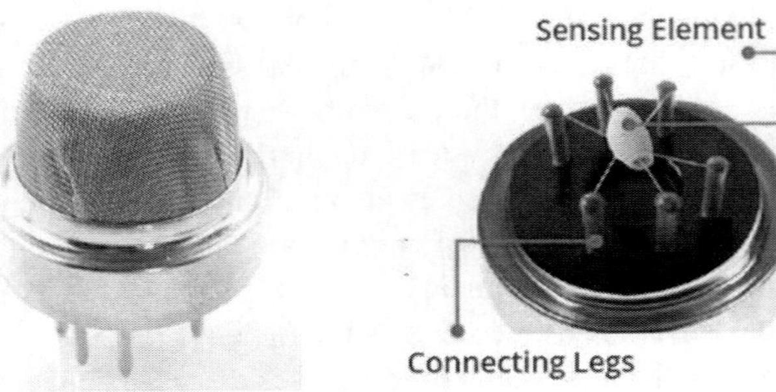

Figure 14. Commercially available gas sensors and its components.

2.1. Metal Oxides for Gas Sensing

Due to their low production costs, high sensitivity, simplicity of use, and capacity to detect numerous gases, Metal-Oxide Semiconductors (MOS), a type of chemiresistive sensor, have attracted a lot of attention recently for environmental monitoring, automobile emission tracking, and food security testing. Metal oxides are among the most widespread, diverse, and potentially largest categories of metals due to their extensive structural, physical, and chemical characteristics and functions. Metal oxides employed as detecting layers in chemo-resistive systems most often include binary oxides like SnO_2, ZnO, TiO_2, and others. The findings of a search inquiry on semiconductor metal oxides, which comprised both n-type and p-type semiconductor metal oxides, employed as sensing components for chemo-resistive gas sensors are shown in Figure 147.

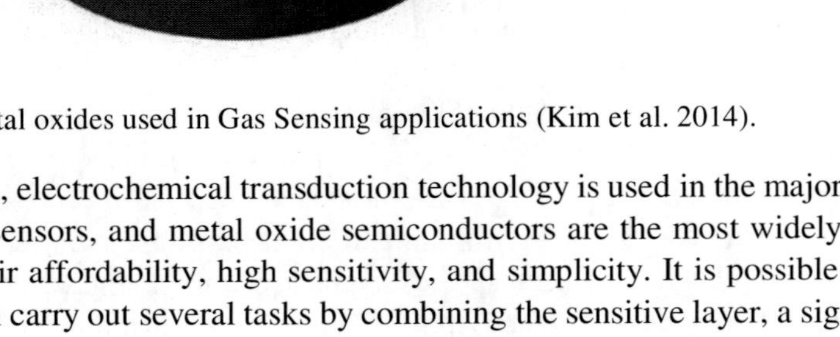

Figure 147. Metal oxides used in Gas Sensing applications (Kim et al. 2014).

Nowadays, electrochemical transduction technology is used in the majority of commercial chemical gas sensors, and metal oxide semiconductors are the most widely used sensor type because of their affordability, high sensitivity, and simplicity. It is possible to create a single device that can carry out several tasks by combining the sensitive layer, a signal converter, and control electronics. Despite the fact that MOS gas sensors offer simple operating ideas, the microscopic-level gassing mechanism is rather intricate.

The interaction of the target analyte with the metal oxide on the surface of MOS-based gas sensors is the basis for their operation. Metal oxide-based gas sensors are classified into two types: n-type (where an electron acts as the principal carrier) and p-type (in which the majority carrier is a hole). Reducing gases such as NH_3, CO, H_2, HCHO, and others enhance the conductivity of n-type semiconductors while decrease the conductivity of p-type semiconductors; conversely, oxidising gases such as NO_2, O_3, Cl_2, and others have the reverse effect. The density of conductive electrons in the polycrystalline sensor element changes when a gas adheres to or contacts the surface of a semi-conducting material:

- Oxygen pre-adsorption on the semi-conductive surface of the material.
- Specific gas adsorption
- Oxygen-adsorbed gas reaction
- Desorption on the surface of the reacted gas

The electron mobility between the semiconductor and the gas is what really contributes to the gas sensitivity of the gas sensor. The conductivity of the semiconductor gas sensors rises when the adsorbed oxygen molecules serve as acceptors in a reaction with the gas; it falls when they act as donors. The fundamental idea is based on a decrease in electron concentration caused by the early, reversible interaction of ambient oxygen with oxide lattice vacancies.

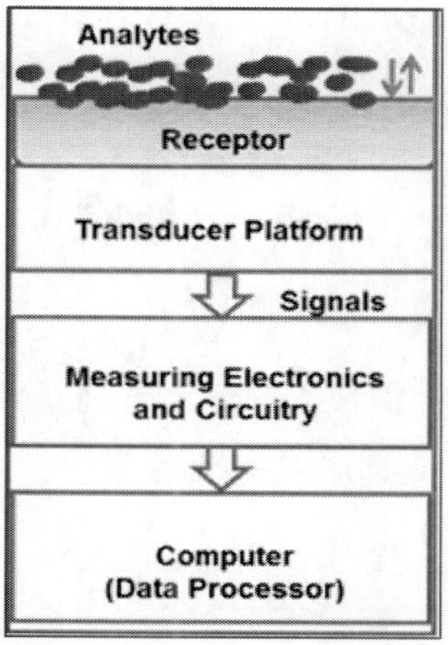

Figure 148. The Process flow in a Gas Sensing device.

The phenomenon of chemical sensing may be restricted to the sensing layer surface or may occur in the entire quantity of the delicate coating. Surface interaction needs the active species to be adsorbed on the surface or interface (gas/solid or liquid/solid); while the interaction in quantity indicates that the species is adsorbed and divided between the sample stage and the bulk of the sensitive material. Once the sensor detects an analyte by generating chemical or electrical signals, the transducer at the receptor captures the signal and modulates it to a useful form for further processing by the electronic circuitry. Figure 148 shows the process of gas sensing and data processing in a gas sensor device.

2.1.1. Mechanism Responsible for Enhanced Sensing Performance

It is essential to show how metal oxide gas detectors operate in order to develop and produce novel gas sensing materials that perform brilliantly. Metal-oxide-based gas sensors primarily detect changes in the equilibrium of surface-chemisorbed oxygen brought on by interaction with the target gas molecule. Based on the type of conduction, metal oxide semiconductors are divided into n- and p-types, each of which has different sensing characteristics for the same gas detection. The resistance of n-type semiconductors decreases while that of p-type semiconductors rises as a consequence of the gas species acting as donors for gas reduction. The metal-oxide semiconductor-based sensing technique is now explained by two different theories. One is the ionosorption model, while the other is the oxygen vacancy model. The first model takes into account changes in the surface's electrical potential brought on by gas adsorption, ionisation, and redox reaction. The second model explains how variations in the number of subsurface or surface oxygen vacancies, as well as their reduction-reoxidation process, affect oxygen stoichiometry. It is concerned with the interaction of gas molecules with oxygen vacancies.

As shown in Figure 149, three factors are taken into account while determining the sensing characteristics in order to allow awareness of the sensing mechanism. These elements include utility factor, transducer function, and receptor function. The responses of each subunit to oxygen and the target gases in the environment are related to the receptor function. The quantity of oxygen adsorption used in this method would undoubtedly change the detecting characteristics, which are reliant on the distinct surface area of the sensing materials. It is now possible to significantly change the specific surface area of sensing materials due to differences in particle size, pore size, and shape. The method of free carrier movement between adjacent crystals determines how each particle's reaction is transformed into the response of the whole device in terms of the transducer function.

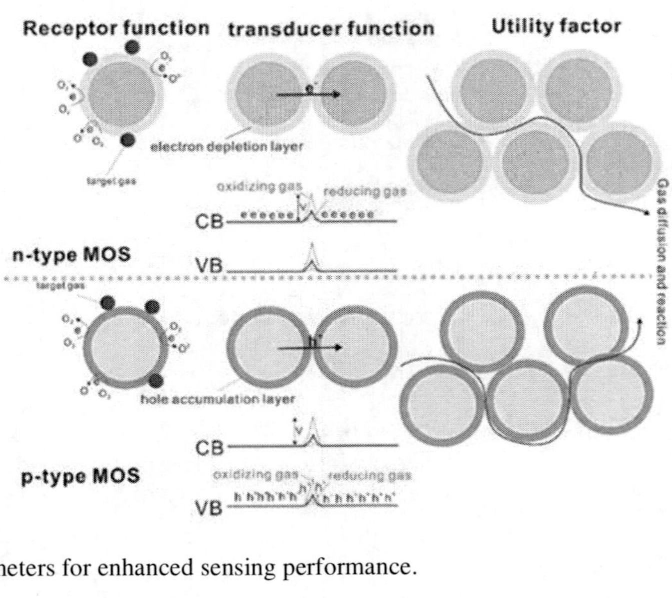

Figure 149. Parameters for enhanced sensing performance.

When oxygen is adsorbed onto the airborne granules that comprise n-type semiconductors, it pulls electrons from those grains to generate oxygen species (O, O^{2-}, O-, and so on), resulting in an electron-depleted layer that elevates the surface potential. As a result, between neighbours, a potential barrier that stops electrons from flowing from one grain to another is generated. Following gas reduction, oxygen species are absorbed, resulting in a decrease in potential barrier height, which is helpful for electron flow between grains. Free carrier transport is influenced not only by the potential barrier generated by gas adsorption, but also by trapping states in grain limits and defect states in semiconductors. The utility factor, on the other hand, is associated with response attenuation as a result of diffusion effects and reactive target gas interaction. As a result, in order to get excellent sensing capabilities from metal-oxide-based sensors, the morphological structure and defect structure should be considered during sensing material production.

2.2. Carbon Nanotubes for Gas Sensing

Carbon nanotubes have sparked the most research interest since Iijima's discovery in 1991, owing to its peculiar form, appearance, and properties. In-depth research has been conducted on all aspects of their production, properties (such as electrical, mechanical, thermal, and optical properties), and applications in a variety of disciplines. Theoretical and modelling work has also been done to better comprehend this nanoscale material and associated phenomena. Carbon nanotubes are a member of the fullerene structural family. Single-walled carbon nanotubes (SWCNT) and multi-walled carbon nanotubes are the two types of nanotubes (MWCNT). An SWCNT may be thought of as a seamless, several nanometer-diameter-thick sheet of graphite that has been rolled up to a thickness of one atom, with a length of one to one hundred microns. MWCNTs are made of many graphite layers that are coiled into tubes and have a common central axis. Due to their unique structural characteristics, carbon nanotubes have unique electrical, physical, and chemical capabilities. Because of the C-C connection, CNTs are the strongest and the most stiff fibres that are currently known.

Depending on the tube diameter and chirality, carbon nanotubes can have elastic or semiconducting electrical properties (the way the graphite sheet is folded to produce the tube). An integer pair is frequently used to denote chirality (n, m). Others are semiconductive, but those with the formula $n - m = 3j$ are elastic. These -electron wires are suitable building blocks for nanoelectronic devices because they have quantized electronic states and coherence lengths that are longer than those attainable with conducting polymers. When combined with recognition components, CNTs are excellent for use in sensing applications since they can be organised into nanowire networks. Because of the almost one-dimensional construction of nanotubes, they have extremely anisotropic dielectric characteristics, allowing them to transmit huge currents with little heating impact.

CNTs are suitable for gas or chemical detection in a variety of methods due to their electrical characteristics, which are very sensitive to their local chemical environment when exposed to certain gases. Many studies have been done on CNT-based gas sensing devices as well as on the theoretical study of gas adsorption and collision effects on nanotubes. When exposed to gas molecules, which may operate as electron donors or acceptors, nanotubes and molecules transfer charges, which causes changes in the electronic characteristics of CNTs. Numerous research has demonstrated that CNTs and functionalized CNTs are utilised to detect low concentration gas species such as NO_2, NH_3, H_2, H_2S, SO_2, NO, NO_2, CO, chemical vapours, etc. CNTs and functionalized CNTs have shown good potential in gas sensing solutions.

2.2.1. Sensing Mechanism in CNT

Contrary to conducting polymers, whose functionality may be determined by molecular processes, it is crucial to characterise the characteristics of CNTs beyond local molecule configurations, as is done in solid-state physics. CNTs' boundary orbitals are better described by band structures than by individual molecule orbitals because of their extended structure. Therefore, when trying to predict or characterise CNT-based sensing processes, chemical knowledge is typically inadequate. CNT-based sensor responses may be ascribed to impacts that take place within the tubes themselves (intra-CNT), impacts that result from points of contact between the tubes (inter-CNT), or effects that result from contact between the tubes and electrodes. The intensity of these multiple processes may be strongly influenced by the analyte, the concentrations of defects in CNTs, and the design of the device.

- *Intra-CNT:* Analyte and specific nanotubes or bundles of nanotubes may associate in ways that allow for intra-CNT sensing. Typically, this involves modifications to the number or mobility of charge carriers as well as the emergence of wall defects in nanotubes. The conductance of the CNT would change (increase or decrease) as a result of charge transfer, either directly or indirectly via interactions with analytes. The physical adsorption of oxygen molecules on the surface of CNTs at normal temperature results in their p-doping. Therefore, adding more p-dopants improves hole conductivity and lowers resistance, but adding more n-type dopants has the opposite effect. Using current-voltage (I-V) characteristics, photoemission spectroscopy (PES), and Raman spectroscopy, charge transfer between analyte and tube may be experimentally quantified for each SWCNT.
- *Inter-CNT:* Mechanisms at the interface between tubes may have a big impact on the electrical characteristics of the whole network in systems constructed of a network of CNTs. Even little spacing variations between two CNTs have a large impact on the contact resistance because the probability of charge tunnelling rapidly decreases with increasing distance. Intertube conduction routes may be changed by separating analytes into interstitial tube gaps or by expanding the supporting matrix or wrapping. An analyte may also cause a molecular or polymer looping in CNTs to disintegrate. Swelling of a matrix material decreases the conductance of CNT networks overall by minimising the tunnelling gaps.
- *Schottky Barrier Modulation:* The effectiveness of the device is sometimes affected by changes to the metal electrode and CNT (Schottky barrier) junction in addition to intra- and inter-CNT effects. Several organisations examine the sensing behaviour with and without passivation of the CNT / electrode contacts in an effort to distinguish between earlier processes and effects at the electrode/CNT interface. Passive layers were only created preferentially on the length of CNTs in these experiments that were not in contact with the electrodes or the whole device. The bottlenecks in the conductive channel are crucial to the sensing system. In instruments with less conductive, defect-rich CNTs, intra-tube impacts predominate the reaction, whereas processes that affect the junctions of the electrode tube dominate the sensing action for exceptionally conductive CNTs.

CNT-based sensor investigations have been pushed into numerous applications, including passive environmental sensing, due to the promise of low-powered, cost-effective sensors.

Sensor selectivity and sensitivity issues arise often when trying to identify substances that might be hazardous to both human and environmental health in liquid and gaseous mediums. Unmodified CNTs have been shown to interact with a broad range of gaseous analytes, and work on functionalizing CNTs is still underway. Unlike gas sensors, detecting the liquid phase generates more shear strain on a stationary CNT screen, bringing new stability concerns to light.

However, there are a number of issues that prohibit the use of CNT as the sensing material directly in practical systems. The first is that the sensor's efficiency, including sensitivity, detection limit, response or recovery time, selectivity, and stability, should be significantly increased in order to match the technological requirements of practical applications. Previous research has shown that pure semi-conductive SWCNTs are far more effective in sensing gases than metallic SWCNTs and MWCNTs, exhibiting higher sensitivity and faster response times. Semi-conductive SWCNTs are therefore preferable for high-quality gas sensors, however the processes for synthesising pure semi-conductive SWCNTs or eliminating metallic SWCNTs from the SWCNT are quite costly. Conditions on the CNT surface, such as oxygen or water molecules, may potentially have an impact on how the sensor responds to the target gases. CNT-based gas sensors face a significant extra hurdle due to the unpredictable nature of CNTs and device-to-device variability caused by the manufacturing procedures of sensors.

2.3. CNT- Metal Oxide Nanocomposite Sensors

Commercially available metal oxide gas sensors need high operating temperatures, which means they use a lot of power and have very little selectivity. When using metal oxide sensors at high temperatures, structural changes or coalescence in the gas-sensitive film's microstructure are likely to take place. The oxygen vacancies considerably increase in mobility as the temperature rises, and the conductive mechanism transitions to a mixed ionic-electronic state. It is well recognised that the migration of oxygen vacancies is what causes long-term drift in metal oxide gas sensors. Therefore, running the sensors at low enough temperatures to ensure that no discernible structural alteration ever happens, so long as gas reactions proceed at an acceptable rate, might be a tactic to prevent long-term changes in their responsiveness.

Carbon nanotubes have gained greater attention owing to their distinctive properties and have emerged as the most promising materials for very sensitive gas sensors, while traditional metal oxide sensing technologies must overcome the problem of inadequate sensitivity at ambient temperature. At room temperature, alcohol, ammonia, carbon dioxide, and nitrogen oxide have been shown to exhibit electrical properties and to be sensitive to very minor gas characteristics, while other compounds need to be heated by a separate heater for typical performance. MWCNTs are a sort of potential sensing material as a result. The field of gas sensors is now paying more and more attention to carbon nanotubes. Due to the significant surface area that may be exposed to gases, they have a special feature that makes them desirable as active materials for gas sensors. The electrical conductance of semi-conducting CNT changes when exposed to toxic gases like NO_2, NH_3, CO_2, benzene, or ethanol, according to experimental reports, even when the device is operated at low temperatures. This results in a significant decrease in the power consumption of the sensing device.

Combining semiconducting metal oxides with carbon nanomaterials has a synergistic effect that can increase sensing effectiveness because (i) conductivity is significantly increased and (ii) numerous active sites are formed at the interfaces, including functional oxygen groups, vacancies, and faults. Due to the swift carrier transport kinetics, composites of semiconducting metal oxide nanoparticles and carbon nanotubes exhibit improved sensitivity and quick reaction or recovery at room temperature. Pre-concentration operations are not necessary because of its high sensitivity, which adds to the benefits of cheap cost, light weight, and simple design. Additionally, MWCNTs perform better than typical sensing materials in terms of high adsorptive capacity, enhanced surface-to-volume ratio, and short reaction time, resulting in significant alterations in electrical properties including capacity and resistance. MWCNTs also offer superior corrosion resistance when used as a sensing material and a greater bandwidth compared to metal oxide semiconductors, which call for ad-hoc circuits, microfabrication processes, and power sources.

When operated at high temperatures, meal oxide sensors are prone to microstructure alterations (Korotcenkov 2005). Therefore, given that gas interactions occur at a suitable pace, operating the sensors at low enough temperatures to guarantee that detectable structural changes never occur may be a tactic to avoid long-term changes in their responsiveness. MWCNTs may be used to boost the sensitivity of traditional metal oxide sensing materials at room temperature. Due to their distinctive features, MWCNTs have drawn attention and have emerged as the most promising materials for extremely sensitive gas sensors.

2.4. Characteristics of Gas Sensors

Any gas sensor's principal objective is to provide accurate, real-time information on the chemical composition of the nearby environment. In order to avoid influencing the sample, a gas detection equipment should preferably function constantly and reversibly. The ideal gas sensor should function with perfect and immediate selectivity with respect to a specific target gas existing in the environment, be inexpensive, risk-free, and work with these characteristics. Any necessary gas concentration ought to provide a detectable signal. Easy manufacturing, quick response, low measurement uncertainty, insensitivity to relative temperature and humidity, strong contamination and poisoning resistance, and low noise are crucial gas sensor parameters for the gas sensor industry. A number of parameters, which are given below, are used to define the sensor's performance.

The ability of a single sensor to distinguish the target analyte from other species present in the sample is referred to as selectivity. Cross-reactivity, or cross-signaling between sensors and analytes, causes selectivity loss and signal compounding. On the assumption that different analytes and the sensor do not have synergistic cooperative effects, selectivity is assessed separately from the signals from the analyte and probable interfering chemicals. The calibration curves for each are then compared (signals vs. concentration). The capacity of the sensor to produce repeated outputs over time for the same setting is referred to as stability. To evaluate the stability of the detectors, several measurements over a long period of time or several cycles of analyte exposure should be made. In a more systematic way, complete calibration curves are accumulated over time for numerous equipment. During a sensor's functioning, stability problems may occur for a number of reasons, such as drift, hysteresis, and irreversibility. When developing sensors for highly reactive analytes, these problems are very common. Drift is a different measure that is characterised as the signal's change over time regardless of outside inputs. It continues to be a problem for CNT-based sensors. Physical changes to selectors, receptors, and the conditions surrounding their interactions with CNTs are common sources of drift, as are minute changes in nanotube positions that affect CNT-CNT junctions and parasitic electrical effects like ion electromotility in response to extremely low applied potentials. The sensor layer may be perturbed by gas movement, pressure variations, and heat reorganisations, which might result in signal drift. Hysteresis is the term for the variation in sensor output while approaching an analyte concentration from an increasing and lowering range. A factor is also the sensor's degree of irreversibility, or its background signal both before and after analyte exposure. Response and recovery times are crucial factors in determining how effective chemical sensors are. The recovery time is the time it takes for the sensor to go back to 10% of its peak value after being exposed to a certain analyte concentration, whereas the response time is the time it takes for the sensor to reach 90% of its stable or maximum value. The degree to which the yield may be restored after exposure is correlated with reversibility and recovery time.

2.5. Mechanically Flexible Sensors

Due to their prospective uses, mechanically flexible gas sensors are much sought after in wearable electronic devices. Stretchable and flexible sensors may be created using semiconducting metal oxide nanosensors, which may have better carrier mobility and mechanical resilience. Semiconducting metal oxide-based gas sensors that are wearable and mechanically flexible have lately become a focus of research. In contrast to those using rigid substrates, the wearable systems require substrates that are flexible, light weight, transparent, portable, small in size, and inexpensive. As flexible substrates for gas sensors, a lot of plastic substrates have been employed because of their excellent dielectric qualities and heat and chemical resilience.

Flexible sensors should be developed to replace rigid ones in order to overcome the implementation barrier of smart devices, where flexibility, carrier mobility, size reduction, or manufacturing restrictions limit their serviceability. The improved transparency of sensors in electronic equipment enables the proper functioning of touch screen displays, flat panel displays, picture recorders, optical communication devices, and solar cells. However, owing to the easily made rifts in the sensor platform, stiff transparent sensors are often negatively damaged by mechanical deformation, which significantly restricts the applications of transparent sensors. Mechanical flexibility and transparency are thus essential properties to consider when designing sensors for mobile and wearable electronics, which have already prompted significant research efforts in this field.

Recently, there has been a lot of development done on a variety of flexible transparent electronic systems, such as touch displays, electronic skins, environmental sensors, and wearable platforms for investigating transpiration. The following essential characteristics should be present in a perfect smart sensor that can be integrated into electronics: (i) flexibility and transparency; (ii) operation in the ambient at room temperature; (iii) absence of an external stimulus for quick response or recovery; (iv) low detection limit; high sensitivity and reproducibility; (v) low cost; and (vi) environmental friendliness. However, there isn't a sensing substance that can satisfy all of these requirements. Selectivity, reversibility, and/or adaptability are often lacking in sensors. It is planned that future research would concentrate on innovative manufacturing methods, new composite materials, and substrates to prevent the use of gas sensors in e-textiles, smart packaging, and other unexpected areas. This is because flexible gas sensors are becoming more popular.

3. Nanosensors for Gas Detection Based on Various Materials

The review of the literature in the subject of nanosensors based on metal oxide semiconductors, carbon nanotubes, and their hybrid materials for the detection of dangerous gases aims to identify research gaps in this area. For each of the methods employed in nanostructure-based sensors, several exemplary examples are chosen, and the review emphasises the novel features that emerge when these materials are used. The study also includes information on the designs, manufacturing processes, and characteristics of flexible sensors created to detect chemical analytes as well as on Internet of Things-based real-time environmental monitoring systems.

3.1. Gas Sensors Based on Metal Oxide Nanostructures and Carbon Nanotubes

Semiconducting oxide-based gas detection has had a considerable effect so far. However, the requirement to work at high temperatures frequently places a limit on this form of detection. These factors have led to the modification of several structural properties, such as crystal form and orientation. Nanostructured materials (also known as nanomaterials) can be used to lower operating temperatures, use less energy, and function more safely. Therefore, a lot of research focuses on making metal oxides smaller as nanoparticles and nanowires. Thin films consisting of certain organic polymers, metal nanoparticles, and most recently carbon nanotubes were also used in this process. When the aforementioned materials get the stimulus from the sensing layer after the molecule or ion-recognition event, they all undergo changes in their electrical (such as conductivity or impedance) and optical (such as molar absorption coefficient or refractive index), among other characteristics.

Metal oxides with nanostructures are a typical kind of material used to create gas sensors. Some of these substances are semi-conductive, allowing them to change their electrical conductivity as the environment around them changes. As a consequence, the vast majority of these devices are conductive nanosensors. The area to volume ratio is very helpful with nanoscale materials. As a consequence, the sensor is more sensitive for the same chemical composition the smaller the nanomaterials are.

Numerous methods for depositing oxides have been devised in order to increase the sensor layer's surface area and, thus, its sensitivity. (Khmissi, El Sayed, and Shaban 2016) used screen-printed technique to deposit La and Ni oxides in the form of nanopowders free of intragranular pores, which are perfect for dense film fabrication. In this technique, a paste made of oxides, agglomerate erate, and an organic substance (ethyl cellulose and aterpineol) is printed on an aluminium substrate. On aluminium substrates with tin oxide doped with Ta and Nb, (Shaban, Mustafa, and El Sayed 2016) employed the technique described by Khimissi to detect CO.

Another method used by (Shukla et al. 2004) to deposit MgO nanoparticle films is the sol-gel approach. The precursor solution for magnesium acetate was made by combining acetic acid and pure metal, which was subsequently oxidised on a pyrex surface. This device may be used to measure the humidity response by altering the electrical or optical characteristics of the sensor material. Adsorption of water molecules may lead to H+ transfer, which can change the refractive index as well as the electrical resistance. The sol-gel method was also used by (Aziz, Saber Abbas, and Wan Baharom 2013) to create NiO nanocrystals on SiO_2. In the film's IR spectra, the bands in the 1070, 800, and 450 cm-1 areas are more absorbent as the deposition temperature rises. They explained this connection by noting how the size of the NiO crystals grew as the deposition temperature increased. Experiments could not establish the sensitivity of the nanocomposite material to CO because of the reduction in anionic oxygen density on the NiO surface during the catalytic oxidation of CO.

(Lundin and Sarakinos 2012) generated SnO_2 films doped with Pt, SiO_2, or CaO using high impulse magnetron sputtering. This process was carried out in a vacuum chamber with the substrate and a blank made of the depositing substance. The material's electrons, ions, and neutral species were discharged from the blank using electricity. Gravity enabled the random dispersion of the deposition particles across the whole surface, ensuring increased strength and sensitivity to the gaseous environment. The operating temperature of the gadget was controlled by a RuO_2 resistor and Au electrodes.

Getting different sensor component shapes is another technique to expand the surface area of the sensing element. To make Si nanowires, for instance, (Shahrjerdi et al. 2011) used restricted lateral preferential epitaxial growth. On top of the first SiO_2 layer, a Si layer and then a second SiO_2 layer were added. The Si layer was then split apart to make a hollow that would act as a mould for the nanowires. The material's conductivity decreased by 9% when exposed to an argon gas mixture that included 20% oxygen. Despite the quick reaction time, the desorption approach was preferred by increasing the temperature (80-90ºC under vacuum).

SnO_2 nanowires were created using porous anodic alumina templates by (Xie et al. 2016). The conductance of the nanowires improved with variations in CO content after they were positioned on a Si/SiO_2 nanowires substratum with Au and Ti electrodes. When sensors were made out of a single nanowire, the electrical current for CO level was lowered by more than 1%.

Self-activated all-graphene NO_2 sensors were developed by (Kim et al. 2015), and they have great transparency, adaptability, and low power consumption. By increasing bias voltage and enhancing gas detecting properties such quick response and reversible sensing behaviour without internal heating, the all-graphene sensor demonstrated self-activation. In addition, a sensor constructed of simply grapheme may be totally transparent and flexible, yielding very precise sensing performance under heavy bending pressure. They claimed that the considerable device efficacy obtained by simple manufacturing procedures considerably increased the possibility for gas sensor applications for next-generation systems, such as wearable electronics.

Reduced grapheme oxide (RGO), which is cheap, simple to manufacture on a large scale, and contains a number of functional groups and defects, has a great amount of promise. It is also simple to functionalize, enabling the production and tweaking of the bandgap. Therefore, it makes logical that scientists are considering RGO as a possible gas sensor (X. Li et al. 2016), for instance, used hydrothermal processing to successfully produce distinctive RGO-decorated TiO_2 microspheres for sensitive room temperature gas sensors. To short-circuit the electrical conducting pathways inside the composite, they coated the surface of the TiO_2 spherical particles with RGO nanosheets. As a consequence, at room temperature, the sensor showed high sensitivity and selectivity for various ammonia levels between 5 and 50 ppm. However, it has been shown that the sensor's ammonia response and recovery times need to be improved. On the other side, certain sensors are often less reliable. In addition to the unique qualities of grapheme and nanoparticles, additional synergistic properties that are desired and helpful for gas sensing applications also appear when metal oxides and grapheme are combined to generate blended nanostructures. The enhanced specific surface area of this composite sensor is one of its main benefits. It has been shown that a metal oxide and graphite combination offers high gas sensitivity and selectivity.

Doping the oxides boosted the analytes' potency and sensitivity, which has been well researched. Numerous writers have discovered that the change in electrical conductance is higher when modest quantities of oxides or transition metals are introduced to the sensing component. In their 2010 study, (Van Hieu, Thuy, and Chien 2008) employed nanosized strong substitutes made from a mixture of hematite, anatase, and zinc oxide using TiO_2, Fe_2O_3, and ZrO_2. In order to increase these solutions' sensitivity to ethanol (100 ppm in air), ceramic substrates doped with SnO_2, Ti, and ZrO were used. The change in the resistance of the sensing material was still the chosen response mechanism. Electronic noses have also been used to cheat.

For usage in gas sensor applications, several metal oxides in a variety of geometric shapes, including nanotubes, films, nanorods, nanotubes, nanofibers, and nanowires, have been created. Metal oxide nanostructures like ZnO, SnO_2, and Cu_2O have been extensively employed in sensing applications because to their large specific surface area, mechanical flexibility, high sensitivity to different gases, and chemical durability. Due to the device's reducing size, the shrunk sensor performance's rapid responsiveness, and the improved sensitivity, selectivity, and practicality, gas sensor technology has recently gained a lot of interest. Furthermore, it is critical to investigate and develop innovative materials, structures, and geometries for gas sensor applications.

Tin oxide's two-dimensional nanosheets were modified into three-dimensional hierarchical nanoflowers by (T. Li et al. 2016), who also studied the gas sensing property. They discovered that the ultra-thin nanosheet-assembled SnO_2 nanoflowers' blooming was diminished to a semi-blooming condition, and that the semi-blooming nanoflowers-based sensor had improved ethanol gas detection characteristics. This problem was mostly brought on by the many nanoscale or microreaction chambers' limited impact on sustaining a full response to the detection of oxygen and ethanol. (Ren et al. 2015) used a nickel-tin-sulfur framework together with heat treatment and pickling to create a porous flower-like tin oxide structure. The porous NiO/SnO_2 flower-like hybrid was produced by the heat-induced disintegration of this flower-like nickel-tin-sulfur precursor.

Carbon nanotubes are a very sensitive material because they can easily interact with different gases and change their conductivity in the presence of several analytes at room temperature, even if these analytes have different chemical behaviours. It is easy to deposit carbon nanotubes and is generally cheap to generate them. As a consequence, a range of substrates with minimal material requirements may be employed. However, their great sensitivity suggests that they are non-selective, which limits the use of them for sensing to actual samples. To obtain selectivity, carbon nanotubes must first be functionalized. There are two types of nanotubes: single-walled and multi-walled. Separate nanotubes naturally align when they are joined by Van Der Waals forces, or more specifically when they are pi stacked. Due to their structural makeup, CNTs have intrinsic electrical, physical, and chemical characteristics. CNT is presently acknowledged as the simplest and strongest fibre in the device because of the C-C bond. In terms of temperature, CNT has outstanding thermal stability in both vacuum and air. The electrical characteristics of the CNT may be either metal or semiconductor, depending on the tube's diameter and chirality. Because of their geometric one-dimensional organisation, nanotubes' dielectric properties are very anisotropic, giving them a modest heating effect at high currents.

A single-walled carbon nanotube-polyaniline composite sensor was created by (Yoo et al. 2015), who also tested how it reacted to DMMP gas in room-temperature air. Experiments showed that these sensors have better sensitivity and faster response times when compared to pure SWCNT network sensors and polyaniline sensors (reaction and reaction times were 27.1% and 5.5s at 10 ppm DMMP). Excellent carrier gathering and quick carrier conveyance may be achieved when SWCNT and polyaniline are combined.

To increase the sensitivity of CNT-based gas sensors to hazardous gases NH_3 and NO_2, (Mishra and Islam 2015) presented a way for fabricating nanofiber composite films of SWCNT-Al_2O_3 utilising gel casting methods. A motor-driven device with a scraper was used to make the film by pouring and rolling the SWCNT powder distributed in the sol-gel alumina solution at a predetermined pace. Enhanced gas stability was achieved in the final experiment at 15% for NH_3 and 32% for NO_2 at 1 ppm.

To further improve the adsorption capacity of CNTs, purification and activation or functionalization are necessary, and this may be accomplished by oxidation treatment. For instance, this was done because purification was required to get rid of contaminants including catalyst particles, soot, amorphous carbon, and other forms that are frequently found combined with CNTs. By adhering to CNT surfaces, impurities can reduce adsorption and hence alter the analyte-adsorption on the external surface of the nanotube. Improving CNT adsorption capacity with oxidation treatment using an oxidising agent such as nitric acid, hydrogen peroxide, sodium permanganate, or a mixture of nitric acid and sulfuric acid is most likely due to an increase in CNT hydrophilicity. CNT oxidation treatment results in the formation of new functional groups on the surface of the nanotubes.

The primary benefits and cons of employing nanostructured materials for sensing applications are listed in Table 5. It is often difficult to compare the different sensor types since not all quality indicators have been assessed and published for all of the sensors. In order to promote gas adsorption, these nanostructures are made to provide higher-quality properties for nanoparticles and thin films. However, the results suggest that this may not always be the case. To improve selectivity and alter the surface or internal organisation of nanostructures or carbon nanotubes, functionalization techniques are used. It is possible to create and use a wide range of nanocomposite films as active components in gas sensors, including those made of blended metal oxides, polymers linked with metals or metal oxides, and carbon nanotubes mixed with polymers, metals, or metal oxides.

Table 5. Nanostructured materials used for gas sensing

Nanomaterial	Advantages	Disadvantages
Metal oxides	• High sensitivity and low response times. • High area/volume ratio. • Metal nanoparticle doping techniques improve the stability and sensitivity of the sensors. • Lower working temperatures (than bulk materials) • Annealing techniques can improve the outcomes.	• Post-treatments including calcination and annealing, as well as the utilisation of high temperatures and sophisticated film deposition processes, are necessary for the nanostructures. • Additional characterisation procedures are needed to learn more about the materials' size and shape.
Metal nanoparticles	• Less material is needed for the sensing layer. Certain nanoparticles function as catalysts and quicken the detection process. • The immobilisation of the sensing layer can be achieved by specific interactions with biomolecules (e.g., Au–cysteine).	• Annealing and sophisticated deposition processes are necessary for nanoparticles. • Characterization methods are pricey.
Metal complexes	• Specific reactions serve as the basis for the detection. As a result, these devices are selective and frequently reversible.	• It costs money to synthesise the complexes.
Polymers	• Response times are quick and repeatability is good with a thin polymer layer.	• Low selectivity and stability
Carbon nanotubes (CNTs)	• Raw CNTs need further purification operations because they have faster reaction and recovery periods, high sensitivity, reversibility, and stability. • Lack of selectivity for raw CNTs; availability of CNTs via a variety of methods (synthesis or deposition); cost of CNTs in relation to other materials; responses at room temperature • The repeatability of the sensors is decreased by networks of CNTs, and their sensitivity is increased by ageing and heat processes. • Working temperatures are lower when metal oxides are placed as nanoparticles on the surface of the CNTs. • Smaller and with lower ionisation voltages are gas ionisation sensors. Consequently, there is less operating risk, expense, and power usage.	• Additional purification processes are required • Lack of selectivity for raw CNTs • CNTs are relatively expensive in comparison with other materials • Networks of CNTs decrease the repeatability of the sensors

3.2. Gas Sensors Based On CNT-Metal Oxide Nanocomposites

In an effort to lower the operating temperature of various materials that may be utilised as the sensing component and to increase the sensitivity of the instruments, some researchers have combined CNTs with oxides or metal particles. Since the CNT structure is not changed, functionalization is not convalent. As opposed to unmodified nanotubes, the technique adds nanoparticles or films on the nanotube surface, changing the analyte's absorption mechanism. In this type of sensor, CNTs then act as the transducer. One of the first studies to use MWNTs obtained using microwave plasma enhanced chemical vapour deposition on silicon substrates to construct nano FETs was done by (Hinkov et al. 2014). A thin palladium layer that was sputter-deposited at 1000 A covers the top section of the MWNTs and serves as an electrode door. The Pd layer's atoms act as catalysts to split the H2 molecules apart. Atomic hydrogen that has been adsorbed on Pd nanoparticles delivers electrons to the CNTs, lowering the amount of hole carriers and, thus, the conductivity of the device.

CNTs may be added to electrical circuits using either pure CNTs or as-grown CNT films, which are the two main production techniques. With sparse CNT deposition, bundling, and inter-CNT impacts, the previous approach—which typically involves the deposition of high-temperature chemical vapour on electrodes with pre-patterned patterns—can be avoided. This method produces solid electrode-CNT connections. For as grown single-CNT devices, the angle of CNT growth between electrodes is challenging to regulate, and the device yield is poor, especially for longer channel lengths. Solution-based functionalization systems and purification methods that may enhance specific diameters or chiralities can be useful for device manufacturing procedures that rely on the deposition of pre-formed CNTs. Purified CNTs can be further refined or treated after deposition to increase the chemitransistor's sensitivity, response time, and transistor on-off ratio.

A method for producing metal oxide-decorated electrospun carbon nanofibers was reported by (L. Zhang et al. 2012). The fact that the metal oxidised carbon nanofibers were obtained all at once is a noteworthy advantage of the technique adopted. In order to adorn the surface of the carbon nanofibre with metal oxides, ZnO and SnO_2, structured nanofibers (PAN) and shell (PVP) were produced. Due to the incompatibility of the two polymers, the core-shell structure was created using single nozzle co-electrospinning. The two polymers' highly different intrinsic physical properties, notably their viscosities, led to their compatibility with the mix solution. The PVP phase was subsequently broken down, the metal precursors were transformed into nano modules of metal oxide, and the PAN was transformed by heat treatment into CNFs of about 40 nm. Precursor concentration in the PVP solution might regulate the metal oxide nanomodules' decorative shape.

(Krishna Kumar and Ramaprabhu 2006) used a predetermined chemical procedure to produce Pt-decorated CNTs that are very hydrogen sensitive. By selectively electroplating carbon monoxide, nitrogen dioxide, methane, hydrogen sulphide, ammonia, and hydrogen with Pt, Pd, Au, Sn, Mg, Fe, Ni, and Zn, (Star et al. 2006) adorned CNTs. The fundamental concept was to employ nano-clusters, which, when adsorbing a target molecule, contribute or take a significant amount of charge to affect electron transport in the nanotube (small size is necessary to optimise the impact of adsorbents on metal clusters).

By altering the electrical conductivity of these hybrid nanomaterials, the effective electronic interaction between metal nanoparticles and the CNT facilitates the detection of gases. This was done by (Zanolli et al. 2011) employing plasma-treated, metal-decorated multiwall carbon nanotube mats to specifically recognise benzene. The molecular benzene detection threshold was found to be less than 50 ppb even under conditions of changing humidity and the presence of interfering species such carbon monoxide, hydrogen sulphide, and nitrogen dioxide.

Sputtering was first used to decorate CNTs with metal nanoparticles by (Penza et al. 2009), (e.g., Au, Pt, Pd, Ru or Ag). The side walls of carbon nanotubes are functionalized by the plasma employed in the sputtering of magnetrons. The method has been successfully used to the *in situ* functionalization of CVD-grown CNT forests on gas sensor transducers, removing the necessity of transferring treated CNTs to obtain a functional unit.

Although less studied than metal decorating, it is also feasible to increase gas sensitivity by coating the sidewalls of carbon nanotubes with metal oxide nanoparticles. When using a suitable ratio between the quantity of plasma treated CNTs and metal oxide, (Leghrib et al. 2010) showed that hybrid sensors operated at room temperature demonstrate superior efficiency in the identification of nitrogen dioxide in the ppb range and carbon monoxide in the small ppm variety. Additionally, these improved sensors show a much reduced cross-sensitivity of moisture.

At low working temperatures, the vapours of NO_2, CO, NH_3, and ethanol might damage sensors modified with SnO_2 or TiO_2. The primary conducting pathways were generated by nanotubes in metal oxide matrices, which drastically altered the conductivity of the composite material during gas absorption. The gas molecules' binding energy to the surface of the CNT had a significant role in determining the sensor retrieval time. By (Liu et al. 2006), a sensor based on SnO_2-coated MCNT was presented. This sensor demonstrated an excellent reaction to oil gas and ethanol vapours and recovered at 335°C in a matter of seconds. As gas concentration grew, the sensors' response did as well. The system's unique electron transport mechanism properties explained the system's high sensitivity and low resistance. SnO_2 granules in MWCNT with low resistance are permeable to electrons.

MWCNT/SnO_2 nanocomposite films were created by (Aroutiounian et al. 2012) utilising the sol gel technique. Figure 150 displays the response of the nanocomposite MWCNT/SnO_2/Pd sensor to 5000 ppm hydrogen and isobutene in a solution sensitised with 0.03M $Ru(OH)Cl_3$ at 12°C.

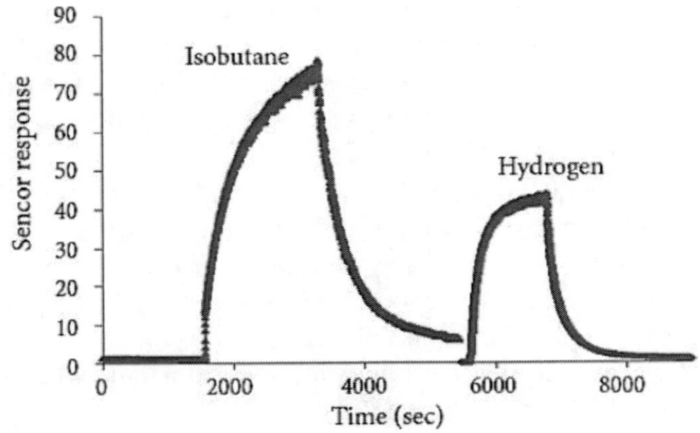

Figure 150. Sensor response of MWCNT/SnO_2 nanocomposite to isobutene and hydrogen (Aroutiounian et al. 2012).

Such hybrid films contain two depletion layers because a metal oxide film largely has n-type semiconducting properties, whereas MWCNTs primarily have p-type semiconducting properties. The first depletion zone is on the metal oxide's surface, while the second is at the point where MWCNTs and metal oxide nanoparticles meet. Hybridized gas detectors' gas sensitivity is boosted by the development of nanochannels, p-n or heterojunctions, which results in lower work function (barrier height) or greater metal oxide sensitive layer conductivity and better gas sensor performance at low working temperatures.

It is crucial to concentrate on band bending on both the metal oxide nanoparticle and CNT surface in order to understand a general mechanism behind the sensing layer's response to specific target molecules. Band bending may provide several strategies for adsorbing target molecules depending on the properties (p-type or n-type) of the two surfaces. Figure 151 depicts a mechanism in which oxidising or reducing molecules, such NO_2 or NH_3, bond to the surface of p- or n-type materials. It is derived from a research by (Rigoni et al. 2016). The terms E_C and E_V denote the borders of the conduction and valence bands, E_F the Fermi energy, E_i the intrinsic energy, and E_A and E_D the bulk energies of the donors and acceptors, respectively, in semiconductors of type and p-type. The molecules NO_2 and NH_3 are used to represent surface acceptors and donors, respectively.

It is crucial to concentrate on the hybrid system, where a nano-heterojunction is frequently observed between the metal oxide nanoparticles and the CNT surface due to the metal oxide nanoparticles' reduced dimensions, once the source of band bending for potential surface adsorption cases has been determined. Additionally, this nano-heterojunction typically has a p-n junction, where metal oxide nanoparticles serve as an n-doped material and CNT in air functions as p-doped owing to surface oxygen adsorption. Depending on the size of the MO nanoparticle depletion zone caused by the target molecule absorption, a depletion region may also be produced on the CNT side of the junction. The load transfer along the CNTs may be significantly impacted by this depletion zone, which may also squeeze the conductivity through the CNTs, preventing the passage of the hole carriers and increasing the strength.

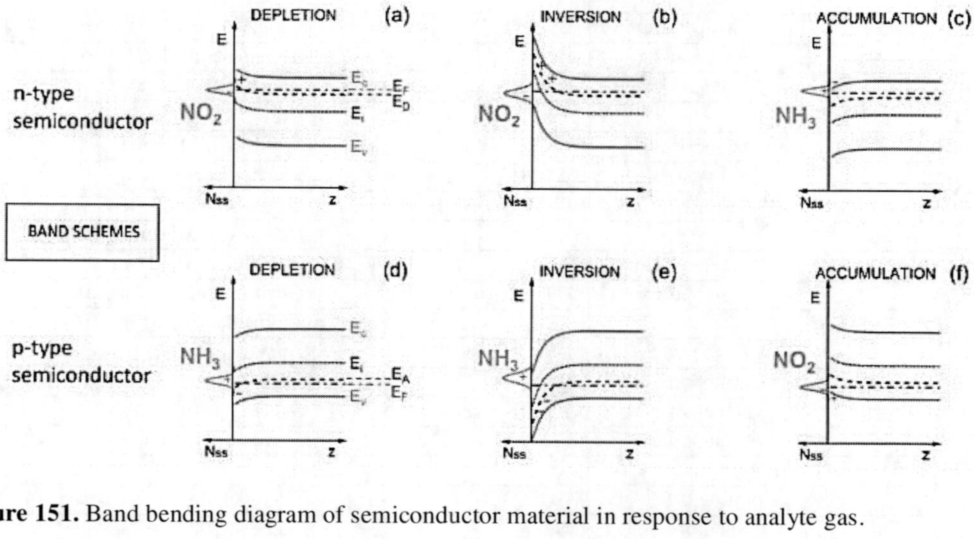

Figure 151. Band bending diagram of semiconductor material in response to analyte gas.

Table 6. List of carbon – metal oxide nanostructure materials researched

Type of Metal oxide in the hybrid sensor	Metal oxide morphology	Type of carbon nanostructure in the hybrid sensor	Author & Year
For Detecting NOx Gas			
SnO_2	-	SWCNT	(Inaba et al. 2021)
SnO_2	Nanoparticle	CNT	(Marichy et al. 2013)
SnO_2	Nanoparticle	MWCNT	(Choi et al. 2010)
SnO_2	Nanopowder	MWCNT	(Sharma, Tomar, and Gupta 2012)
SnO_2	Nanocluster	MWCNT	(Leghrib et al. 2010)
SnO_2	Nanoparticle	SWCNT	(Mubeen et al. 2013)
SnO_2	Nanoparticle	rGO	(Z. Zhang et al. 2011)
SnO_2	Nanocrystals	rGO	(Mao, Cui, Lu, et al. 2012)
WO_3	Nanopowder	MWCNT	(Espinosa et al. 2007)
WO_3	Film	MWCNT	(Su and Pan 2011)
WO_3	Nanonundle	SWCNT	(J. S. Lee et al. 2013)
WO_3	Nanorod	rGO	(Wongchoosuk et al. 2013)
ZnO	Nanoparticle	MWCNT	(Vyas et al. 2012)
Cu_2O	Nanowire	MWCNT	(Deng et al. 2012)
NiO	Nanosheet	CNT	(Hoa et al. 2013)
For Detecting NH_3 Gas			
SnO_2	-	SWCNT	(Ghaddab et al. 2010)
SnO_2	Nanoparticle	MWCNT	(Van Hieu, Thuy, and Chien 2008)
SnO_2	Nanoparticle	Carbon Nanoflakes	(S. K. Lee, Chang, and Kim 2014)
MnO_2	-	MWCNT	(Seredych and Bandosz 2012)
ZnO	Film	Carbon nanofiber	(Tulliani et al. 2011)
ZnO	Nanorod	CNT	(Mbuyisa et al. 2012)
In_2O3	Nanotube	MWCNT	(Du et al. 2007)
TiO_2	Nanoparticle	MWCNT	(Sánchez and Rincón 2012a)
TiO_2	Film	MWCNT	(Sánchez and Rincón 2012b)
For Detecting CO Gas			
SnO_2	Nanoparticle	SWCNT	(Y. Zhang et al. 2013)
SnO_2	Nanocluster	MWCNT	(Leghrib et al. 2010)
SnO_2	Nanopowder	MWCNT	(Luo, Zhu, and Wang 2012)
WO_3	Nanopowder	MWCNT	(A. Yang et al. 2007)
ZnO	Nanoparticle	MWCNT	(Khanderi et al. 2009)
For Detecting CO_2 Gas			
SnO_2	Nanocrystal	SWCNT	(M. Yang et al. 2010)
SnO_2	Nanoparticle	SWCNT	(Mubeen et al. 2013)
SnO_2	Nanofiber	MWCNT	(Gong, Sun, and Chen 2008)
SnO_2	-	SWCNT	(Mao, Cui, Yu, et al. 2012)
SnO_2	Nanoparticle	rGO	(Russo et al. 2012)
WO_3	Nanoparticle	SWCNT	(Ghasempour and Zad 2009)
WO_3	Film	MWCNT	(Wongchoosuk et al. 2010)
ZnO	Nanoparticle	MWCNT	(Anand et al. 2014)
MnO_2	Nanoparticle	SWCNT	(Jung, Yoon, and Lee 2013)
TiO_2	Nanoparticle	SWCNT	(Santangelo et al. 2010)
NiO	Nanoparticle	MWCNT	(Dhall and Jaggi 2016)
For Detecting Ethanol Gas			
SnO_2	Film	MWCNT	(Luo et al. 2013)
SnO_2	Nanotube	CNT	(Jia et al. 2009)

Type of Metal oxide in the hybrid sensor	Metal oxide morphology	Type of carbon nanostructure in the hybrid sensor	Author & Year
SnO_2	Nanoparticle	CNT	(Mendoza et al. 2014)
SnO_2	-	MWCNT	(Tyagi et al. 2017)
SnO_2	Nanoparticle	MWCNT	(Chen, Zhu, and Wang 2006)
SnO_2	Mesoporous nanoparticle	rGO	(Chang et al. 2013)
SnO_2	Nnaocrystals	CNT	(Feng et al. 2013)
ZnO	Nanoparticle	MWCNT	(Yuan and Shi 2013)
ZnO	Nanowire	Carbon Nanofiber	(Afzali, Abdi, and Arzi 2014)
ZnO	Nanoparticle	SWCNT	(Tian et al. 2014)
NiO	Nanoparticle	MWCNT	(S. J. Li et al. 2014)
Fe_2O_3	Nanoparticle	SWCNT	(Liang et al. 2014)

(Van Hieu, Thuy, and Chien 2008) created a SnO_2/MWCNT composite NH_3 sensor that operates at room temperature using common thin-film microelectronic technology. With NH_3 gas concentration changing from 60 to 800 ppm, the optimal composite sensor at ambient temperature demonstrated a substantially higher reactivity and faster reaction and regeneration (less than 5 min). An increase in sensor resistance might be overestimated since the composite sensing layer will continue to be a p-type semiconductor.

Numerous CNTs/MO hybrid have so far been investigated for the detection of various gases and compounds. Table 6 lists a wide variety of carbon-metal oxide hybrid nanostructures that have been produced and evaluated for use in gas, chemical, and biological sensing, including CNT and rGO metal oxide hybrids.

In addition to combining these outstanding qualities, the resultant carbon-metal oxide nanostructure hybrid materials also exhibit several novel hybridization traits. These composites are novel materials with unique functions and qualities that are not simply the sum of their separate components. At ambient temperature, several of the carbon-metal oxide nanostructure hybrids that have been described exhibit higher sensitivity to the target gas than comparable metal oxide or carbon detectors. According to the data thus far, enhancing the reaction may depend significantly on the kind of junctions formed at the places of contact and the establishment of a p-n heterojunction between a generally p-type carbon support and an n-type metal oxide. The properties required to increase sensor performance can therefore be improved utilising carbon-metal oxide hybrids to produce devices with high sensitivity to target molecules and low or room temperature gas reaction for nanostructured conductive carbon. Furthermore, the possibility of printing the sensor on flexible substrates should not be understated.

3.3. Gas Sensors on Flexible Substrates

Due to its potential uses in wearable and mobile electronics, RFID tags, and intelligent packaging for the surveillance of perishable commodities, flexible gas sensing technologies are of major importance to the scientific community. Intelligent flexible sensors can be cheap, portable, durable, and even stretchable. Additionally, they may be made translucent and extremely thin so that they can be easily hidden in specific spaces, such museums and galleries. Since the first array of 100m crystalline silica-based photovoltaic cells was incorporated into a plastic substrate in 1960, flexible electronics have come a long way. The production of automated tape bonding interconnects has utilised plastic substrates for the first time in commerce. Philips Research introduced the first polymer-based transistors in 2001. In the same year, Siemens introduced polymer-based organic field effect transistors. A conventional chemical gas sensor consists of a substrate, a sensing layer, and a few electrodes. To accomplish the device's flexibility, all of these components must be capable of being bent to some extent without causing damage or a decrease in the sensor's response. This necessitates a careful selection of materials and special deposition techniques. Modern production techniques for bendable electronics sensors include wafer thinning, printing, and direct manufacture of flexible substrates.

The use of printed, inexpensive copper acetate-based sensors to monitor the quality of raw broiler meat is described by (Koskela et al. 2015). Hydrogen sulphide (H_2S), a byproduct of microbial activity, is found using the sensor. The concept of the sensor platform is also presented, which combines the sensor with a printed planar coil antenna, a printed condenser, and a wirelessly readable printed resonance circuit. Because the sensor is ideal for large-scale manufacture, it may be sufficiently affordable to be incorporated into a low-cost retail food packaging. (Mattana et al. 2013) use flexible plastic strips to weave temperature and humidity sensors into fabrics. The researchers get temperature and humidity sensors that are easily woven into cloth using a standard loom.

The subject of flexible or bendable sensors and electronics is expanding quickly, and printing technologies support and revolutionise it by providing an affordable method of processing diverse electronic materials at temperatures suitable for processing plastic substrates. Printing technologies are particularly tempting for efficient manufacturing because of their streamlined production processes, reduced material waste, low production costs, and simple patterning techniques. Due to the unique properties of printed electronics, scientists have been able to investigate novel possibilities for the processing of materials and to develop sensors and devices on even non-planar surfaces, which is otherwise challenging using conventional wafer-based production techniques. Electronic structures that are conformable and sensitive can be wrapped around the body of a robot or artificial hands thanks to printed electronics on flexible substrates. Additionally, the use of printed electronics on polymer substrates has created new opportunities for creating low-cost electronics in areas larger than those now served by commercially available conventional wafers. Study in this field is gradually advancing toward a merger of well-established microelectronics with centuries-old printing processes, in keeping with the Electronics Industry Roadmap.

In order to print electronics and sensors conventionally, functional inks or an equivalent must be added after placing pre-patterned module components in contact with flexible (or non-flexible) substrates. Contact printing and non-contact printing are the two primary techniques often used to perform the printing or coating scheme, as seen in Figure 152.

Touch printing involves physically bringing the patterned bodies with inked surfaces into contact with the substrate. In a non-contact method, the substrate holder is moved in a predetermined order to define the structures while the solution is given through slots or nozzles. Engraving, offset printing, flexographic printing, and roll-to-roll printing are examples of contact-based printing methods. The most popular non-contact printing techniques include inkjet printing, slot-die coating, and screen printing. Due to its several advantages—including ease of use, affordability, speed, flexibility to industrial processes, decreased material waste, improved pattern resolution, and ease of control by adjusting a few process parameters—non-contact printing techniques have grown in popularity. A lot of interest has recently been shown in novel polymetric stamp printing techniques as transfer printing, micro-contact printing, and nanoprinting, especially for flexible inorganic mano-crystalline semiconductor circuits.

(Lakafosis et al. 2010) highlight how RFID-enabled sensor nodes and inkjet-printed electronics methods on flexible and paper substrates are being used to develop the first RFID-enabled wireless network sensor infrastructure. (Sung, De La Fuente Vornbrock, and Subramanian 2010) used gravure printing to create flexible organic light-emitting diodes based on tiny molecules. Using a slot-die roll-coating process, (Sandstrom et al. 2012) reported an all-solution manufacture of an alternate emissive system, a light-emitting electrochemical cell. Due to the use of air-stable materials and an emissive layer that is micrometres thick, the created flexible devices are appealingly durable and fault-tolerant.

In order to distinguish between different gaseous species, (Courbat 2010) investigated the combination of three different plastic foil transducers. The initial stage was making a selective colorimetric ammonia gas sensor right away using PET or PEN foil as an optical waveguide. As a consequence, processing has been simplified using large-scale manufacturing-compatible additive manufacturing techniques. Second, the impact of downsizing on the detecting capabilities of drop-coated metal-oxide gas sensors produced on polyimide foil was investigated. By streamlining the technique, they were able to reduce the power usage by taking advantage of the substrate's poor heat conductivity.

A method for producing nanoparticle silicon-based inks was described by (Britton and Härting 2006). It began with bulk silicon that had been milled and reduced to a fine powder with a maximum nanoparticle diameter of 50 nm. This powder was then suspended in ethanol with the addition of polymeric binders. This method was later patented and was used to screen print dense film, silicone-based resistors at room temperature on top of low-temperature substrates like paper.

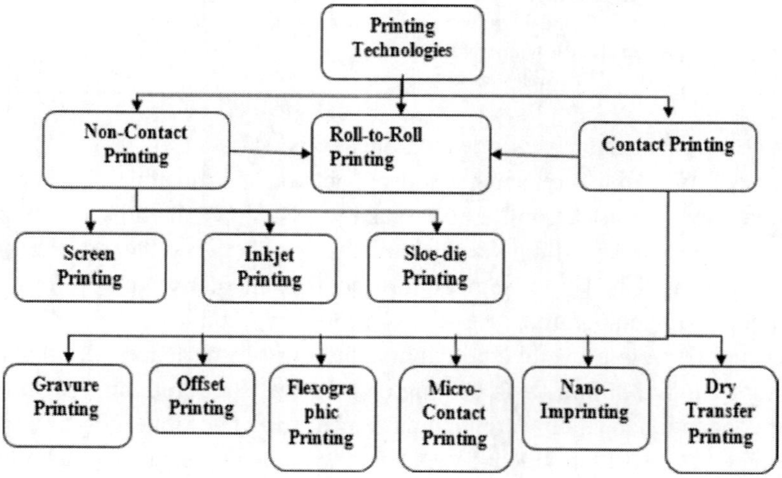

Figure 152. Classification of printing technologies on flexible substrates.

The most common flexible substrates used in flexible electronics, particularly gas sensors, are made of plastic. They are relatively affordable, roll-to-roll manufactured, and very adaptable. The most common substrates used for gas sensing applications include polyethylene terephthalate (PET), polyethylene naphthalate (PEN), parylene, nylon, polystyrene, and polyimide (PI). Because paper is a flexible substrate, inexpensive electronic devices may be made out of a variety of cheap, easily available, recyclable, and eco-friendly materials. Additionally, it allows for the fabrication of devices that can be bent, folded, unfolded, and utilised in ways that rigid substrates could not. Electronics made of paper are more portable, lighter, roll-to-roll compatible, and easily cut with scissors without causing damage. As flexible substrates, textiles are also an increasing source of worry. However, there aren't many publications that have been published on the incorporation of gas sensors into textiles. Textiles consisting of polyaniline and nylon 6 were produced by (Hong, Oh, and Kang 2004). These fabrics are very selective and ammonia-sensitive. In 2013, (Han et al. 2013) combined a cotton yarn with a sensing layer constructed of a single-walled carbon nanotube. However, the main disadvantage of this tactic is the bending stress that the yarn experiences during the weaving process. The three most common types of flexible substrate possibilities at the moment are ultra-thin glass, metal foil, and plastic (polymer) films. As shown in Table 7, transparent plastic substrates in particular offer great optical transmittance that is equivalent to that of thin glass, good flexibility, and outstanding resistance that is comparable to that of metal foils.

Table 7. Flexible substrates and its properties

Flexible Material	Advantages	Disadvantages
Plastic substrate	Rugged Bendable, rollable Transparent	Poor dimension stability Low process temperature High H_2O and O_2 permeation Low chemical resistance
Thin glass	Conformable Transparent Low H_2O and O_2 permeation	Low mechanical stability Low process temperature
Metal foil	Rugged, conformable Low H_2O and O_2 permeation High process temperature Good dimension stability	Opaque Capacitive effect

For low-power operation in wireless applications, (Briand et al. 2011) concentrate on integrating temperature, moisture, and gas sensors onto plastic substrates. The creation of inkjet silver meander-shaped resistors on flexible plastic foils is described by (Molina-Lopez et al. 2013). (PET). These devices exhibit very linear behaviour between the temperatures of 60° and 108°C and are distinguished by a temperature coefficient of resistance. Another intriguing illustration of a printed temperature sensor was offered by (Aliane et al. 2014). In this case, the authors described flexible polyethylene naphthalate or polyethylene terephthalate foil screen printed resistors. First, a 30 nm-thick coating of gold was sputtered onto the surface to create the electrical connections, which were then patterned using photolithography.

(Crowley et al. 2010) provided details of a completely printed polyaniline-based gas detector with a 2.5 ppm detection threshold and a linear connection between measured current and concentration in the 10-100 ppm range (2010). (Hibbard et al. 2013) described breathing ammonia detectors based on printed polyaniline nanoparticle detectors that operate in the sub ppm concentration range (2013). Simple inkjet printing has been used to create a paper-based, flexible ammonia gas detector using silver and single-walled carbon nanotubes functionalized with poly (m-aminobenzene sulphonic acid).

According to (Yun, Cha, and Gong 2014), moisture detecting polyelectrolyte ink was crosslinked and connected to the gold electrode in screen printed silver paste transducers on Cu, Ni, and Au electroless polyimide. The relative humidity sensor showed good sensitivity, changing the intensity by almost four orders of magnitude for relative humidity ranging from 20% to 95% relative humidity, quick response and retrieval times, high linearity, and minimal hysteresis. An ammonia sensor was reported by (Claramunt et al. 2013) and consisted of a flexible polyimide substrate with an array of interdigitated silver electrodes inkjet-printed on one side and an ubiquitous silver heater on the other. Carbon nanofibers embellished with metal nanoparticles have been applied to interdigitated electrodes as a protective coating. The sensor's operation was made reversible by using the heater to chemically reset the nanofibers.

Several efforts have also been done on printing the gas-sensitive layers themselves. Inkjet printing allows for the storage of several polymeric materials. A printed radio frequency antenna, resistive and FET transducers, and inkjet-printed CNTs were employed by (Hammock et al. 2013) to detect NO_2 and H_2S. On silicon and aluminium substrates, (Shen 2012) inkjet-printed pure and doped SnO_2 films. By printing SnO_2 and doping in succession, inkjet printing allowed for simple doping. But because its printed layers needed to be annealed at 550°C, they couldn't be used with plastic substrates. (Peter, Kneer, and Wollenstein 2011) overcame this issue by developing a titanium-doped chromium oxide ink that didn't require any heat. However, by sintering the printed layer at 400ºC, the adhesion to the silicon substrate and the film stability were improved. However, high-performance polymers like certain polyimides can operate at this temperature.

4. Internet of Things for Environmental Monitoring Applications

Device integration is crucial for the full realisation of independent, multimodal, multifunctional, and multidirectional flexible physical sensing platforms, together with device design and production. Real-Time monitoring applications for different parameters are significantly dependent on the network of sensors used to monitor and transmit recorded data. The Internet of Things recently promises to offer everyday objects the benefits of internet and technology, enabling us to increase our perception and our ability to change our environment.

In the areas of environment, health, safety, and security, the dispersed monitoring of environmental factors is a major problem. The development of intelligent sensors and wireless sensor networks based on silicon technology, aiming at many sorts of applications, is the subject of extensive study. Agricultural and environmental domains are excellent prospects for the introduction of IoT solutions since they take place across large regions and require ongoing monitoring and management. Under the IoT paradigm, inexpensive electronic devices may enhance how people interact with the physical environment, and the Internet's computational power and software resources can offer useful insights. According to (Saville, Hatanaka, and Wada 2016), when the data collected are utilised to provide predictive learning algorithms, IoT provides additional possibilities beyond ground floor automation and makes decision-making for owners, managers, and policy makers easier. It might be used to gather and store environmental data, monitor compliance with local laws and environmental factors, set off alarms, or provide advice to the public and authorities.

An approach for examining the behaviour of environmental noise pollution was put out by (Torres-Ruiz et al. 2016). It comprises of a mobile application named "Noise Monitor" that uses the microphone sensor to detect noise coming from the mobile device. Volunteered Geographic Information, a sizable geospatial collection of urban information about Mexico City, is the source of the information on geo-referenced noise. (Hachem et al. 2015) discuss their approach to enhancing municipally sensed data using information from participatory sensing-based data collected via cellphones. Data assimilation approach is utilised to create more accurate noise maps that incorporate measured and simulated noise concentrations, and the solution for noise-specific applications customises the Urban Civics middleware.

The system suggested by (Dhanaraju et al. 2022) captures different kinds of factors such as temperature, humidity, soil moisture, CO_2, luminosity, water pH and pictures for online crop surveillance. Experimental trails indicate that data about the field setting can be precisely collected by the system prototype. Furthermore, information can be stored in the surveillance centre server of the agricultural setting. Internet remote users can access servers, coordinators and gateways via web browsers, control the entire system, brows and download information. This system basically achieves low-energy consumption as well as expandable design goal and can meet the demand for contemporary surveillance of the agricultural environment.

To minimise negative and hazardous effects, the IoT for air monitoring sector aims to offer frequent or continuous measurements, analyse, and identify environmental parameters or pollutant concentrations. Predicting probable changes in the ecosystem or the entire biosphere is another aspect of it. In this solution, (Watthanawisuth, Tuantranont, and Kerdcharoen 2009) present a real-time micro-climate surveillance technique based on a WSN. Solar-powered temperature and relative humidity sensors (SHT15) supported by ZigBee communication technology were part of the solution. The GEMS proposal by (Luan et al. 2016), which offered a system for monitoring apple orchards based on GPRS technology, is another IoT approach for air monitoring. Monitoring parameters such relative humidity, temperature, and radiation were assessed over the course of two years in fifteen distinct regions of China.

(Khandani and Kalantari 2009)'s description of a design process to identify the spatial sampling of humidity detectors in the wireless sensor network soil is offered as an example of a prediction of environmental conditions. Another example of predicting environmental conditions was given by (Luan et al. 2016), who detailed a plan combining IoT with drought observation and forecasting.

Modern environmental monitoring instruments now have more advanced management and decision-making capabilities. For instance, (Giorgetti et al. 2016) proposed a tailored wireless sensor network-based landslide risk surveillance system that enables swift deployments in hostile environments without human intervention because the system is set up to copy with node failure and poor communication connections that reorganise the network on their own. The adaptive controller in the web service and real-time data architecture created by (Wong and Kerkez 2016) alters the parameters of each sensing node in a wireless sensor network in accordance with a predetermined strategy. (Zheng et al. 2015) provided an IoT management strategy to safeguard environmental quality and the environment while constructing an artificial river where nature and urbanisation meet. The system kept a close eye on important factors including soil, water, atmosphere, and wind across a large region with a high spatial resolution. Intelligent sensors for air humidity, temperature, pressure, and other natural disasters will function as the front-end instruments to collect and analyse environmental contextual data.

Another element influencing the capabilities of built-in issue detection and diagnostics is the worldwide convergence of industrial systems.

As we live in an automated culture, a smart environment would incorporate this trend by connecting computers, smartphones and the internet into out atmosphere. (Yi et al. 2015) described the next generation air pollution monitoring system concept with emphasis on the need to develop low-cost, energy-efficient and high spatiotemporal resolution real-time air pollution surveillance technologies. The sensor static sensor network, community sensor network, vehicle sensor networks are three types of sensor networks based on sensor element carriers in the next-generation air pollution monitoring system.

(Jamil et al. 2015) used wireless sensor network nodes to monitor city-wide air pollution as well as moving public transportation buses and autos. This approach provided writers with information from fixed nodes installed across the municipality to mobile nodes on buses and vehicles used for public transportation. Air pollution particle data such as gases, smoke, and other pollutants are gathered using sensors aboard public transportation buses, and the data is processed when the buses and vehicles return to their origin after passing through stationary nodes across the city. It is an important and essential task to preserve the environment and health of individuals by measuring air pollutants. In urban regions, owing to rise in the amount of factories and automotive cars, emissions of toxic gasses such as CO, NO_2, and CO_2 are also rising, causing serious impacts on the general public's health conditions.

One of the greatest breakthroughs of Internet of Things is bringing together the physical world and the data world. Sensors play a very significant part in bridging the barrier between the outside world and information. Sensors collect data from their environment, raising awareness of context by generating information. Thus the change in their environment can be monitored and if necessary, the corresponding things can make some responses. Radio-Frequency Identification Technology (RFID) is usually regarded as a main enabler for the Internet of Things due to its capacity to track a big amount of unique objects using Electronic Product Codes (EPC). Other types of ubiquitous sensor systems, barcodes, or 2D codes, however, may also be used to facilitate the IoT. Nanotechnology and miniaturization can create integrated intelligence in products that call smart devices themselves. They can process data, configure themselves and make decisions independently. Using IoT, embedded sensor technology can be more advanced to monitor and communicate critical environmental parameters such as temperature, humidity, pressure.

4.1. Internet of Things (IoT)

A fully integrated device with sensing, data analysis, and decision-making modules should be created in order to reduce air pollution. It is possible to improve estimates of the spatial and temporal variability of air pollutants, support studies of environmental modelling, exposure, and health effects, and give the general public immediate access to information by enhancing regulatory networks with portable, inexpensive air pollution sensors that provide real-time data. An autonomous real-time monitoring system for air quality was created in this study effort, and it involves detecting several characteristics, including temperature, humidity, carbon monoxide, and air pressure. Today, the Internet of Things is used extensively across many industries and is crucial to the system for monitoring air quality. Aurdino microcontroller, which uses minimal power and is inexpensive, collects and transmits data from various sensors as part of the Internet of Things and cloud computing convergence. The difficult part is getting accurate measurements of air quality in time and place that are helpful for public health initiatives. The major goal of creating this prototype is to track the gases contributing to environmental pollution and gauge their intensity so that it can be determined which gases are responsible for pollution in a specific location and at what rate the air is impacted.

IoT is an internet-accessible eco system of connected physical objects. The 'thing' in IoT can be an individual with a heart rate monitor or even a motor vehicle with integrated-in sensors, the objects allocated in IP address and intelligent of gathering and transferring data over a network without manual aid or intervention. These gadgets, or things, communicate to a router to provide feedback that they collect from the environment via sensors, or to enable other systems to reach out through actuators. As a result of improvements in computer power, electronics downsizing, and network connections, this technology is implemented in a broad variety of networked goods, systems, and sensors that provide new capabilities that were previously unattainable.

IoT device adoption on a wide scale has the potential to drastically alter many facets of our way of life. We are heading toward a "smart home" concept, with new IoT items including internet-enabled appliances, home automation components, and energy management systems delivering improved security and energy efficiency for customers. The way healthcare services are delivered is changing because to other IoT personal gadgets including networking medical equipment and wearable fitness and health monitors. The use of this technology should benefit the elderly and those with disabilities by enhancing their freedom and quality of life at a fair price.

4.1.1. Working of IoT

The whole mechanism begins with the devices themselves, such as smartphones, digital watches, electronic devices that interact safely with a platform of stuff online. IoT platform gathers and combines information from various devices and platforms and uses analytics to share the most significant information with industry-specific applications. The hardware used in IoT systems involves remote dashboard equipment, control devices, servers, a device for routing or bridging, and sensors. These devices handle important tasks and operations for supporting specific objectives and actions such as system activation, action specifications, safety, communication, and detection. IoT software uses platforms, embedded systems, partner systems, and middleware to address it main fields of networking and action. These individual and master applications are accountable for information collection, integration of devices, real-time analytics, and IoT network implementation and process expansion.

First and foremost, sensors and other devices enable the collection of incredibly little data from their surroundings. The intricacy of all this data collection might vary, ranging from a simple temperature surveillance sensor to a comprehensive full video feed. The information is then sent to a cloud infrastructure, however a transportation method is needed. The sensors can be connected to the cloud via a variety of communication and conveyance media, including cellular networks, satellite networks, Wi-Fi, Bluetooth, wide-area networks (WAN), low-power wide area networks, and many more. Next, the end user will be able to access the data in some way. This can be done by setting alarms on their phones or sending emails or messages to alert them. A user may occasionally have access to an interface that allows them to actively monitor their IoT plan. The programme processes the gathered data once it has been collected and transferred to the cloud.

4.1.2. IoT-Based Air-Quality Solutions

Precise monitoring is a particularly challenging part of air quality control. The air quality monitoring infrastructure in many cities is restricted to large, fixed monitoring stations, making it tough for experts to accuratelyassess the air pollution levels 7 in their cities and their effect on the lives of citizens. The Internet of Things (IoT) advancement has the potential to transform this situation Advances in Low Power Wide Area (LPWA) networks have resulted in the availability of small, portable, low cost, and always closely linked sensors. LPWA sensors can be attached to street furniture, bicycles or even people with low data rates and long range to measure and report air quality more regularly and accurately than ever before. Big data features, such as analytics as well as machine learning, could then be applied to these datasets and related measurements, such as weather and traffic updates, adding context to the data and understanding the causes and uncertainties of air pollution levels. While not meant to completely replace established monitoring networks, governments around the world are investing enormous sums in policies and partnerships to investigate air quality solutions driven by IoT.

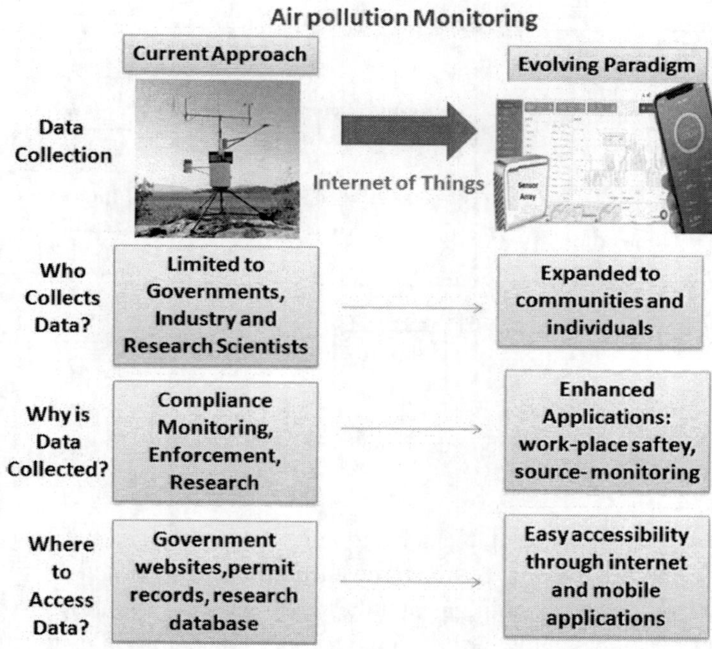

Figure 153. Paradigm shift with the implementation of IoT based system.

In the past, methods for measuring air pollution have typically relied on pricey, sophisticated, and immobile equipment, which has limited who may collect data, why they are gathered, and how they can be accessed. Figure 153 depicts the paradigm change that has occurred as a result of the development of affordable, user-friendly, portable air pollution sensors and the Internet of Things (IoT), which is now able to deliver high-resolution data in real-time. These characteristics provide chances to improve a variety of current air pollution monitoring capabilities and perhaps open the door for new uses.

4.2. Real-Time Air Quality Monitoring System Design

The work includes designing and implementing an embedded system using a network of gas sensors and communication module providing a framework for monitoring pollution from anywhere using IoT. The system block diagram is presented in Figure 154. Four gas sensors are used for continuous monitoring of pollutant gases of the surrounding. The sensors are integrated to a microcontroller and the sensed continuous data is sent to the controller for pollution-level verification. The controller then compares the data to the threshold value and, if the threshold is exceeded, provides an alert through the buzzer system. A wi-fi module connects the system to internet for remote monitoring.

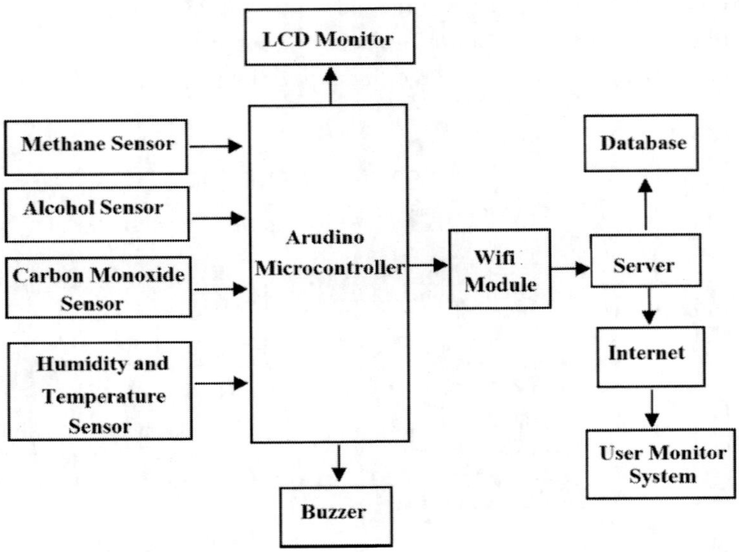

Figure 154. Block diagram of the proposed air quality monitoring system.

4.2.1. Sensor Module

Five air pollution sensors make up the sensing unit. SnO_2, the substance used in the MQ-3's sensitive gas sensor, has a lower conductivity in pure air. The conductivity of the sensor is higher when both the target alcohol gas and the rising gas concentration are present. Convert conductivity change to match the output signal for gas concentration using a simple electrical circuit. The MQ-3 gas sensor has a high alcohol sensitivity as well as great resistance to disturbances from fuel, smoke, and mist. The detector is suitable for several applications and is inexpensive. It may be used to recognise alcohol at various concentrations. The Al_2O_3 micro ceramic tube, Tin Dioxide (SnO_2) sensitive layer, measuring electrode, and heater make up the MQ-7 sensor, which is mounted inside a network crust constructed of stainless steel and plastic. For delicate parts, the heater provides the ideal operating conditions. It employs a cycle approach for high and low temperatures, detecting CO at the latter (heated by 1.5V). Along with the rising CO gas concentration, the sensor's conductivity rises. The other gases are cleansed at high temperature after being adsorbed at low temperature (heated by 5.0V). In air quality control systems, the MQ-135 Gas sensors are used to detect or measure NH_3, NO_x, Alcohol, Benzene, Smoke, and CO_2. When attempting to detect a specific gas, the sensor module MQ-135's Digital Pin, which enables this sensor to work even without a microcontroller, is helpful. The analogue pin can be utilised with the majority of popular microcontrollers because it is powered by TTL and runs on 5V. SnO_2, the sensitive component of the MQ-4 gas sensor, exhibits a decreased conductivity in clean air. The conductivity of the sensor increases along with the growing gas concentration when the target combustible gas is present. Through a simple circuit, users may transform the conductivity shift into output signals for gas concentration. The MQ-4 gas sensor is very sensitive to methane and resistant to influence from other gases including alcohol. A basic and inexpensive digital temperature and humidity sensor is the DHT22. It uses a thermistor and a capacitive humidity sensor to measure the ambient air, and the interface is a digital signal on the data pin. The four gas sensors' technical parameters, as well as the temperature and humidity sensors that were received from their datasheets, are shown in Table 8.

Table 8. Technical specifications of the gas sensors

Sensor	Operating Voltage	Current Consumption	Operating Temperature	Detection Limit
MQ3	5V ± 0.1	150 mA	-10°C~70°C	0.05mg/L—10mg/L Alcohol
MQ7	5V ± 0.1	200 mA	-20°C-50°C	20ppm-2000ppm carbon monoxide
MQ135	5V ± 0.1	150 mA	-10°C-45°C	10 ppm-300 ppm NH3, 10 ppm-1000 ppm Benzene
MQ4	5V ± 0.1	150 mA	-20°C-50°C	200-10000 ppm CH4, natural gas
DHT22	3.3 -5.5V	150 mA	-10°C-65°C	humidity 0 to-100% RH, temperature - 40 to 800C

4.2.2. Microcontroller Unit

Arduino is an open source programmable circuit board which can be implemented into a wide range of simple and complex makerspace projects. This module includes a microcontroller that can be programmed in the physical environment for sensing and controlling items. The arduino can communicate with a wide range of outputs such as LEDs, engines and displays by reacting to sensors and inputs. Because of its flexibility and low cost, Arduino Uno has become a common option for manufacturers and manufacturers seeking interactive hardware projects. The current Arduino Uno model includes a USB interface, 14 digital I/O ports, and 6 analogue input pins for connecting to external electrical circuits. 14 I/O ports may be utilised to output PWM to six pins. Designers can observe and feel the real world of external technological gadgets thanks to it.

The board includes 14 digital I/O and 6 analog pins that allow external link with any board circuit. These pins give the external devices the flexibility and ease of use that it can be connected via these pins. To connect the equipment to the board, there is no hard and fast interface needed. The 6 analog pins are labeled as A0 to A5 and have a 10-bit resolution. However, these pins measure from 0 to 5V but can be configured to the high range using the function analog reference() and the pin AREF. 13 KB of flash memory is used to store the number of code instructions. Only 5V is required to turn the board on which the USB port or external adapter can be used directly, but it can support external power source up to 12V which can be adjusted and limit 5V or 3.3V based on project requirements. Arduino UNO has the capacity to interface with other boards, microcontrollers and computers. The Atmega 328 on the board uses pins such as Rx and Tx to provide serial communication.

4.2.3. Output Display and Alerting Unit

The sensor output can be read from an LCD display and once the gas concentration goes above the threshold value, a buzzer gives audio alert. An LCD is a type of electronic display module that use liquid crystal to create a visual picture. A very fundamental module that is frequently used in electrical circuits is the 16x2 built-in LCD display. The 16x2 display format converts 16 characters per line onto 2 of those lines. This LCD uses a 5x7 pixel matrix to show each letter. A mechanical, electromechanical, or piezoelectric audio signalling device is a buzzer or beeper. Alarm systems, timers, and user input confirmation, such as a mouse click or keystroke, are common uses for buzzers and beepers. This type of buzzer is used in gas alarms, burglar alarms, etc. The buzzer gives the output sound in this detector when any of the sensor detects the target gas above the threshold level. Thus the buzzer serves as the system's audio output.

4.2.4. Wi-Fi Module and Web Server

The gadget uses an ESP8266 WiFi Module to provide on-board network connectivity. The Espressif system created the Wi-fi enabled system on chip module known as ESP8266. It uses a 32-bit RISC processor with an 80 MHz clock speed based on the Tensilica Xtensa L106 (or overclocked to 160 MHz). There are 96 KB of data RAM, 64 KB of instruction RAM, and 64 KB of boot ROM. Serial peripheral interface allows access to external flash memory. A series of AT instructions must be used by the microcontroller in order to connect with the ESP8266 module. Also it uses UART to connect e ESP8266 module with specified Baud rate. Users can access the data anywhere, anytime, with the help of the Internet, Aurdino Uno is interfaced with different sensors (Temperature & Humidity, MQ3, MQ4, MQ7, MQ135 Gas Sensors) and data is collected and stored in a Web server in real time.

4.3. Software Design

An intelligent object, the building block of the Internet of Things, is just another n ame for an internet-connected embedded system. Arduino IDE Software is the programming software for Arduino microcontroller. A software architecture with three layers—perception layer, gateway/processing layer, and remote environment—is created to synchronise the collection of data from the sensor kits. Remote Environment represents authorized users who can use the Internet via 3G/4G or Wi-Fi network toaccess the data in Thingspeak website. Perception layer is a module for the hardware interface or the collection of data from the sensors. The Gateway's primary function for the proposed architecture is to provide internet-wide data translation services. The gateway's main component is a cloud-based server. Thingspeak server is an API and open data platform for IoT, that allows user to monitor, collect, analyze, visualize, store, and act on sensor data.

4.3.1. Arduino IDE Software

The Arduino IDE is an integrated development environment for Arduino hardware that aids in programming the Arduino microcontrollers to connect with sensors and other components and carry out operations utilising library functions on the local and global domain. Arduino Uno is programmed using Java based Arduino Software, a cross-platform application called IDE. The base set out AVR microcontroller Atmega 328 comes with built-in bootloader that allows to upload the program to the board using a distinct burner.

4.3.2. Thingspeak: A Sensor-Cloud Application

ThingSpeak is an IoT analytics service platform that gives consumers access to real-time cloud data streams for collection, analysis, and evaluation. The data that the gadgets upload to ThingSpeak is immediately viewable. Real-time data processing and analysis are made possible by the ability to execute MATLAB code in ThingSpeak as it arrives. IoT devices that need analytics are routinely developed and validated using ThingSpeak. It is an open source Internet of Objects programme that has an open API for storing and retrieving data from device assets or objects via a local area network (LAN) or the Internet using HTTP applications. It also has a social network for device assets with relevant status updates. Thingspeak supports several operations for processing numerical data, such as averaging, time-scaling, rounding, median, and summing. The Thingspeak programme includes graphing with JavaScript, key management for read/write APIs, and time zone management.

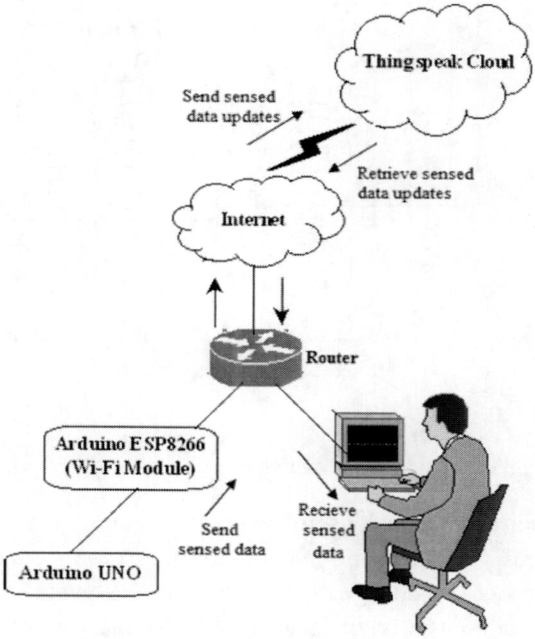

Figure 155. Software architecture of the air quality monitoring system.

Thingspeak utilises an internet connection to act as a data carrier between the connected things and the cloud in order to receive, save/store, analyse, monitor, and interact with sensed data from the connected sensor to the host microcontroller, such as an Arduino, TI CC3200 module, Raspberry-pi, etc. The Thingspeak allows the creation of captivating location-based tracing apps, sensor-based logging applications, and "social networks" of constantly updated items and things. An option is to use the Internet to remotely operate "Home Automation" devices linked to the public network from a person's home. The distinguishing element of the Thingspeak software is the word "Channel," which includes data fields, location fields, and status fields for different sensed information. Data may be applied or alternatively processed and shown once channels have been built. The software architecture of the system for monitoring air quality is shown in Figure 155.

4.4. Working on the Air Quality Monitoring System

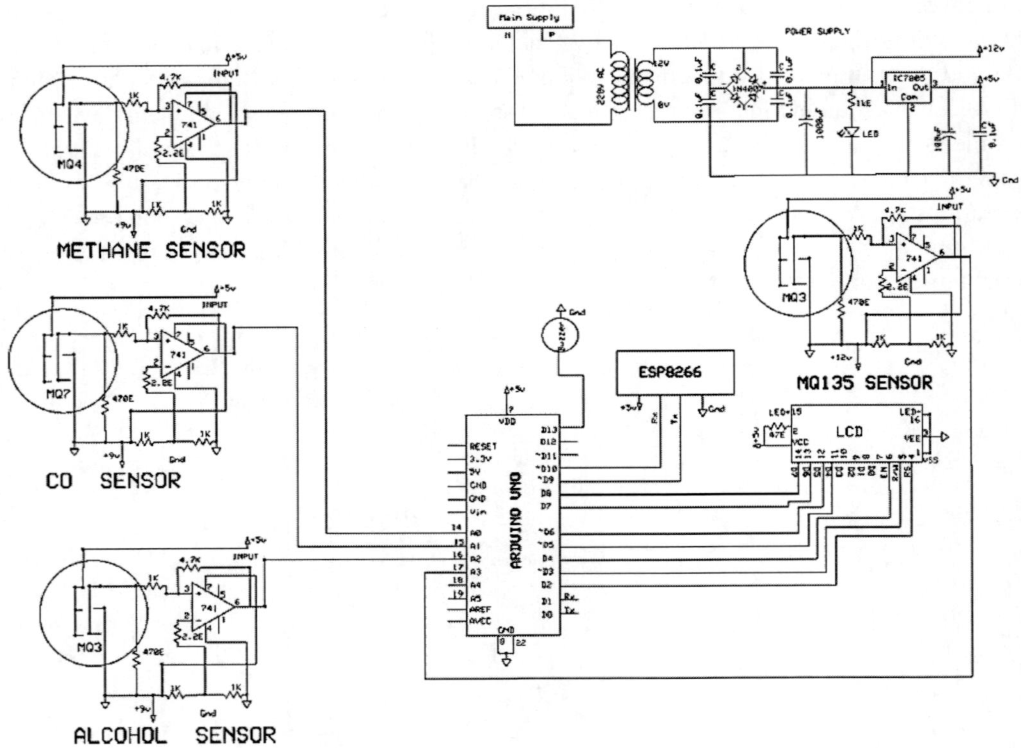

Figure 156. Circuit diagram of the hardware implementation.

Arduino Uno is the important and most useful part of the system. Arduino controls all the output devices and simultaneously, it reads and manipulates the sensor inputs. The gas sensors used are electrochemical sensors that change resistance to varying gases concentrations. To form a voltage divider circuit, each sensor is connected in series with a variable resistor and the variable resistor is used to change sensitivity. After heating, when one of the target gaseous elements comes into contact with the sensor, the resistance of the sensor changes. The resistance variation changes the voltage across the sensor, and a microcontroller can read this voltage. By knowing the reference voltage and the resistance of the other resistor, the voltage value can be used to find the resistance of the sensor. For the different type of gas, the sensor has different sensitivity. Figure 156 shows the circuit diagram of the hardware implemented.

The primary function of the system is to monitor the concentration of toxic gasses in the atmosphere and when any of the sensed parameters exceed their corresponding threshold value, send a alert to the user through the buzzer. The values are recorded continuously and transmitted in intervals of 3, 5 or 10 min as specified by the user to an online database. In the Aurdino module, the pins connected to the sensors are initialized. Using the record temperature, humidity and gas concentration values, a 'reference value' is calculated and this value is stored in the database. The appropriate threshold values are calculated from the reference value. Figure 157 shows the flowchart for the systemoperation.

Gas sensors collect data from the surrounding environment for gas concentration sensing and continually check it with the threshold of that particular sensor. If the threshold is found to exceed, buzzer turns on and time and gas-concentration values are continually recorded in the database. Similarly, the system senses, records and alerts the temperature and humidity values in the same manner.

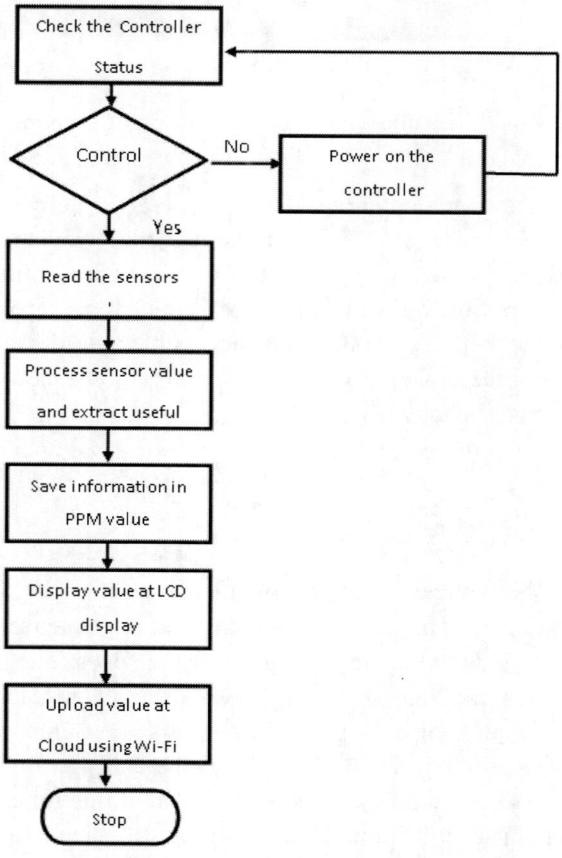

Figure 157. Flowchart of the system operation.

4.5. System Integration and Testing

Once the system is designed and error free coding is established, the user has to load sensor libraries in Arduino IDE Software and execute the program through the microcontroller to view the sensed data on LCD. The sensor values are converted to corresponding PPM readings in the embedded programming, and the LCD shows the concentration of each sensed gas in parts per million. The Figure 158 shows the hardware system with sensors and controller the sample LCD reading for the system.

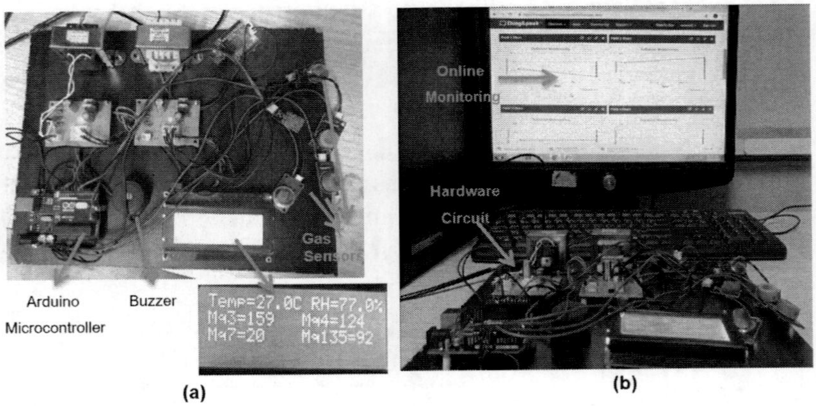

Figure 158. (a) Hardware implementation and LCD output display (b) Entire system connected to Internet.

The system connects to the internet once the user enters the network credentials in the IDE software and runs the application. At that point, the user may view the gas sensor data on the Thingspeak homepage. The system was placed to the test outside in the city of Coimbatore on several days, and the output readings in Thingspeak cloud were recorded. Any ambient data may be plotted locally and kept in a cloud that is accessible over the internet. As long as there is an internet connection, the device may be duplicated and installed at many sites to gather data and analyse their trends affordably.

Conclusion

Modern chemical sensors have been available on the market for many years, but the materials, manufacturing processes, and sensor form factors utilised to create these conventional sensors are frequently incompatible with the creation of flexible devices. The preparation of a sensing film with high selectivity and sensitivity to a desired target gas, long-term stability, good repeatability, quick reaction, compact size, and low power consumption is the key to success in constructing a gas sensor device. The bulk of commercially available gas sensors are still based on pure metal oxide gas sensors, as is well known at this time. Such gas sensors have been utilised effectively in several applications, but they still have issues with poor selectivity, high working temperatures, and high power consumption. The fundamental component of a gas sensor is the sensing film, and improving the material properties is crucial for improved sensor performance. The active and competitive research topic over the past several years has been nanostructuring of the sensor layers. Growing well-known gas-sensitive materials with better controlled morphology, crystalline structure, and chemical composition is made feasible by nanoscience and nanotechnology in particular. Conventional chemical sensors, for instance, are also heavy and bulky, making it impossible to utilise them for applications that need flexibility and wearability. Therefore, it is necessary to create sensors that are compact and flexible to enable seamless integration with curved surfaces like the human body for those working in hazardous situations. To obtain higher sensitivity at room temperature, composite materials based on metal oxide and multiwalled carbon nanotube were synthesized, as they often show synergistic effects compared to their singular components. The morphological and compositional properties were extensively studiedusing various characterization methods.

One of the main concerns is the long-term durability and stability of the sensors for implementation under different environmental circumstances, such as different humidity concentrations and presence of interfering gases, which could significantly influence the sensing characteristics of many gas sensors. The sensing performance of nanosensor reported in the results section is based on measurements in artificial environment of individual devices, that do may not be reproduced in field conditions. The carrier gas is nitrogen in this case, but for practical applications there need to be studies with introduction of air, humidity or interfering gases. Nevertheless, owing to the resemblance in functional groups and the property of different gases, selectivity and reliability in a complicated setting are still significant impediments in real-world applications to the realization of gas sensors.

There are promising futures for the use of transparent and flexible electronic sensors in smart electronics. There is still much to learn about the creation of fully flexible gas sensors with the benefits of simple design, straightforward operation, and inexpensive manufacturing costs. Flexible batteries and circuit boards, which are currently in the development stage, might be used to create a fully flexible gadget. All components with various capacities must be successfully merged into a single frame and then modified for large-area integration in order to apply the flexible sensor created in this research effort as fully integrated and functioning sensing platforms. The investigations done show a potential development in the sensing technology employing CNT-metal oxide nanomaterials and flexible substrates, however it is yet unknown how to build a stand-alone flexible device in terms of power supply, operation, and mobility.

References

Afzali, Parvaneh, Yaser Abdi, and Ezatollah Arzi. 2014. "Directional Reduction of Graphene Oxide Sheets Using Photocatalytic Activity of ZnO Nanowires for the Fabrication of a High Sensitive Oxygen Sensor." *Sensors and Actuators B: Chemical* 195 (May): 92–97. https://doi.org/10.1016/J.SNB.2013.12.105.

Aliane, A., V. Fischer, M. Galliari, L. Tournon, R. Gwoziecki, C. Serbutoviez, I. Chartier, and R. Coppard. 2014. "Enhanced Printed Temperature Sensors on Flexible Substrate." *Microelectronics Journal* 45 (12): 1621–26. https://doi.org/10.1016/J.MEJO.2014.08.011.

Anand, Kanika, Onkar Singh, Manmeet Pal Singh, Jasmeet Kaur, and Ravi Chand Singh. 2014. "Hydrogen Sensor Based on Graphene/ZnO Nanocomposite." *Sensors and Actuators B: Chemical* 195 (May): 409–15. https://doi.org/10.1016/J.SNB.2014.01.029.

Aroutiounian, V. M., V. M. Arakelyan, E. A. Khachaturyan, G. E. Shahnazaryan, M. S. Aleksanyan, L. Forro, A. Magrez, K. Hernadi, and Z. Nemeth. 2012. "Manufacturing and Investigations of I-Butane Sensor Made of SnO_2/Multiwall-Carbon-Nanotube Nanocomposite." *Sensors and Actuators B: Chemical* 173 (October): 890–96. https://doi.org/10.1016/J.SNB.2012.04.039.

Aziz, Madzlan, Saad Saber Abbas, and Wan Rosemaria Wan Baharom. 2013. "Size-Controlled Synthesis of SnO2 Nanoparticles by Sol–Gel Method." *Materials Letters* 91 (January): 31–34. https://doi.org/10.1016/J.MATLET.2012.09.079.

Barsan, Nicolae, and Udo Weimar. 2001. "Conduction Model of Metal Oxide Gas Sensors." *Journal of Electroceramics* 7 (3): 143–67. https://doi.org/10.1023/A:1014405811371/METRICS.

Briand, Danick, Alexandru Oprea, Jérôme Courbat, and Nicolae Bârsan. 2011. "Making Environmental Sensors on Plastic Foil." *Materials Today* 14 (9): 416–23. https://doi.org/10.1016/S1369-7021(11)70186-9.

Britton, D. T., and M. Härting. 2006. "Printed Nanoparticulate Composites for Silicon Thick-Film Electronics." *Pure and Applied Chemistry* 78 (9): 1723–39. https://doi.org/10.1351/PAC200678091723/MACHINEREADABLECITATION/RIS.

Chang, Yanhong, Yunfeng Yao, Bin Wang, Hui Luo, Tianyi Li, and Linjie Zhi. 2013. "Reduced Graphene Oxide Mediated SnO_2 Nanocrystals for Enhanced Gas-Sensing Properties." *Journal of Materials Science & Technology* 29 (2): 157–60. https://doi.org/10.1016/J.JMST.2012.11.007.

Chen, Yujin, Chunling Zhu, and Taihong Wang. 2006. "The Enhanced Ethanol Sensing Properties of Multi-Walled Carbon/SnO_2 Core/Shell Nanostructures." *Nanotechnology* 17 (12): 3012. https://doi.org/10.1088/0957-4484/17/12/033.

Choi, Kwang Yong, Joon Shik Park, Kwang Bum Park, Hyun Jae Kim, Hyo Derk Park, and Seong Dong Kim. 2010. "Low Power Micro-Gas Sensors Using Mixed SnO_2 Nanoparticles and MWCNTs to Detect NO_2, NH3, and Xylene Gases for Ubiquitous Sensor Network Applications." *Sensors and Actuators B: Chemical* 150 (1): 65–72. https://doi.org/10.1016/J.SNB.2010.07.041.

Claramunt, S., O. Monereo, M. Boix, R. Leghrib, J. D. Prades, A. Cornet, P. Merino, C. Merino, and A. Cirera. 2013. "Flexible Gas Sensor Array with an Embedded Heater Based on Metal Decorated Carbon Nanofibres." *Sensors and Actuators B: Chemical* 187 (October): 401–6. https://doi.org/10.1016/J.SNB.2012.12.093.

Courbat, Jérôme. 2010. "Gas Sensors on Plastic Foil with Reduced Power Consumption for Wireless Applications." https://doi.org/10.5075/EPFL-THESIS-4710.

Crowley, Karl, Aoife Morrin, Roderick L. Shepherd, Marc In Het Panhuis, Gordon G. Wallace, Malcolm R. Smyth, and Anthony J. Killard. 2010. "Fabrication of Polyaniline-Based Gas Sensors Using Piezoelectric Inkjet and Screen Printing for the Detection of Hydrogen Sulfide." *IEEE Sensors Journal* 10 (9): 1419–26. https://doi.org/10.1109/JSEN.2010.2044996.

Deng, Suzi, Verawati Tjoa, Hai Ming Fan, Hui Ru Tan, Dean C. Sayle, Malini Olivo, Subodh Mhaisalkar, Jun Wei, and Chorng Haur Sow. 2012. "Reduced Graphene Oxide Conjugated Cu2O Nanowire Mesocrystals for High-Performance NO_2 Gas Sensor." *Journal of the American Chemical Society* 134 (10): 4905–17. https://doi.org/10.1021/JA211683M.

Dhall, Shivani, and Neena Jaggi. 2016. "Hydrogen Gas Sensing Characteristics of Multiwalled Carbon Nanotubes Based Hybrid Composites." *Journal of Electronic Materials* 45 (1): 695–702. https://doi.org/10.1007/S11664-015-4176-8/METRICS.

Dhanaraju, Muthumanickam, Poongodi Chenniappan, Kumaraperumal Ramalingam, Sellaperumal Pazhanivelan, and Ragunath Kaliaperumal. 2022. "Smart Farming: Internet of Things (IoT)-Based Sustainable Agriculture." *Agriculture 2022, Vol. 12, Page 1745* 12 (10): 1745. https://doi.org/10.3390/AGRICULTURE12101745.

Du, Ning, Hui Zhang, Bindi Chen, Ma Xiangyang, Zhihong Liu, Jianbo Wu, and Deren Yang. 2007. "Porous Indium Oxide Nanotubes: Layer-by-Layer Assembly on Carbon-Nanotube Templates and Application for Room-Temperature NH3 Gas Sensors." *Advanced Materials* 19 (12): 1641–45. https://doi.org/10.1002/ADMA.200602128.

Espinosa, E. H., R. Ionescu, E. Llobet, A. Felten, C. Bittencourt, E. Sotter, Z. Topalian, P. Heszler, C. G. Granqvist, J. J. Pireaux, and X. Correig. 2007. "Highly Selective NO_2 Gas Sensors Made of MWCNTs and WO3 Hybrid Layers." *Journal of The Electrochemical Society* 154 (5): J141. https://doi.org/10.1149/1.2667855.

Feng, Xiaoyuan, Jian Jiang, Hao Ding, Ruimin Ding, Dan Luo, Jianhui Zhu, Yamin Feng, and Xintang Huang. 2013. "Carbon-Assisted Synthesis of Mesoporous SnO_2 Nanomaterial as Highly Sensitive Ethanol Gas Sensor." *Sensors and Actuators B: Chemical* 183 (July): 526–34. https://doi.org/10.1016/J.SNB.2013.04.006.

Fratoddi, Ilaria, Iole Venditti, Cesare Cametti, and Maria Vittoria Russo. 2015. "Chemiresistive Polyaniline-Based Gas Sensors: A Mini Review." *Sensors and Actuators B: Chemical* C (220): 534–48. https://doi.org/10.1016/J.SNB.2015.05.107.

Ghaddab, B., F. Berger, J. B. Sanchez, and C. Mavon. 2010. "Detection of O3 and NH3 Using Tin Dioxide/Carbon Nanotubes Based Sensors: Influence of Carbon Nanotubes Properties onto Sensor s Sensitivity." *Procedia Engineering* 5 (January): 115–18. https://doi.org/10.1016/J.PROENG.2010.09.061.

Ghasempour, R., and A. Iraji Zad. 2009. "Hybrid Multiwalled Carbon Nanotubes and Trioxide Tungsten Nanoparticles for Hydrogen Gas Sensing." *Journal of Physics D: Applied Physics* 42 (16): 165105. https://doi.org/10.1088/0022-3727/42/16/165105.

Giorgetti, Andrea, Matteo Lucchi, Emanuele Tavelli, Marco Barla, Giovanni Gigli, Nicola Casagli, Marco Chiani, and Davide Dardari. 2016. "A Robust Wireless Sensor Network for Landslide Risk Analysis: System Design, Deployment, and Field Testing." *IEEE Sensors Journal* 16 (16): 6374–86. https://doi.org/10.1109/JSEN.2016.2579263.

Gong, Jianwei, Jianren Sun, and Quanfang Chen. 2008. "Micromachined Sol–Gel Carbon Nanotube/SnO_2 Nanocomposite Hydrogen Sensor." *Sensors and Actuators B: Chemical* 130 (2): 829–35. https://doi.org/10.1016/J.SNB.2007.10.051.

Hachem, Sara, Vivien Mallet, Raphael Ventura, Animesh Pathak, Valerie Issarny, Pierre Guillaume Raverdy, and Rajiv Bhatia. 2015. "Monitoring Noise Pollution Using the Urban Civics Middleware." *Proceedings - 2015 IEEE 1st International Conference on Big Data Computing Service and Applications, BigDataService 2015*, August, 52–61. https://doi.org/10.1109/BIGDATASERVICE.2015.16.

Hammock, Mallory L., Alex Chortos, Benjamin C. K. Tee, Jeffrey B. H. Tok, and Zhenan Bao. 2013. "25th Anniversary Article: The Evolution of Electronic Skin (E-Skin): A Brief History, Design Considerations, and Recent Progress." *Advanced Materials* 25 (42): 5997–6038. https://doi.org/10.1002/ADMA.201302240.

Han, Jin Woo, Beomseok Kim, Jing Li, and M. Meyyappan. 2013. "A Carbon Nanotube Based Ammonia Sensor on Cotton Textile." *Applied Physics Letters* 102 (19). https://doi.org/10.1063/1.4805025/26696.

Hibbard, Troy, Karl Crowley, Frank Kelly, Frank Ward, John Holian, Alan Watson, and Anthony J. Killard. 2013. "Point of Care Monitoring of Hemodialysis Patients with a Breath Ammonia Measurement Device Based on Printed Polyaniline Nanoparticle Sensors." *Analytical Chemistry* 85 (24): 12158–65. https://doi.org/10.1021/AC403472D.

Hieu, Nguyen Van, Luong Thi Bich Thuy, and Nguyen Duc Chien. 2008. "Highly Sensitive Thin Film NH3 Gas Sensor Operating at Room Temperature Based on SnO_2/MWCNTs Composite." *Sensors and Actuators B: Chemical* 129 (2): 888–95. https://doi.org/10.1016/J.SNB.2007.09.088.

Hinkov, Ivaylo, Samir Farhat, Cristian P. Lungu, Alix Gicquel, François Silva, Amine Mesbahi, Ovidiu Brinza, Cornel Porosnicu, Alexandru Anghel. 2014. "Microwave Plasma Enhanced Chemical Vapor Deposition of Carbon Nanotubes." *Journal of Surface Engineered Materials and Advanced Technology* 4 (4): 196–209. https://doi.org/10.4236/JSEMAT.2014.44023.

Hoa, Le Thuy, Huynh Ngoc Tien, Van Hoang Luan, Jin Suk Chung, and Seung Hyun Hur. 2013. "Fabrication of a Novel 2D-Graphene/2D-NiO Nanosheet-Based Hybrid Nanostructure and Its Use in Highly Sensitive NO_2 Sensors." *Sensors and Actuators B: Chemical* 185 (August): 701–5. https://doi.org/10.1016/J.SNB.2013.05.050.

Hoang, T., N. Chu, and Trung Tran. 2017. "The Environmental Pollution In Vietnam: Source, Impact And Remedies." *International Journal of Scientific & Technology Research*.

Hong, Kyung Hwa, Kyung Wha Oh, and Tae Jin Kang. 2004. "Polyaniline–Nylon 6 Composite Fabric for Ammonia Gas Sensor." *Journal of Applied Polymer Science* 92 (1): 37–42. https://doi.org/10.1002/APP.13633.

Inaba, Masafumi, Takenori Oda, Masaki Kono, Nisarut Phansiri, Takahiro Morita, Shota Nakahara, Michihiko Nakano, and Junya Suehiro. 2021. "Effect of Mixing Ratio on NO_2 Gas Sensor Response with SnO2-Decorated Carbon Nanotube Channels Fabricated by One-Step Dielectrophoretic Assembly." *Sensors and Actuators B: Chemical* 344 (October): 130257. https://doi.org/10.1016/J.SNB.2021.130257.

Jamil, Muhammad Saqib, Muhammad Atif Jamil, Anam Mazhar, Ahsan Ikram, Abdullah Ahmed, and Usman Munawar. 2015. "Smart Environment Monitoring System by Employing Wireless Sensor Networks on

Vehicles for Pollution Free Smart Cities." *Procedia Engineering* 107: 480–84. https://doi.org/10.1016/J.PROENG.2015.06.106.

Jia, Yong, Lifang He, Zheng Guo, Xing Chen, Fanli Meng, Tao Luo, Minqiang Li, and Jinhuai Liu. 2009. "Preparation of Porous Tin Oxide Nanotubes Using Carbon Nanotubes as Templates and Their Gas-Sensing Properties." *Journal of Physical Chemistry C* 113 (22): 9581–87. https://doi.org/10.1021/JP9001719.

Jung, Daewoong, Youngsam Yoon, and Gil S. Lee. 2013. "Hydrogen Sensing Characteristics of Carbon-Nanotube Sheet Decorated with Manganese Oxides." *Chemical Physics Letters* 577 (July): 96–101. https://doi.org/10.1016/J.CPLETT.2013.05.047.

Khandani, Salma Kalantari, and Mehdi Kalantari. 2009. "Using Field Data to Design a Sensor Network." *Proceedings - 43rd Annual Conference on Information Sciences and Systems, CISS 2009*, 219–23. https://doi.org/10.1109/CISS.2009.5054720.

Khanderi, Jayaprakash, Rudolf C. Hoffmann, Aleksander Gurlo, and Jörg J. Schneider. 2009. "Synthesis and Sensoric Response of ZnO Decorated Carbon Nanotubes." *Journal of Materials Chemistry* 19 (28): 5039–46. https://doi.org/10.1039/B904822G.

Khmissi, H., A. M. El Sayed, and Mohamed Shaban. 2016. "Structural, Morphological, Optical Properties and Wettability of Spin-Coated Copper Oxide; Influences of Film Thickness, Ni, and (La, Ni) Co-Doping." *Journal of Materials Science* 51 (12): 5924–38. https://doi.org/10.1007/S10853-016-9894-7.

Kim, Yeon Hoo, Sang Jin Kim, Yong Jin Kim, Yeong Seok Shim, Soo Young Kim, Byung Hee Hong, and Ho Won Jang. 2015. "Self-Activated Transparent All-Graphene Gas Sensor with Endurance to Humidity and Mechanical Bending." *ACS Nano* 9 (10): 10453–60. https://doi.org/10.1021/ACSNANO.5B04680/SUPPL_FILE/NN5B04680_SI_001.PDF.

Korotcenkov, Ghenadii. 2005. "Gas Response Control through Structural and Chemical Modification of Metal Oxide Films: State of the Art and Approaches." *Sensors and Actuators B: Chemical* 107 (1): 209–32. https://doi.org/10.1016/J.SNB.2004.10.006.

Koskela, J., J. Sarfraz, P. Ihalainen, A. Määttänen, P. Pulkkinen, H. Tenhu, T. Nieminen, A. Kilpelä, and J. Peltonen. 2015. "Monitoring the Quality of Raw Poultry by Detecting Hydrogen Sulfide with Printed Sensors." *Sensors and Actuators B: Chemical* 218 (October): 89–96. https://doi.org/10.1016/J.SNB.2015.04.093.

Krishna Kumar, M., and S. Ramaprabhu. 2006. "Nanostructured Pt Functionlized Multiwalled Carbon Nanotube Based Hydrogen Sensor." *Journal of Physical Chemistry B* 110 (23): 11291–98. https://doi.org/10.1021/JP0611525.

Lakafosis, Vasileios, Amin Rida, Rushi Vyas, Li Yang, Symeon Nikolaou, and Manos M. Tentzeris. 2010. "Progress towards the First Wireless Sensor Networks Consisting of Inkjet-Printed, Paper-Based RFID-Enabled Sensor Tags." *Proceedings of the IEEE* 98 (9): 1601–9. https://doi.org/10.1109/JPROC.2010.2049622.

Lee, Jun Seop, Oh Seok Kwon, Dong Hoon Shin, and Jyongsik Jang. 2013. "WO3 Nanonodule-Decorated Hybrid Carbon Nanofibers for NO_2 Gas Sensor Application." *Journal of Materials Chemistry A* 1 (32): 9099–9106. https://doi.org/10.1039/C3TA11658A.

Lee, Soo Keun, Daeic Chang, and Sang Wook Kim. 2014. "Gas Sensors Based on Carbon Nanoflake/Tin Oxide Composites for Ammonia Detection." *Journal of Hazardous Materials* 268 (March): 110–14. https://doi.org/10.1016/J.JHAZMAT.2013.12.049.

Leghrib, Radouane, Roman Pavelko, Alexandre Felten, Alexey Vasiliev, Carles Cané, Isabel Gràcia, Jean Jacques Pireaux, and Eduard Llobet. 2010. "Gas Sensors Based on Multiwall Carbon Nanotubes Decorated with Tin Oxide Nanoclusters." *Sensors and Actuators B: Chemical* 145 (1): 411–16. https://doi.org/10.1016/J.SNB.2009.12.044.

Lelieveld, J., J. S. Evans, M. Fnais, D. Giannadaki, and A. Pozzer. 2015. "The Contribution of Outdoor Air Pollution Sources to Premature Mortality on a Global Scale." *Nature 2015 525:7569* 525 (7569): 367–71. https://doi.org/10.1038/nature15371.

Li, Su Juan, Ning Xia, Xia Lei Lv, Meng Meng Zhao, Bai Qing Yuan, and Huan Pang. 2014. "A Facile One-Step Electrochemical Synthesis of Graphene/NiO Nanocomposites as Efficient Electrocatalyst for Glucose and Methanol." *Sensors and Actuators B: Chemical* 190 (January): 809–17. https://doi.org/10.1016/J.SNB.2013.09.047.

Li, Tianming, Wen Zeng, Huiwu Long, and Zhongchang Wang. 2016. "Nanosheet-Assembled Hierarchical SnO_2 Nanostructures for Efficient Gas-Sensing Applications." *Sensors and Actuators B: Chemical* 231 (August): 120–28. https://doi.org/10.1016/J.SNB.2016.03.003.

Li, Xiaogan, Yangyang Zhao, Xueyan Wang, Jing Wang, Alexander M. Gaskov, and S. A. Akbar. 2016. "Reduced Graphene Oxide (RGO) Decorated TiO_2 Microspheres for Selective Room-Temperature Gas Sensors." *Sensors and Actuators B: Chemical* 230 (July): 330–36. https://doi.org/10.1016/J.SNB.2016.02.069.

Liang, Shiming, Junwu Zhu, Chao Wang, Songtao Yu, Huiping Bi, Xiaoheng Liu, and Xin Wang. 2014. "Fabrication of α-Fe_2O_3@graphene Nanostructures for Enhanced Gas-Sensing Property to Ethanol." *Applied Surface Science* 292 (February): 278–84. https://doi.org/10.1016/J.APSUSC.2013.11.130.

Liu, Yan Li, Hai Feng Yang, Yu Yang, Zhi Min Liu, Guo Li Shen, and Ru Qin Yu. 2006. "Gas Sensing Properties of Tin Dioxide Coated onto Multi-Walled Carbon Nanotubes." *Thin Solid Films* 497 (1–2): 355–60. https://doi.org/10.1016/J.TSF.2005.11.018.

Luan, Qingzu, Xiaoyi Fang, Caihua Ye, and Yonghong Liu. 2016. "An Integrated Service System for Agricultural Drought Monitoring and Forecasting and Irrigation Amount Forecasting." *International Conference on Geoinformatics* 2016-January (January). https://doi.org/10.1109/GEOINFORMATICS.2015.7378617.

Lundin, Daniel, and Kostas Sarakinos. 2012. "An Introduction to Thin Film Processing Using High-Power Impulse Magnetron Sputtering." *Journal of Materials Research 2012 27:5* 27 (5): 780–92. https://doi.org/10.1557/JMR.2012.8.

Luo, Liqiang, Fang Li, Limei Zhu, Yaping Ding, Zhao Zhang, Dongmei Deng, and Bo Lu. 2013. "Nonenzymatic Glucose Sensor Based on Nickel(II)Oxide/Ordered Mesoporous Carbon Modified Glassy Carbon Electrode." *Colloids and Surfaces B: Biointerfaces* 102 (February): 307–11. https://doi.org/10.1016/J.COLSURFB.2012.08.003.

Luo, Liqiang, Limei Zhu, and Zhenxin Wang. 2012. "Nonenzymatic Amperometric Determination of Glucose by CuO Nanocubes-Graphene Nanocomposite Modified Electrode." *Bioelectrochemistry (Amsterdam, Netherlands)* 88 (December): 156–63. https://doi.org/10.1016/J.BIOELECHEM.2012.03.006.

Mao, Shun, Shumao Cui, Ganhua Lu, Kehan Yu, Zhenhai Wen, and Junhong Chen. 2012. "Tuning Gas-Sensing Properties of Reduced Graphene Oxide Using Tin Oxide Nanocrystals." *Journal of Materials Chemistry* 22 (22): 11009–13. https://doi.org/10.1039/C2JM30378G.

Mao, Shun, Shumao Cui, Kehan Yu, Zhenhai Wen, Ganhua Lu, and Junhong Chen. 2012. "Ultrafast Hydrogen Sensing through Hybrids of Semiconducting Single-Walled Carbon Nanotubes and Tin Oxide Nanocrystals." *Nanoscale* 4 (4): 1275–79. https://doi.org/10.1039/C2NR11765G.

Marichy, Catherine, Patricia A. Russo, Mariangela Latino, Jean Philippe Tessonnier, Marc Georg Willinger, Nicola Donato, Giovanni Neri, and Nicola Pinna. 2013. "Tin Dioxide-Carbon Heterostructures Applied to Gas Sensing: Structure-Dependent Properties and General Sensing Mechanism." *Journal of Physical Chemistry C* 117 (38): 19729–39. https://doi.org/10.1021/JP406191X/SUPPL_FILE/JP406191X_SI_001.PDF.

Mattana, Giorgio, Thomas Kinkeldei, David Leuenberger, Caglar Ataman, Jinyu J. Ruan, Francisco Molina-Lopez, Andrés Vásquez Quintero, G. Nisato, Gerhard Troster, Danick Briand, Nico F. de Rooij. 2013. "Woven Temperature and Humidity Sensors on Flexible Plastic Substrates for E-Textile Applications." *IEEE Sensors Journal* 13 (10): 3901–9. https://doi.org/10.1109/JSEN.2013.2257167.

Mbuyisa, Puleng, Sunil P. Bhardwaj, Federica Rigoni, Elvio Carlino, Stefania Pagliara, Luigi Sangaletti, Andrea Goldoni, Muzi Ndwandwe, and Cinzia Cepek. 2012. "Controlled Synthesis of Carbon Nanostructures Using Aligned ZnO Nanorods as Templates." *Carbon* 50 (15): 5472–80. https://doi.org/10.1016/J.CARBON.2012.07.034.

Mendoza, Frank, Dionne M. Hernández, Vladimir Makarov, Emmanuel Febus, Brad R. Weiner, and Gerardo Morell. 2014. "Room Temperature Gas Sensor Based on Tin Dioxide-Carbon Nanotubes Composite Films." *Sensors and Actuators B: Chemical* 190 (January): 227–33. https://doi.org/10.1016/J.SNB.2013.08.050.

Mishra, Prabhash, and S. S. Islam. 2015. "Highly Efficient Ceramic-SWCNT Based Free Standing Films for Gas Sensing Application." *Materials Science in Semiconductor Processing* 40 (December): 276–84. https://doi.org/10.1016/J.MSSP.2015.05.047.

Molina-Lopez, F., A. Vásquez Quintero, G. Mattana, D. Briand, and N. F. De Rooij. 2013. "Large-Area Compatible Fabrication and Encapsulation of Inkjet-Printed Humidity Sensors on Flexible Foils with Integrated Thermal Compensation." *Journal of Micromechanics and Microengineering* 23 (2): 025012. https://doi.org/10.1088/0960-1317/23/2/025012.

Mubeen, Syed, Min Lai, Ting Zhang, Jae Hong Lim, Ashok Mulchandani, Marc A. Deshusses, and Nosang V. Myung. 2013. "Hybrid Tin Oxide-SWNT Nanostructures Based Gas Sensor." *Electrochimica Acta* 92 (March): 484–90. https://doi.org/10.1016/J.ELECTACTA.2013.01.029.

Penza, M., R. Rossi, M. Alvisi, M. A. Signore, G. Cassano, D. Dimaio, R. Pentassuglia, E. Piscopiello, E. Serra, and M. Falconieri. 2009. "Characterization of Metal-Modified and Vertically-Aligned Carbon Nanotube Films for Functionally Enhanced Gas Sensor Applications." *Thin Solid Films* 517 (22): 6211–16. https://doi.org/10.1016/J.TSF.2009.04.009.

Peter, C., J. Kneer, and J. Wollenstein. 2011. "Inkjet Printing of Titanium Doped Chromium Oxide for Gas Sensing Application." *Sensor Letters* 9 (2): 807–11. https://doi.org/10.1166/SL.2011.1619.

Ren, Haibo, Wei Zhao, Liyou Wang, Si Ok Ryu, and Cuiping Gu. 2015. "Preparation of Porous Flower-like SnO_2 Micro/Nano Structures and Their Enhanced Gas Sensing Property." *Journal of Alloys and Compounds* 653 (December): 611–18. https://doi.org/10.1016/J.JALLCOM.2015.09.065.

Rigoni, F., G. Drera, S. Pagliara, E. Perghem, C. Pintossi, A. Goldoni, and L. Sangaletti. 2016. "Gas Sensing at the Nanoscale: Engineering SWCNT-ITO Nano-Heterojunctions for the Selective Detection of NH3 and NO_2 Target Molecules." *Nanotechnology* 28 (3): 035502. https://doi.org/10.1088/1361-6528/28/3/035502.

Russo, Patrícia A., Nicola Donato, Salvatore Gianluca Leonardi, Seunghwan Baek, Donato E. Conte, Giovanni Neri, and Nicola Pinna. 2012. "Room-Temperature Hydrogen Sensing with Heteronanostructures Based on Reduced Graphene Oxide and Tin Oxide." *Angewandte Chemie International Edition* 51 (44): 11053–57. https://doi.org/10.1002/ANIE.201204373.

Sánchez, M., and M. E. Rincón. 2012a. "Effect of Multiwalled Carbon Nanotube Functionalization on the Gas Sensing Properties of Carbon Nanotube–Titanium Dioxide Hybrid Materials." *Diamond and Related Materials* 21 (January): 1–6. https://doi.org/10.1016/J.DIAMOND.2011.09.010.

———. 2012b. "Effect of Multiwalled Carbon Nanotube Functionalization on the Gas Sensing Properties of Carbon Nanotube–Titanium Dioxide Hybrid Materials." *Diamond and Related Materials* 21 (January): 1–6. https://doi.org/10.1016/J.DIAMOND.2011.09.010.

Sandstrom, Andreas, Henrik F. Dam, Frederik C. Krebs, and Ludvig Edman. 2012. "Ambient Fabrication of Flexible and Large-Area Organic Light-Emitting Devices Using Slot-Die Coating." *Nature Communications 2010 3:1* 3 (1): 1–5. https://doi.org/10.1038/ncomms2002.

Santangelo, S., G. Messina, G. Faggio, A. Donato, L. De Luca, N. Donato, A. Bonavita, and G. Neri. 2010. "Micro-Raman Analysis of Titanium Oxide/Carbon Nanotubes-Based Nanocomposites for Hydrogen Sensing Applications." *Journal of Solid State Chemistry* 183 (10): 2451–55. https://doi.org/10.1016/J.JSSC.2010.08.018.

Saville, Ramadhona, Katsumori Hatanaka, and Masaaki Wada. 2016. "ICT Application of Real-Time Monitoring and Estimation System for Set-Net Fishery." *OCEANS 2015 - MTS/IEEE Washington*, February. https://doi.org/10.23919/OCEANS.2015.7404524.

Seredych, Mykola, and Teresa J. Bandosz. 2012. "Manganese Oxide and Graphite Oxide/MnO_2 Composites as Reactive Adsorbents of Ammonia at Ambient Conditions." *Microporous and Mesoporous Materials* 150 (1): 55–63. https://doi.org/10.1016/J.MICROMESO.2011.09.010.

Shaban, Mohamed, Mona Mustafa, and A. M. El Sayed. 2016. "Structural, Optical, and Photocatalytic Properties of the Spray Deposited Nanoporous CdS Thin Films; Influence of Copper Doping, Annealing, and Deposition Parameters." *Materials Science in Semiconductor Processing* 56 (December): 329–43. https://doi.org/10.1016/J.MSSP.2016.09.006.

Shahrjerdi, D., B. Hekmatshoar, S. W. Bedell, M. Hopstaken, and D. K. Sadana. 2011. "Low-Temperature Epitaxy of Compressively Strained Silicon Directly on Silicon Substrates." *Journal of Electronic Materials 2011 41:3* 41 (3): 494–97. https://doi.org/10.1007/S11664-011-1807-6.

Sharma, Anjali, Monika Tomar, and Vinay Gupta. 2012. "Room Temperature Trace Level Detection of NO_2 Gas Using SnO_2 Modified Carbon Nanotubes Based Sensor." *Journal of Materials Chemistry* 22 (44): 23608–16. https://doi.org/10.1039/C2JM35172B.

Shen, Wenfeng. 2012. "Properties of SnO_2 Based Gas-Sensing Thin Films Prepared by Ink-Jet Printing." *Sensors and Actuators B: Chemical* 166–167 (May): 110–16. https://doi.org/10.1016/J.SNB.2012.01.019.

Shukla, S. K., G. K. Parashar, A. P. Mishra, Puneet Misra, B. C. Yadav, R. K. Shukla, L. M. Bali, and G. C. Dubey. 2004. "Nano-like Magnesium Oxide Films and Its Significance in Optical Fiber Humidity Sensor." *Sensors and Actuators B: Chemical* 98 (1): 5–11. https://doi.org/10.1016/J.SNB.2003.05.001.

Star, Alexander, Vikram Joshi, Sergei Skarupo, David Thomas, and Jean Christophe P. Gabriel. 2006. "Gas Sensor Array Based on Metal-Decorated Carbon Nanotubes." *Journal of Physical Chemistry B* 110 (42): 21014–20. https://doi.org/10.1021/JP064371Z/SUPPL_FILE/JP064371ZSI20060821_010618.AVI.

Su, Pi Guey, and Te Tsun Pan. 2011. "Fabrication of a Room-Temperature NO_2 Gas Sensor Based on WO3 Films and WO3/MWCNT Nanocomposite Films by Combining Polyol Process with Metal Organic Decomposition Method." *Materials Chemistry and Physics* 125 (3): 351–57. https://doi.org/10.1016/J.MATCHEMPHYS.2010.11.001.

Sung, Donovan, Alejandro De La Fuente Vornbrock, and Vivek Subramanian. 2010. "Scaling and Optimization of Gravure-Printed Silver Nanoparticle Lines for Printed Electronics." *IEEE Transactions on Components and Packaging Technologies* 33 (1): 105–14. https://doi.org/10.1109/TCAPT.2009.2021464.

Tian, Hailin, Huiqing Fan, Hui Guo, and Na Song. 2014. "Solution-Based Synthesis of ZnO/Carbon Nanostructures by Chemical Coupling for High Performance Gas Sensors." *Sensors and Actuators B: Chemical* 195 (May): 132–39. https://doi.org/10.1016/J.SNB.2014.01.016.

Torres-Ruiz, Miguel, Juan H. Juárez-Hipólito, Miltiadis Demetrios Lytras, and Marco Moreno-Ibarra. 2016. "Environmental Noise Sensing Approach Based on Volunteered Geographic Information and Spatio-Temporal Analysis with Machine Learning." *Lecture Notes in Computer Science (Including Subseries Lecture Notes in Artificial Intelligence and Lecture Notes in Bioinformatics)* 9789: 95–110. https://doi.org/10.1007/978-3-319-42089-9_7/COVER.

Tulliani, Jean Marc, Alessio Cavalieri, Simone Musso, Eloisa Sardella, and Francesco Geobaldo. 2011. "Room Temperature Ammonia Sensors Based on Zinc Oxide and Functionalized Graphite and Multi-Walled Carbon Nanotubes." *Sensors and Actuators B: Chemical* 152 (2): 144–54. https://doi.org/10.1016/J.SNB.2010.11.057.

Turner, Michelle C., Zorana J. Andersen, Andrea Baccarelli, W. Ryan Diver, Susan M. Gapstur, III C. Arden Pope, Diddier Prada, Jonathan Samet, George Thurston, and Aaron Cohen. 2020. "Outdoor Air Pollution and Cancer: An Overview of the Current Evidence and Public Health Recommendations." *CA: A Cancer Journal for Clinicians* 70 (6): 460–79. https://doi.org/10.3322/CAAC.21632.

Tyagi, Punit, Anjali Sharma, Monika Tomar, and Vinay Gupta. 2017. "A Comparative Study of RGO-SnO_2 and MWCNT-SnO_2 Nanocomposites Based SO_2 Gas Sensors." *Sensors and Actuators B: Chemical* 248 (September): 980–86. https://doi.org/10.1016/J.SNB.2017.02.147.

Vyas, Rishi, Sarla Sharma, Parul Gupta, Arun K. Prasad, A. K. Tyagi, K. Sachdev, and S. K. Sharma. 2012. "CNT-ZnO Nanocomposite Thin Films: O_2 and NO_2 Sensing." *Advanced Materials Research* 585: 235–39. https://doi.org/10.4028/WWW.SCIENTIFIC.NET/AMR.585.235.

Watthanawisuth, Natthapol, Adisorn Tuantranont, and Teerakiat Kerdcharoen. 2009. "Microclimate Real-Time Monitoring Based on ZigBee Sensor Network." *Proceedings of IEEE Sensors*, 1814–18. https://doi.org/10.1109/ICSENS.2009.5398587.

Wong, Brandon P., and Branko Kerkez. 2016. "Real-Time Environmental Sensor Data: An Application to Water Quality Using Web Services." *Environmental Modelling & Software* 84 (October): 505–17. https://doi.org/10.1016/J.ENVSOFT.2016.07.020.

Wongchoosuk, Chatchawal, Anurat Wisitsoraat, Ditsayut Phokharatkul, Mati Horprathum, Adisorn Tuantranont, and Teerakiat Kerdcharoen. 2013. "Carbon Doped Tungsten Oxide Nanorods NO2 Sensor Prepared by Glancing Angle RF Sputtering." *Sensors and Actuators B: Chemical* 181 (May): 388–94. https://doi.org/10.1016/J.SNB.2013.01.066.

Wongchoosuk, Chatchawal, Anurat Wisitsoraat, Ditsayut Phokharatkul, Adisorn Tuantranont, and Teerakiat Kerdcharoen. 2010. "Multi-Walled Carbon Nanotube-Doped Tungsten Oxide Thin Films for Hydrogen Gas Sensing." *Sensors 2010, Vol. 10, Pages 7705-7715* 10 (8): 7705–15. https://doi.org/10.3390/S100807705.

Xie, Yadian, Duygu Kocaefe, Chunying Chen, and Yasar Kocaefe. 2016. "Review of Research on Template Methods in Preparation of Nanomaterials." *Journal of Nanomaterials*. Hindawi Limited. https://doi.org/10.1155/2016/2302595.

Yang, An, Xiaoming Tao, Rongxing Wang, Shuncheng Lee, and Charles Surya. 2007. "Room Temperature Gas Sensing Properties of Sn O_2/Multiwall-Carbon-Nanotube Composite Nanofibers." *Applied Physics Letters* 91 (13). https://doi.org/10.1063/1.2783479/334163.

Yang, Myung, Dai Hong Kim, Won Sik Kim, Tae June Kang, Byung Yang Lee, Seunghun Hong, Yong Hyup Kim, and Seong Hyeon Hong. 2010. "H2 Sensing of SnO_2 Coated Single Wall Carbon Nanotube Network Sensors." *Nanotechnology* 21 (21): 215501. https://doi.org/10.1088/0957-4484/21/21/215501.

Yi, Wei Ying, Kin Ming Lo, Terrence Mak, Kwong Sak Leung, Yee Leung, and Mei Ling Meng. 2015. "A Survey of Wireless Sensor Network Based Air Pollution Monitoring Systems." *Sensors 2015, Vol. 15, Pages 31392-31427* 15 (12): 31392–427. https://doi.org/10.3390/S151229859.

Yoo, Ran, Jeongmin Kim, Min Jung Song, Wooyoung Lee, and Jin Seo Noh. 2015. "Nano-Composite Sensors Composed of Single-Walled Carbon Nanotubes and Polyaniline for the Detection of a Nerve Agent Simulant Gas." *Sensors and Actuators, B: Chemical* 209 (March): 444–48. https://doi.org/10.1016/J.SNB.2014.11.137.

Yuan, Wenjing, and Gaoquan Shi. 2013. "Graphene-Based Gas Sensors." *Journal of Materials Chemistry A* 1 (35): 10078–91. https://doi.org/10.1039/C3TA11774J.

Yun, Sang Woo, Jae Ryung Cha, and Myoung Seon Gong. 2014. "Water-Resistive Humidity Sensor Prepared from New Polyelectrolyte Containing Both Photo-Curable 4-Styrylpyridinium Function and Thiol Anchor." *Sensors and Actuators B: Chemical* 202 (October): 1109–16. https://doi.org/10.1016/J.SNB.2014.06.065.

Zanolli, Zeila, Radouane Leghrib, Alexandre Felten, Jean Jacques Pireaux, Eduard Llobet, and Jean Christophe Charlier. 2011. "Gas Sensing with Au-Decorated Carbon Nanotubes." *ACS Nano* 5 (6): 4592–99. https://doi.org/10.1021/NN200294H/SUPPL_FILE/NN200294H_SI_001.PDF.

Zhang, Jun, Xianghong Liu, Giovanni Neri, and Nicola Pinna. 2016. "Nanostructured Materials for Room-Temperature Gas Sensors." *Advanced Materials (Deerfield Beach, Fla.)* 28 (5): 795–831. https://doi.org/10.1002/ADMA.201503825.

Zhang, Lifeng, Xiaoxu Wang, Yong Zhao, Zhengtao Zhu, and Hao Fong. 2012. "Electrospun Carbon Nano-Felt Surface-Attached with Pd Nanoparticles for Hydrogen Sensing Application." *Materials Letters* 68 (February): 133–36. https://doi.org/10.1016/J.MATLET.2011.10.064.

Zhang, Yang, Shumao Cui, Jingbo Chang, Leonidas E. Ocola, and Junhong Chen. 2013. "Highly Sensitive Room Temperature Carbon Monoxide Detection Using SnO_2 Nanoparticle-Decorated Semiconducting Single-Walled Carbon Nanotubes." *Nanotechnology* 24 (2). https://doi.org/10.1088/0957-4484/24/2/025503.

Zhang, Zhenyu, Rujia Zou, Guosheng Song, Li Yu, Zhigang Chen, and Junqing Hu. 2011. "Highly Aligned SnO_2 Nanorods on Graphene Sheets for Gas Sensors." *Journal of Materials Chemistry* 21 (43): 17360–65. https://doi.org/10.1039/C1JM12987B.

Zheng, Rubin, Tianhai Zhang, Zhifang Liu, and Haowei Wang. 2015. "An EIoT System Designed for Ecological and Environmental Management of the Xianghe Segment of China's Grand Canal." *Https://Doi.Org/10.1080/13504509.2015.1124470* 23 (4): 372–80. https://doi.org/10.1080/13504509.2015.1124470.

Chapter 10

Integration of Nanosensors with Electronic Gadgets

Learning Outcomes

- The reader could comprehend the importance and uses of combining nanosensors with electrical devices.
- The learner will be able to name the criteria for choosing appropriate electronic devices, taking into account elements such sensor interface compatibility, computing power, wireless connection, power management, and data security.
- Insights into the integration of nanosensors with the Internet of Things (IoT) ecosystem will be given in this chapter, covering the significance of sensing and data collecting, networking and communication, intelligent decision-making, miniaturisation, form factor adaptation, and scalability.
- It is possible to study the integration strategies and difficulties related to integrating nanosensors, including sensor integration strategies, interface and communication protocols, power supply concerns, size and form factor restrictions, and privacy and security issues.
- Discovered here are new developments in integration methods, miniaturisation, flexible electronics, sensor fusion, and multi-modal integration, among other future ideas on nanosensor integration.
- The learner gains a thorough awareness of the technical considerations involved in incorporating nanosensors into electronic devices, allowing the creation of cutting-edge and intelligent devices for a variety of applications.

Abstract

A ground-breaking topic with enormous promise for improving sensing capacities and allowing enhanced functionality is the use of nanosensors into electronic systems. Real-time data collection and processing may be accomplished by seamlessly integrating nanosensors with electronic devices, such as smartphones, wearables, or environmental monitoring systems. This opens the door to applications in healthcare, environmental monitoring, and other fields. The chapter examines the importance, difficulties, and potential directions of incorporating nanosensors into electrical devices. It offers a thorough study of the criteria for choosing suitable devices, the integration of nanosensors with the Internet of Things (IoT) ecosystem, integration strategies, and new trends. The necessity of seamless integration for improved sensing capabilities and real-time data collecting is highlighted in this chapter. It covers a range of topics, including sensor interface compatibility, computing power, wireless connection, power management, and data security, to be taken into account when choosing suitable devices. The importance of sensing and data collecting, networking and communication, intelligent decision-making, miniaturisation, and scalability are highlighted as they relate to the integration of nanosensors with the IoT ecosystem. Advances in integration methods, miniaturisation, sensor fusion, and multi-modal integration are some of the other topics covered in this chapter, along with integration difficulties and future prospects. Overall, this chapter is an invaluable resource for academics and industry experts who are interested in adding nanosensors to electronic devices to increase functionality and user experience.

1. Introduction

In general, nanosensors are highly sensitive and selective sensors that use nanomaterials or nanostructures to detect and quantify target analytes. By putting these nanomaterials into integrated systems or platforms, integrated nanosensors extend this idea and provide improved functionality and integration into a range of applications. These sensors are often made of nanomaterials, such as nanoparticles, nanowires, or nanotubes, which, because of their tiny size, display distinctive characteristics and behaviours. Integrated nanosensors are made to be small, very sensitive, and capable of monitoring in real time or very close to it. They may be incorporated into a wide range of platforms and systems, including industrial applications, biomedical equipment, environmental monitoring systems, and electronics. These devices may now be equipped with nanosensors to obtain accurate and thorough data about the immediate environment or particular chemicals. Nanosensor integration has various benefits over using conventional sensors.

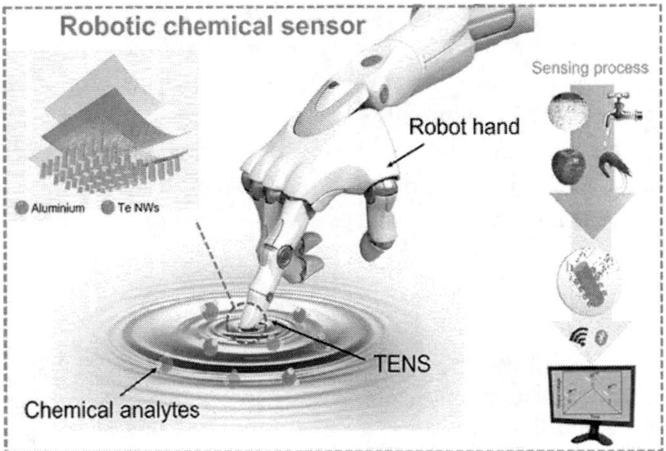

Figure 159. Tribolelectric nanosensor integrated with robotics to sense the chemical analytes (Roy Barman et al. 2023).

Depending on the particular application and needs, there are a variety of ways to combine nanosensors with electrical devices. Below, a few typical strategies are outlined. Thin film deposition methods like physical vapour deposition (PVD) and chemical vapour deposition (CVD) may be used to create nanosensors. With these methods, sensing materials are deposited in thin layers on the surface of electrical components like silicon chips or flexible substrates. During the manufacturing process, nanosensors may be incorporated directly into electrical devices. The creation of nanosensors, for instance, may be integrated into the manufacturing process of wearables or smartphones, enabling the seamless integration of sensing capabilities. Nanosensors may be created using methods like inkjet printing or screen printing on electrical devices. With the use of this method, the device may have localised sensing capabilities by having nanoscale sensing materials deposited onto certain regions of the device. The technology of microelectromechanical systems (MEMS) may be used to incorporate nanosensors into electrical devices. On a tiny scale, MEMS devices integrate mechanical and electrical elements. MEMS devices may be equipped with nanosensors, enabling the detection of a range of physical and chemical characteristics. Nanosensors are capable of wireless communication with electrical devices. Wi-Fi or Bluetooth are two examples of wireless communication technologies that may be used to send sensor data. This makes it possible for electrical equipment to remotely monitor and operate the nanosensors. The precise integration technique will depend on the kind of nanosensor, the intended application, and the electrical device being utilised, it is vital to note. Depending on the intended sensing capabilities and the compatibility with the electronic device, several technologies and materials may be used.

The integrated nanosensors' tiny size and high surface area-to-volume ratio enable them to detect and quantify analytes with great sensitivity. Nanosensors may be made to bind to targets selectively by functionalizing their surfaces with certain chemicals or coatings, providing highly targeted detection. Extremely compact integrated nanosensors make it possible to create wearable and portable gadgets that are simple to deploy and integrate into a variety of applications. The embedded nanosensors can monitor the target analytes continuously and in real-time, allowing for quick reaction and intervention as required. Nanosensors with various analyte detection and measurement capabilities are conceivable, offering a plethora of data in a single instrument. The illustration of an integrated triboelectric nanosensor with robotics that is capable of sensing the chemical analytes is shown in Figure 159.

Numerous industries, including healthcare, environmental monitoring, food safety, agriculture, and security, use integrated nanosensors. They may be used to find and keep an eye on diseases, pollutants, gases, poisons, biomarkers, and many other interesting compounds. In general, integrated nanosensors provide a potential route for very sensitive and selective sensing in a variety of fields, opening up new possibilities for monitoring, diagnostics, and control in a variety of applications.

2. Significance and Applications of Integrated Nanosensors

The potential for revolutionising the capabilities and functions of electronic devices via the integration of nanosensors is enormous. Nanosensors may improve the sensing capabilities of electronic devices, allowing them to detect and quantify a broad variety of physical, chemical, and biological factors thanks to their small size and high sensitivity. Electronic devices may develop more intelligence, responsiveness, and environmental adaptability by using nanosensors. Numerous applications, such as environmental sensing, smart agriculture, industrial automation, and healthcare monitoring are made possible by this combination. In the end, the incorporation of nanosensors with electrical devices opens the door to the creation of sophisticated, miniature devices with enhanced performance and unique functions.

Monitoring the environment and human health may benefit significantly from the integration of nanosensors with electrical devices. Due to their great sensitivity, nanosensors can detect toxins or pollutants in the environment at extremely low quantities. This makes it possible to find and monitor dangerous compounds in the air, water, or soil early on.

Nanosensors built into electronic devices may provide real-time information on environmental factors including radiation levels, water quality, and air quality as illustrated in Figure 160. This quick input makes it possible to respond and intervene in time to reduce environmental dangers. These sensors are often small and simple to incorporate into tiny electrical systems. The ability to deploy portable environmental monitoring devices at different places for extensive data collecting and analysis is made possible by this miniaturisation. Nanosensors and electrical devices may be used to provide monitoring systems that are affordable. When compared to conventional sensing technologies, nanosensors often need less resources and materials, which lowers production costs and enables scale deployment.

For instance, nanosensors incorporated into electronic devices have showed considerable promise in the monitoring of air quality. Tin oxide nanowire gas sensors were successfully integrated into cellphones for real-time air quality monitoring in a research. This demonstrates the nanosensors' great sensitivity and precision in identifying dangerous gases including nitrogen dioxide and volatile organic compounds (VOCs) (Lahlalia et al. 2019). The nanosensors may also be combined with electrical devices to track various aspects of water quality. In a study, it was shown how to create graphene-based nanosensors to detect heavy metal ions in water. The detection of minute levels of heavy metals was made possible by the nanosensors' exceptional sensitivity and selectivity (Zhang et al. 2018).

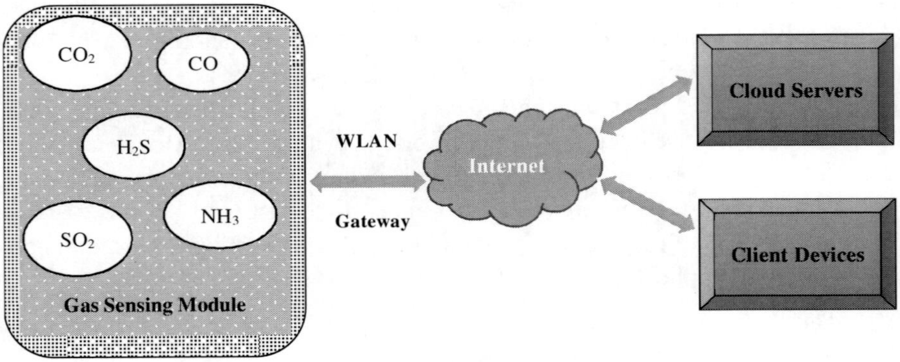

Figure 160. Illustration of integrated gas sensing nanosensors.

Real-time health monitoring for humans may be provided through nanosensors built into wearable technology or implanted inside the body. With the use of these sensors, personalised healthcare and early illness detection are made possible. These sensors can measure vital signs, biomarkers, and disease indicators. Remote patient health monitoring is made possible by the integration of nanosensors with electronic devices. As a result, less frequent hospital visits may be necessary, patient comfort may increase, and healthcare providers may be able to deliver timely treatments based on the data gathered. Non-invasive or slightly invasive monitoring solutions may be provided via nanosensors. They may, for instance, find biomarkers in physiological fluids or exhaled air instead of invasive procedures like blood sample. The comfort and compliance of the patient are improved. By continually monitoring health indicators, identifying early warning signals, and giving feedback for dietary modifications or medication adherence, nanosensors may help with illness prevention and control. This proactive strategy may result in improved disease management and patient outcomes.

For instance, wearable technology's embedded nanosensors allow for continuous health monitoring. Real-time monitoring of heart rate, respiration rate, and skin temperature was made possible by the incorporation of flexible, graphene-based nanosensors into wearable patches for vital sign monitoring (Huang et al. 2019). This technology may one day be used for remote patient monitoring and the early diagnosis of medical problems. Nanosensors incorporated into electronic devices may help in the detection and treatment of diseases. The development of gold nanoparticle-based nanosensors for disease biomarker detection demonstrated high sensitivity and specificity in detecting biomarkers related to cancer and infectious diseases, offering potential applications in early diagnosis and personalised treatment (Tai et al. 2022). In general, integrating nanosensors with electronic devices in environmental and healthcare monitoring provides enhanced sensitivity, real-time monitoring, portability, cost-effectiveness, and personalizability. These developments help to create monitoring systems that are more effective and precise, which improves environmental management and healthcare delivery.

3. Electronic Gadgets for Integration

The integration of nanosensors with electronic devices is essential for providing increased sensing capabilities for a variety of applications. These devices act as platforms for nanosensors, hosting and interacting with them to enable data gathering, processing, and communication. Electronic gadgets, such as cellphones, wearable technology, and specialised monitoring systems, offer the framework required to fully realise the promise of nanosensors in a range of industries, such as environmental monitoring and healthcare as mentioned in Figure 161.

Monitoring of the environment and healthcare systems gains several advantages from the use of nanosensors with electronic devices. First of all, it makes it possible to detect and measure several characteristics with increased sensitivity and precision. Nanosensors provide great sensitivity and selectivity for detecting target compounds or biomarkers because of their nanoscale size and unique material characteristics. The ability to identify minute changes in health indicators or tiny levels of contaminants in the environment enables for more accurate monitoring and the early identification of possible problems.

Real-time monitoring capabilities are another benefit of combining nanosensors with electrical devices. Users may get real-time insights into their surroundings or health state by continually collecting data from the nanosensors and analysing it inside the electronic gadget. The ability to make educated judgements and quickly take the necessary action is given to people, healthcare professionals, and environmental authorities by this timely information. For instance, individuals may avoid polluted regions or take measures when air quality deteriorates via real-time air quality monitoring utilising nanosensors built into cellphones.

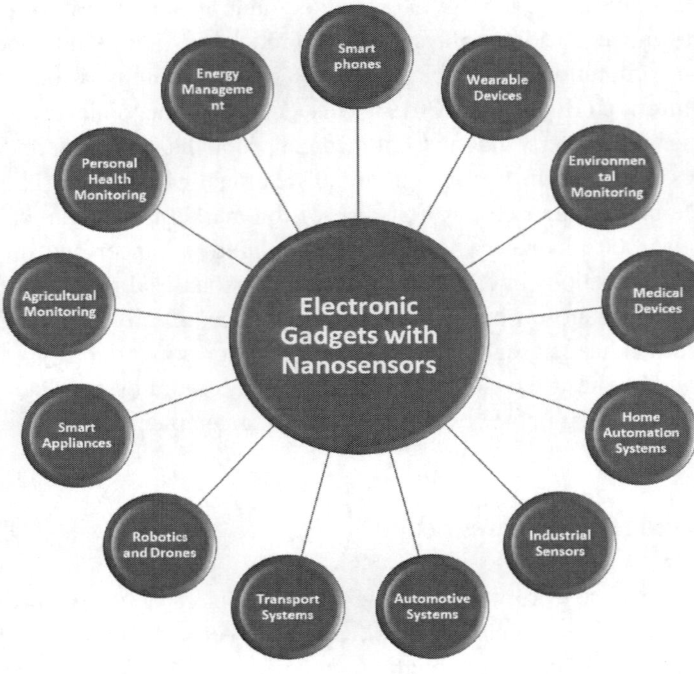

Figure 161. Different types of electronic gadgets integrated with the nanosensors.

Monitoring systems may become more portable and accessible thanks to the incorporation of nanosensors with electronic devices. Nanosensors may be readily included into small, portable devices because to miniaturisation and improvements in electronic device technology. This mobility makes it simple to install monitoring systems in a variety of locations, including distant or less resourced places. For instance, wearable technology with built-in nanosensors may offer continuous health monitoring for people wherever they are in their everyday life. For on-site inspections and quick reaction to environmental events, portable environmental monitoring systems incorporating nanosensors may be used.

Monitoring of the environment and human health benefits significantly from the incorporation of nanosensors with electronic devices. These benefits consist of better mobility, real-time monitoring capabilities, and increased sensitivity. We can accomplish more precise and timely monitoring and improve environmental management and personalised healthcare by using the potential of nanosensors built into electronic gadgets. The incorporation of nanosensors with electronic devices is anticipated to further revolutionise monitoring systems as technology develops, creating new opportunities for proactive healthcare and environmental sustainability.

3.1. Overview of Electronic Gadgets

One of the most often used electronic devices for incorporating nanosensors is the smartphone. Smartphones may be used as flexible platforms for hosting nanosensors due to their powerful processing capabilities, built-in sensors (such as an accelerometer, gyroscope, and GPS), and wireless communication. Nanosensors built into smartphones may gather and analyse sensor data, display information in real time, and send the gathered data to distant servers for analysis by taking use of the processing power of these devices. Numerous applications, including tracking personal health and monitoring air and water quality, are made possible by this interface.

Another class of electronic gadgets that are well-suited for nanosensor integration is wearable technology, which includes smartwatches, fitness trackers, and biosensors. These gadgets are designed to be worn on the body, which makes them perfect for applications including ongoing health monitoring and individualised treatment. Users may effortlessly monitor their vital signs, exercise levels, sleep patterns, and other health factors in real-time by incorporating nanosensors into wearable devices. The wearable gadget serves as the interface for data visualisation and communication with other devices or healthcare specialists while the nanosensors offer the essential sensing capabilities.

Nanosensors may be used in specialised monitoring systems and devices, in addition to smartphones and wearables. These systems have been specifically designed for tasks including environmental monitoring, workplace safety, and medical diagnostics. Nanosensors, for example, may be used in environmental monitoring stations to assess air pollutants, radiation levels, or water quality factors. To give thorough environmental insights, these devices often include cutting-edge data recording capabilities, networking choices, and analysis tools. Specialised medical tools and gadgets may use nanosensors to detect biomarkers, track medication levels, or carry out fast diagnostic tests in healthcare, resulting in more effective and precise healthcare delivery.

Electronic devices include a broad variety of medical tools and equipment used in healthcare monitoring. These consist of pulse oximeters, blood pressure monitors, blood glucose monitors, and other devices. These devices often include sensors that assess certain health indicators and give useful information for illness management, diagnosis, and therapy monitoring. Electronic components and sensors are used in advanced medical equipment including imaging systems (MRI, CT scan), diagnostic analyzers, and laboratory analyzers to produce high-quality pictures or carry out precise laboratory tests for accurate medical evaluation.

To allow remote data gathering and analysis, remote monitoring systems integrate electronic devices with communication technology. These systems may integrate a number of tools, including wireless sensors, data recorders, and communication modules, to track environmental factors or patient health in distant places. Stakeholders may access and analyse the data in real-time by transmitting it from these devices to a central server or cloud platform. Systems for remote monitoring are used in telemedicine, patient monitoring, and environmental surveillance, allowing quick and effective interventions.

Both environmental and medical monitoring depend on electronic devices. These gadgets, which range from smartphones and wearables to specialised monitoring stations and medical equipment, incorporate sensors, data processing tools, and networking choices to allow real-time monitoring, data analysis, and communication. Electronic devices are used in monitoring systems to improve data accuracy, portability, and accessibility, which helps to improve environmental management and personalised healthcare.

3.2. Selection Criteria for Compatible Gadgets

Environmental and medical monitoring have been transformed by the incorporation of nanosensors with electronic devices, which provides improved sensing capabilities and real-time data processing. Several important factors must be taken into consideration when choosing appropriate electrical devices for nanosensor integration. Compatibility with the sensor interface, adequate processing power and data storage, wireless connectivity options, effective power management, user-friendly interfaces and displays, appropriate size and form factor, cost-effectiveness, data security and privacy measures, durability and reliability, scalability and interoperability, compatibility with operating systems and software, battery life and energy efficiency, and sensor data processing and analysis are a few of these. Researchers and developers can make well-informed decisions by carefully considering these criteria to enable seamless integration, dependable performance, usability, and cost-effectiveness of electronic devices and nanosensors, ultimately resulting in effective and efficient environmental and healthcare monitoring systems.

3.2.1. Compatibility with Sensor Interface

It is essential to establish compatibility between the sensor's interface and the hardware and software of the electronic device when integrating nanosensors with that device. The electrical or communication protocols used by various nanosensors may need to be matched with the capabilities of the device. To communicate with the electronic device, for instance, certain nanosensors may need analog-to-digital converters or particular communication protocols (such I2C, SPI). The integration of a smartphone with a microsensor system for gas sensing applications highlighted how crucial it is to choose smartphones with the right hardware interfaces for smooth communication between the nanosensors and the device (Gomes et al. 2019).

3.2.2. Processing Power and Data Storage

For the electronic device to handle the sensor data produced by the nanosensor, it needs have enough processing power and data storage capacity. For applications that call for on-device data processing and real-time monitoring, this criterion becomes more crucial. A wearable electronic device for monitoring glucose was integrated with nanosensors in a study paper, which emphasised the need for a device with sufficient processing power and memory to manage continuous glucose readings and carry out essential data analysis (Gia et al. 2017).

3.2.3. Wireless Connectivity

When sending data from nanosensors to external devices or cloud platforms for analysis, wireless communication is often essential. Depending on the needs of the application, the electronic device should have the appropriate wireless communication capabilities, such as Wi-Fi, Bluetooth, or cellular connection. Real-time data transfer, remote monitoring, and easy integration with other systems or networks are all made possible via wireless communication. In order to promote effective glucose monitoring, a research on the integration of nanosensors with smartphones for fluorescence-based glucose sensing highlighted the significance of wireless smartphone capabilities for remote monitoring and data transfer (Ngo et al. 2020).

3.2.4. Power Management

To guarantee dependable and continuous functioning, the electronic device should allow effective power management. Nanosensors coupled with electronic devices may have specialised power needs. The electronic device need to have a sufficient battery, power-saving features, or the potential to generate energy from the environment. The device's battery life is extended through effective power management, guaranteeing continuous monitoring. The integration of a nanosensor system with a self-powered wearable electronic device brought attention to the significance of power management tactics, such as energy harvesting and power optimisation approaches, for long-term, self-sustaining operation (Ball et al. 2018).

3.2.5. User Interface and Display

To offer real-time feedback and visualise the sensor data, the electronic device should have a user-friendly interface and display capabilities. Users may interact with the nanosensor-integrated device via the user interface, which also gives them clear information. This is especially important for applications where users must evaluate and react to the data that has been acquired. The combination of nanosensors with smartphones to monitor water quality has highlighted the need of an intuitive user interface and a graphical representation of water quality parameters for simple interpretation, allowing users to make knowledgeable choices (Lakshmikantha et al. 2021).

3.2.6. Size and Form Factor

The electronic device's size and form factor are essential for smooth integration with nanosensors. To fit the nanosensor and make its integration into diverse applications easier, the device should have a small footprint and suitable dimensions. The electronic device should be portable, flexible, and pleasant to wear, for example, in wearable applications. The work demonstrated the integration of a wearable electronic device with nanosensors for biomarker detection, emphasising the necessity for a lightweight and flexible device to provide pleasant and unobtrusive wearable integration (Sempionatto et al. 2022).

3.2.7. Cost and Accessibility

Cost is a crucial factor to take into account, particularly in situations with limited resources or large-scale installations. To achieve general acceptance and affordability, the gadget should be inexpensive and simple to use. This criteria becomes especially important when several devices are required for an application or when contemplating deployment in less developed economic regions. Nanosensors and inexpensive Arduino microcontrollers were used to monitor air pollution, demonstrating how accessible and cost-effective Arduino-based platforms are for environmental sensing applications, making them appropriate for wide-scale use (Kelechi et al. 2021).

3.2.8. Data Security and Privacy

Data security and privacy issues are crucial when combining nanosensors with electrical devices. To safeguard the private information that the nanosensors gathered, the electronic device has to have strong security measures and encryption capabilities. When personal health information is incorporated in healthcare monitoring apps, this is very important. In order to protect patient privacy and prevent unauthorised access to the collected data, a study highlighted the necessity of secure authentication protocols and encryption mechanisms when integrating nanosensors with wearable devices for healthcare monitoring (Ullah et al. 2021).

3.2.9. Durability and Reliability

In particular, whether employed in demanding environmental circumstances or long-term monitoring situations, electronic devices used for nanosensor integration should be dependable and long-lasting. They have to be resilient enough to tolerate physical strain, temperature changes, and contact with chemicals or moisture. The use of ruggedized electronic equipment for environmental monitoring in challenging outdoor locations emphasises the need of equipment toughness and dependability to enable continuous functioning and precise data collecting under testing circumstances (Almuslem, Shaikh, and Hussain 2019).

3.2.10. Scalability and Interoperability

Scalability and interoperability should be taken into account when choosing electrical devices for nanosensor integration, especially for large-scale deployments or connection with pre-existing monitoring infrastructure. To provide easy integration with other devices or network systems, the device should be compatible with industry-standard protocols and communication interfaces. The incorporation of nanosensors into electronic devices for smart city applications highlighted the need for interoperability standards, such as the Internet of Things (IoT) protocols, to guarantee smooth data exchange and communication across various platforms and devices (Alam, Poniszewska-Maranda, and Maranda 2021).

3.2.11. Battery Life and Energy Efficiency

The battery life and energy efficiency of electronic devices used for nanosensor integration should be adequate to sustain continuous monitoring and reduce the need for frequent recharging or battery replacement. Battery life may be increased with the use of effective power management systems, low power components, and optimisation techniques. The significance of energy-efficient design was emphasised in a low-power microcontroller platform for wearable gas sensing applications, guaranteeing longer field operation (Saoutieff et al. 2021).

3.2.12. Compliance with Regulatory Standards

To guarantee patient safety and data security, electronic devices used in healthcare monitoring applications should adhere to regulatory requirements, such as medical device regulations and privacy rules. The observance of quality and security criteria is shown by compliance with regulations like the General Data Protection Regulation (GDPR) in Europe and ISO 13485 for medical devices. In order to guarantee data privacy, integrity, and interoperability when integrating nanosensors with electronic health record (EHR) systems, a research emphasised the significance of regulatory compliance (Abunadi and Kumar 2021).

The capabilities of environmental and healthcare monitoring systems have substantially increased thanks to the integration of nanosensors with electronic devices. Researchers and developers can choose the best electronic devices for seamless integration with nanosensors by taking into account a variety of selection criteria, including compatibility, processing power, connectivity, power management, user interface, size, cost, security, durability, and scalability. They can also take into consideration operating systems, battery life, data processing, ease of integration, and regulatory compliance. These standards guarantee dependable performance, real-time monitoring, intuitive user interfaces, and affordable solutions. Since accurate and rapid data collection, analysis, and transmission are made possible by the integration of nanosensors with electronic devices, it gives people, environmental agencies, and healthcare experts the ability to make well-informed judgements and take the necessary steps. The integration of nanosensors with electronic devices is set to further revolutionise environmental management and individualised healthcare, leading to a more sustainable and healthier future. This is due to continuous technological breakthroughs and ongoing research.

3.3. Examples of Electronic Gadgets for Integration

Few examples of integrated electronic gadgets with nanosensors are discussed below.

3.3.1. Water Quality Monitor with Nanosensors

The identification of impurities is made possible by the integration of nanosensors into water quality monitoring systems. High sensitivity was shown in the detection of heavy metals, such as lead and copper ions, in water samples when graphene-based nanosensors were combined with a system for monitoring water quality (Irkham et al. 2023). The exact and quick analysis provided by the water quality monitor fitted with these nanosensors ensured the safety of the water for drinking.

3.3.2. Air Pollution Monitor with Nanosensors

Nanosensor-integrated air pollution monitoring systems give real-time data on air quality parameters. The portable air pollution monitor equipped with nanosensors allowed for the detection and measurement of numerous gases, such as nitrogen dioxide (NO_2) and ozone (O_3), as well as particulate matter (PM2.5 and PM10) (Mohamed and Awad 2022). The integrated system provided a convenient and economical way to monitor air quality.

3.3.3. Glucose Monitoring Patch with Nanosensors

Continuous glucose monitoring is possible thanks to wearable patches that include glucose nanosensors. High sensitivity and selectivity in detecting glucose levels in sweat were shown by the incorporation of graphene-based nanosensors into a wearable patch for non-invasive glucose monitoring (Holmes 2016; Xuan, Yoon, and Park 2018). Through this connection, real-time monitoring was made possible, and people with diabetes were given a practical and pleasant choice.

3.3.4. Smart Inhaler with Nanosensors

Nanosensors included into smart inhalers allow for the monitoring of medicine use and inhalation patterns. Nanosensors were included into a smart inhaler for the treatment of asthma. These nanosensors tracked medicine delivery and recognised breathing patterns, giving users feedback on how to use the inhaler correctly (Himes et al. 2019). Through this combination, personalised care was made possible and medication adherence was enhanced.

3.3.5. Biomedical Implants with Nanosensors

Nanosensor-equipped implantable medical devices provide improved monitoring capabilities. Biomarkers were identified and real-time information on the state of the patient's health was supplied through biomedical implants for monitoring cellular and physiological changes (Wang et al. 2021). Early disease identification and personalised healthcare were made possible by this combination.

3.3.6. Food Safety Checker with Nanosensors

Nanosensor-equipped handheld devices make it possible to quickly and accurately check the safety of food. The portable food safety tester's embedded nanosensors quickly and accurately provided examination of food pollutants such heavy metals, pesticides, or allergens (Müller-Maatsch and van Ruth 2021). With this connection, testing for food safety was made portable and simple.

3.3.7. Environmental Monitoring Drone with Nanosensors

Nanosensor-equipped drones provide airborne environmental parameter monitoring. The drones equipped with nanosensors found contaminants and assessed radiation levels as well as variables like temperature and humidity (Srivastava et al. 2022). This connection made it possible to efficiently and broadly monitor remote locations while also providing useful information for environmental management.

3.3.8. Soil Moisture Sensor with Nanosensors

As a soil moisture sensor, an electrical device containing nanosensors may be employed. These nanosensors can precisely assess soil moisture content, assisting farmers in streamlining irrigation procedures and better managing water resources. A research showed how to include nanosensors into a system for measuring soil moisture, allowing for real-time monitoring of soil moisture levels (Bogena, Weuthen, and Huisman 2022).

3.3.9. UV Radiation Monitor with Nanosensors

Nanosensor-equipped electronic devices can measure UV radiation for sun safety applications. These nanosensors are able to monitor and detect UV radiation levels, enabling users to determine their exposure and take the appropriate safety measures.

3.3.10. Wearable Sweat Analysis Patch with Nanosensors

Nanosensor-integrated wearable electronic patches allow for non-invasive investigation of sweat composition. These nanosensors can identify biomarkers in sweat, giving information on electrolyte balance, metabolic activity, and hydration levels.

3.3.11. Smart Agriculture Monitor with Nanosensors

Nanosensor-equipped electronic devices may be utilised for smart agricultural monitoring. Nanosensors provide more effective crop cultivation and resource management by measuring a variety of characteristics, such as temperature, humidity, light intensity, and soil nutrient levels.

3.3.12. Smart Bandage with Nanosensors

Nanosensor-equipped electronic devices may be utilised as smart bandages for monitoring wounds. By detecting markers in wounds like pH, temperature, or bacterial presence, nanosensors placed in the bandage may provide real-time input on the healing process.

These illustrations show the wide variety of electronic devices that have nanosensors built in for use in the environmental and medical fields. These devices' integration enables enhanced monitoring capabilities, real-time data processing, and increased precision when identifying important parameters. These developments support more effective environmental management, individualised healthcare, and improved safety across a range of industries.

4. Internet of Things and Nanosensors

The ideas of the Internet of Things (IoT) and the incorporation of nanosensors are intertwined and closely linked. Figure 162 shows the basic conceptual block diagram of the integration of real world IoT devices with a virtual world over a connecting network where the real-time data is processed. The Internet of Things (IoT) is a network of linked physical things or gadgets that are equipped with connection, software, and sensors to gather and share data. On the other hand, nanosensors are tiny sensors with nanoscale parts that provide very sensitive and focused detecting capabilities. When nanosensors are included into Internet of Things (IoT) devices, they improve the ecosystem's functionality by offering sophisticated sensing, monitoring, and data gathering capabilities.

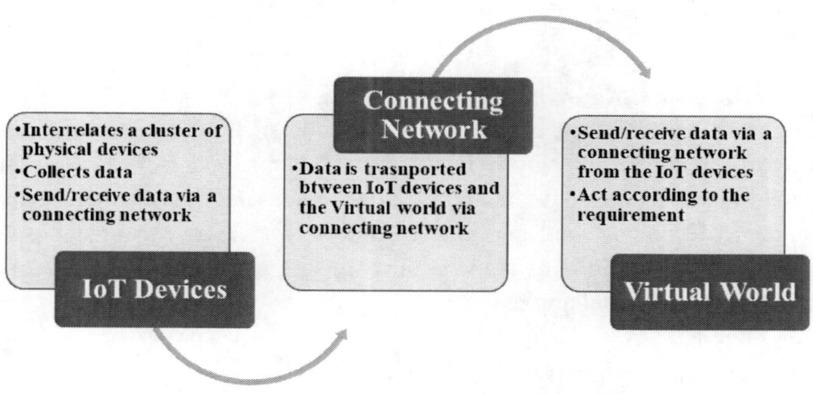

Figure 162. Basic conceptual block diagram of IoT.

4.1. Significance of the Integration of Nanosensors with IoT ecosystem

4.1.1. Sensing and Data Collection

Nanosensors provide real-time and highly accurate data gathering, which is essential to the IoT ecosystem. Numerous physical, chemical, and biological factors, including temperature, humidity, pressure, contaminants, biomarkers, and others, may be detected and measured using these sensors. The IoT network's objects are given the capacity to detect their surroundings precisely and offer real-time data updates by incorporating nanosensors into IoT devices. Environmental monitoring, healthcare, smart homes, industrial automation, and agriculture are just a few of the applications made possible by this data collecting, which also serves as the cornerstone for data-driven decision-making.

4.1.2. Connectivity and Communication

Nanosensors and IoT devices work together seamlessly to provide connection and communication. Nanosensors may be made to interact with cellular networks, Bluetooth, Wi-Fi, Zigbee, and other wireless communication protocols. Nanosensors may send the data they have gathered to other IoT devices or a central server for processing and analysis thanks to their connectivity capabilities. Nanosensors can provide real-time data updates, allow remote monitoring, and streamline communication between various IoT devices by using IoT connection. The IoT ecosystem is formed by a network of linked sensors and gadgets thanks to this connection feature.

4.1.3. Intelligent Decision-making

The IoT ecosystem's intelligence and decision-making powers are improved by the incorporation of nanosensors. In order to get useful insights, the acquired data from nanosensors may be integrated with additional contextual information, such as location data or history data. To evaluate the data and uncover significant patterns, trends, or anomalies, sophisticated analytics and machine learning techniques might be used. Intelligent decision-making is made possible by these insights, which also allow for proactive maintenance, efficient resource allocation, early illness diagnosis, and improved energy efficiency. The creation of intelligent and autonomous systems is made easier by the integration of nanosensors with the IoT ecosystem.

4.1.4. Miniaturization and Form Factor

Nanosensors have the benefit of being able to be made smaller, which makes it possible for them to be easily included into tiny IoT gadgets and items. Nanosensors are appropriate for wearable, smart home, and even implanted medical devices due to their tiny size, which allows the development of compact and unobtrusive IoT devices. The use of nanosensors aids in overcoming the size limitations of conventional sensors and permits the use of sensors in a broader variety of settings and applications.

4.1.5. Scalability and Deployment Flexibility

Nanosensors' incorporation into the Internet of Things environment offers scalability and deployment flexibility. Scalable manufacturing methods may be used to generate nanosensors in large numbers, enabling their cost-effective deployment in a variety of applications. Nanosensors' and IoT networks' scalability allows for the deployment of multiple sensors in a variety of situations, resulting in a dense network of linked sensors. By facilitating thorough monitoring and data collecting, this scalability and deployment flexibility improve insights and decision-making.

The IoT ecosystem benefits from the addition of nanosensors by improving the system's sensing, connection, intelligence, and scalability. Nanosensors support the development of smart and linked IoT applications across a variety of disciplines by enabling real-time data gathering, seamless communication, intelligent decision-making, and miniaturisation. In fields including healthcare, agriculture, environmental monitoring, industrial automation, and smart cities, the combination of IoT with nanosensors creates new possibilities for sophisticated monitoring, automation, and data-driven solutions.

4.2. Architecture of IoT Ecosystem

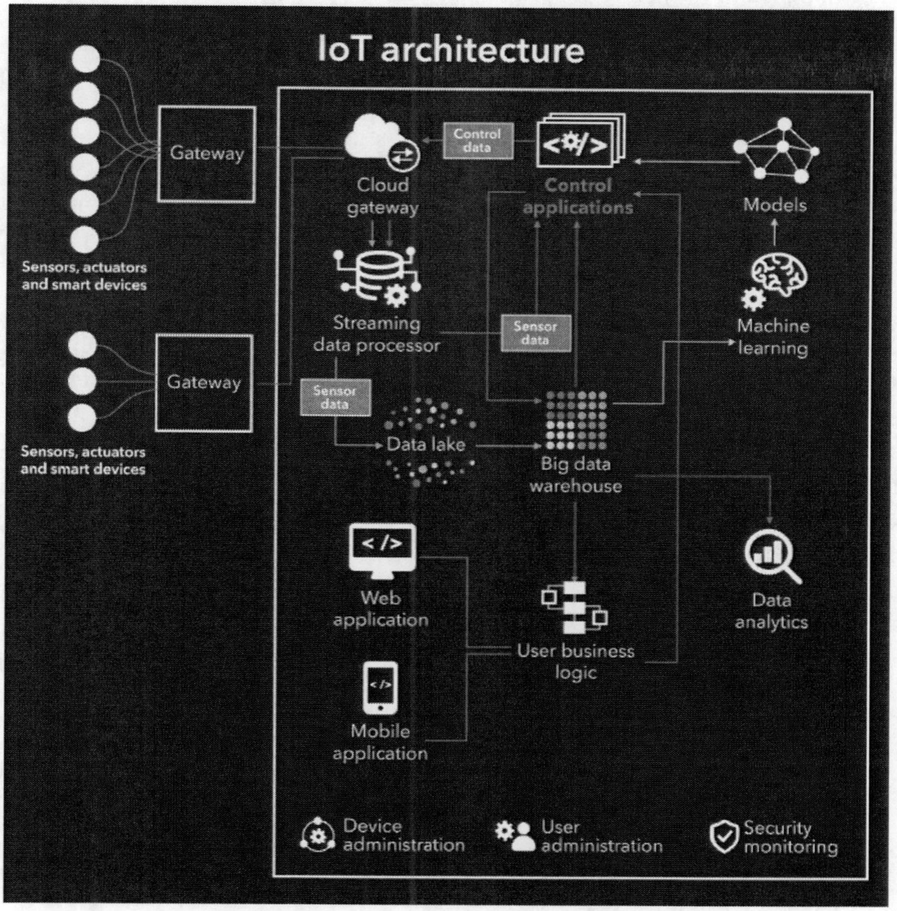

Figure 163. Representation of IoT architecture.

The Internet of Things has applications in practically every conceivable industry, and each application uses a distinct system. However, the underlying building design that makes all IoT systems function is essentially the same. A base layer, or an item or a device equipped with sensors or actuators to perceive and gather data, will be present in every system. An Internet/Edge gateway receives the environmental data that the devices have collected. The Data Acquisition system then receives the raw data and pre-processes it for additional analysis before converting it to digital signals. After being pre-processed and filtered, the raw data is then saved in data warehouses, such as the cloud, where more processing is done to improve and analyse it. Machine learning and other forms of Artificial Intelligence (AI) are used in the data analytics portion to increase accuracy and user experience. The command and result are then sent through different communication protocols to the end-devices or end-users to enable them to carry out the intended job. Since all of these processes are done to enhance and fulfil the user experience, it is the most crucial component of the whole IoT system. As a result, several User Interface approaches are used for improvement. Following is a detailed explanation of each architectural step.

Despite all the architectural components, the overall architecture of the Internet of Things system is composed of only three levels, as follows:

- IoT Device Layer (client side)
- IoT Gateway Layer (server side)
- IoT Cloud Layer (connecting clients and operators)

These three architectural layers has to be addressed in all the architectural stages. The factors like scalability, maintenance and functionality of these layers play a crucial role in determining the successful design of an IoT system.

4.2.1. IoT Architectural Stages

The workflow of how the data is collected stored and processed is shown in the Figure 163 representing the IoT architecture diagram below.

4.2.1.1. Sensors/Actuators

Numerous gadgets claim a spot on the lengthy list of IoT devices. They may be generically categorised as sensors, actuators, and general electrical devices. Any IoT system's initial step is to gather environmental data. By using a variety of sensors and combining them, this is achieved. The deployment of sensors/actuators in the IoT system and their role is illustrated in Figure 164.

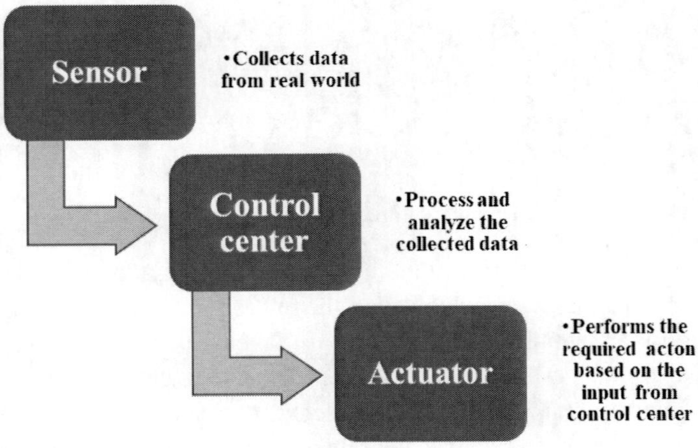

Figure 164. Sensors/Actuator deployment in IoT systems.

The sensors might be a whole device or a solitary sensor, depending on the situation. For instance, a stand-alone proximity sensor can only detect the presence of a person and communicate that information to the data collecting systems. Mobile phones, on the other hand, are classified as common electronic devices since they have many sensors, such as proximity, accelerometer, gyroscope, camera, GPS, orientation, magnetometer, microphone, ambient light detector, fingerprint detector, barcode reader, etc. Mobile phones are an example of a gadget that is more than just a sensor; it is a full system capable of performing a number of additional functions.

The client-side components known as actuators are made to react to the needs of the system. Through the use of an internet connection, they may be managed and turned on remotely. An actuation system must be created that is capable of switching the lights if, for instance, an IoT system is intended to automatically turn on and off the lights in a room when it detects the presence of a person. When a person is detected, the proximity sensor will be positioned to look for them and relay information to the CPU. The actuator will receive orders from the processor to turn on or off the lights in that specific room once the processor has processed the information it has collected. Devices for rehabilitation, controls for streetlights, industrial automation, and more are among the further uses. By linking the whole system through a connecting network, several operations may be accomplished.

4.2.1.2. Internet/Edge Gateways

It is necessary to transport the data gathered by sensors and other IoT devices to the cloud for processing and storage. Similar to this, the actuator must get an action command from the cloud in order to carry out the necessary duties. Through the internet's IoT gateways, this data flow between physical objects and the cloud is made possible and is illustrated in Figure 165. It is necessary to process and analyse the raw data on the cloud. Therefore, before data is saved in the cloud, it must be filtered, preprocessed, and compressed without causing significant loss. The information is then transformed into digitally equivalent sequences that the system can understand and process further. The link allows for bidirectional data transfer and is either wired (LAN, Ethernet) or wireless (Wi-Fi, Bluetooth, RFID). The TCP/IP protocol is often the most widely utilised of the several protocols that the gateways handle, depending on the application and connection. IoT gateways handle massive data transmissions, thus they need to be very secure by using a variety of encryption mechanisms.

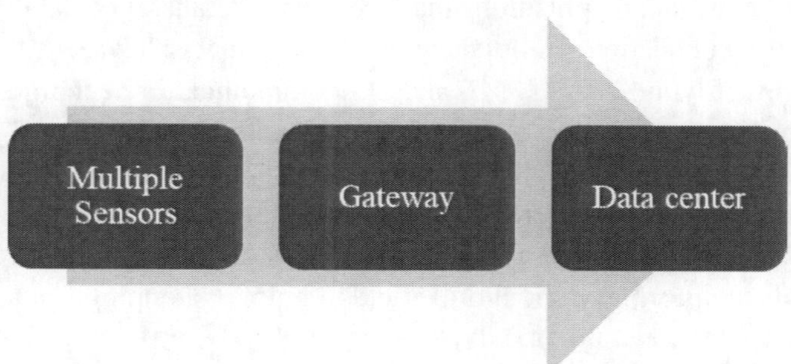

Figure 165. IoT gateway structure.

4.2.1.3. Data Acquisition Systems

Analogue signals will be used to represent the data that sensors and associated equipment collect from the real environment. Because analogue data is more challenging to handle, accuracy and efficiency will suffer. Additionally, it is more prone to deformation. Therefore, it is crucial to transform analogue data into its digital equivalent signals. With better network security, the system's overall accuracy while processing digital information improves. To operate a device in real-time, the data must be processed effectively. Over the course of the time, the sensor will continually collect and deliver data. The system has to be operated based on the data collected from the sensors. To make the system smarter, the data must be gathered, transported, processed, and analysed in a shorter amount of time. The data collection system, where the information acquired is analysed, should demonstrate high level optimised performance with high speed data transfer in order to make intelligent judgements rapidly.

A cloud-based system is used to acquire the data, which includes a number of protocols, sensor data, data storage, information processing, and predictive analysis. The effectiveness of the data collecting system used in the IoT system is the foundation for improving services and user experience.

4.2.1.4. Connecting Protocols

As was already established, the main purpose of an IoT system is to gather, transfer, and receive data through a cloud storage from anywhere in the globe. To get to the cloud system, these transferred data require a way to go. It is claimed that this media uses a connecting protocol. Numerous techniques, including Wi-Fi, cellular, Bluetooth, Zigbee, satellite, low-power wide-area networks (LPWAN), Li-Fi, RFID, etc., may be used to establish a connection.

Data transmission over a long distance must be done without loss or distortion, which might use more power than normal. Any system with a high power need will operate less effectively in general. The trade-off between bandwidth, range, and power consumption remains constant. As a result, creating a proper connection between the devices and the cloud is considerably more difficult and ends up being the most crucial aspect of constructing an IoT system. Low power consumption and wide-ranging, high-bandwidth data transmission are characteristics of the optimal link. Below, several connection techniques are described in further depth.

4.2.1.4.1. Cellular

The most popular way to create a secure, distant connection between the devices and the cloud is over a cellular network. Applications that need for the transfer of data across great distances build this kind of connection. Cellular connection connects devices to the internet utilising mobile phones employing 2G, 3G, 4G, and 5G communication techniques. Huge amounts of data may be sent to the cloud in a shorter period of time with 4G and 5G connection. The disadvantage of this technology is its high power consumption, which might rapidly deplete the battery of the devices despite its great transmission range and wide bandwidth.

Applications for this include mobile-controlled home automation, real-time monitoring, controlling and monitoring agricultural fields, online shopping, purchasing tickets, and entertainment (movie streaming, music).

4.2.1.4.2. Bluetooth

One of the most significant and popular connection options when a short-range data transfer is necessary is Bluetooth communication. Through the use of Bluetooth technology, many devices are linked together to communicate data in a short-range context. It permits wireless communication between many devices. Wireless connection of headsets, speakers, smart watches, printers, file transfers, etc. utilising smart phones and PCs are popular everyday uses of Bluetooth. It has a significant role to play in the area of home automation where mobile phones are used to provide message-controlled device functionality with reduced power consumption (compared to cellular technology).

Bluetooth is a wireless personal area network (WPAN) communication technology that adheres to the IEEE standard 802.15.1. A typical Bluetooth device has an RF transmission range of up to 10 metres. Up to 7 devices may connect over Bluetooth to create a connected environment. The term "piconet" is often used to describe a group of devices linked through Bluetooth. Even in loud settings, it can establish a solid connection and is less prone to distortion across shorter distances. It just transfers data with less bandwidth. The Bluetooth communication now uses Bluetooth-Low-Energy (BLE), a new technology that uses less power than normal. As a result, Bluetooth is often used in indoor IoT systems to link devices. The most recent version of Bluetooth, Bluetooth 5, is said to provide more privacy and security while using less power and taking up less space. With a transmission rate of 48MBps, this Bluetooth version is designed to connect several devices across a distance of 300 metres.

4.2.1.4.3. Zigbee

Similar to Bluetooth, Zigbee is a wireless connection technology. Compared to Bluetooth, Zigbee technology offers several substantial benefits. The main benefit is that a Zigbee can connect to more than 65,000 devices simultaneously, which is a significant advantage. Compared to Bluetooth, it can broadcast over longer distances with comparably less power consumption and improved security. Although it delivers a data rate of roughly 250 kbps slower than Bluetooth, a standard Zigbee has an RF signal range of 10 to 100 metres. It adheres to IEEE standard 802.15.4 and is a WPAN communication protocol. This kind of communication is often used in low-range applications with relatively little data transmission rate, such as home automation.

4.2.1.4.4. Wi-Fi

In a confined space, Wi-Fi is one of the most widely used and desired communication protocols. A few hundred MB per second of high data rate are provided. 802.11n is the IEEE standard for it. Wi-Fi can transmit more data at once because its working frequency range is substantially greater than the ranges utilised by cellular technologies. But this method's severe power consumption and short data transmission range are its biggest drawbacks. Wi-Fi typically has a range of around 100 feet.

4.2.1.4.5. Satellite

The creators of the Internet of Things also employ satellites to build connection. The satellite will be able to give access across the whole world and link a sizable number of sensors and actuators to the cloud. This is used in applications that call for very long-range data transmission. The whole earth is covered by this, the most reliable connection protocol. The two main ways to use a satellite to link a device to the cloud are direct and backhaul. Wherever feasible, the direct connectivity may create a direct connection between the devices and the cloud via cellular data. It will build the connection utilising satellite coverage in the locations where cellular data cannot reach. For instance, ships will utilise cellular data connection while they are close to a shore or port and satellite connectivity when they are in open water. In contrast, the backhaul technology connects to the satellite through a tower and connects to the sensors and actuators via various wired or wireless technologies. These are used in rural locations with little internet access. The main disadvantage of this sort of communication is its high power consumption, which might negatively impact system performance by causing a rise in latency. Despite having tremendous coverage, it necessitates the installation of bigger, more power-hungry equipment, which raises the cost.

4.2.1.4.6. Near Field Communication (NFC)

The NFC devices, such as RFID, Bar Code, QR Code, etc., are used to transfer data across extremely small distances of little more than 4 cm, or a few centimetres. This necessitates the presence of a person in the near field, in contrast to other Wireless communication protocols that were stated before. The scanning of electronic tickets at theatre and metro ticket counters, ID access, automated billing systems, card payment systems, library upkeep, etc., are the greatest examples. The necessary communication protocol will be included into an electronic device, such as a smartphone, in order to create a secure connection between different devices.

4.2.1.4.7. Data center and cloud

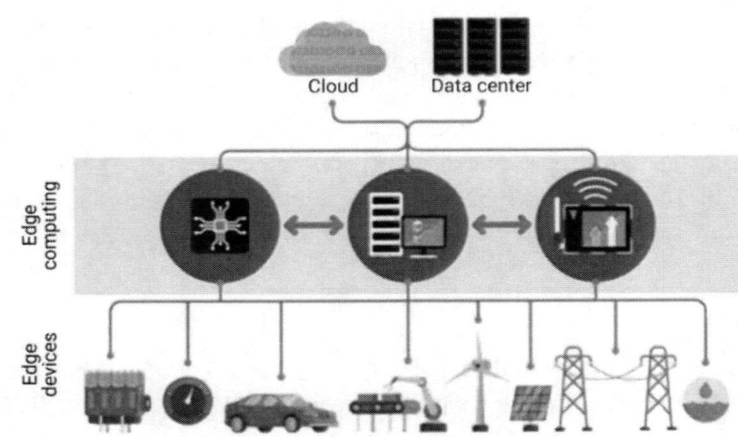

Figure 166. Architecture of data centre and cloud.

The IoT system's brain is comprised of the data centre and cloud which is illustrated in Figure 166. IoT systems produce enormous amounts of data that are gathered in real-time through sensors and actuators. The information obtained has to be kept and utilised in a secure location with maximum effectiveness. When it comes to building and deploying IoT-based applications, the cloud is important. The cloud is used to process and analyse all of the pre-processed raw data that is stored in a data lake. End users may remotely access the analysed data from any location and utilise the forecasts to make wise choices.

A network of high-performance servers is used by the cloud computing used in IoT systems to analyse the data that is acquired quickly from the billions of client side IoT devices (sensors/actuators). Machine learning and other AI methods and algorithms are used to more accurately get deeper insights into the vast amounts of acquired data.

4.2.1.5. Edge Data Analytics

Edge devices, in particular, may prove to be quite advantageous for large-scale IoT projects, even though they are not a necessary element of practically any IoT design. The deployment of edge analytics in IoT systems to control real-time actions is illustrated in Figure 167. Edge systems may provide quicker answers and more flexibility in data processing and analysis due to the IoT cloud platforms' restricted accessibility and data transmission speed. As the speed of data processing is critical in certain industrial applications, edge computing has recently seen a remarkable increase in interest among industrial IoT ecosystems.

Data analysis is the act of transforming analogue data from billions of smart devices and sensors into insightful understandings that can all be understood and utilised for thorough study. This may enhance the entire IoT system's management and effectiveness. Real-time smart data analysis is one of the main advantages of an effective IoT system since it enables engineers to spot anomalies in the data collected and take rapid action to stop an unsuitable situation. If the information is reliably collected at the right moment, network operators may be able to prepare for additional projects. It is simpler and quicker for edge infrastructure to make choices on the IoT system in real-time and to specify intelligent actions quickly since it can be placed more physically near to the source system. In these situations, only the bigger data chunks are sent to the edge analytics, which really need the dominance of the cloud to be processed.

Global businesses are already making use of the immense information gleaned through IoT mobile and online apps. Data is very crucial for any business model, and predictive modelling ensures success in any industry. Companies may predict business trends and be ready for a successful deployment in advance with careful preparation.

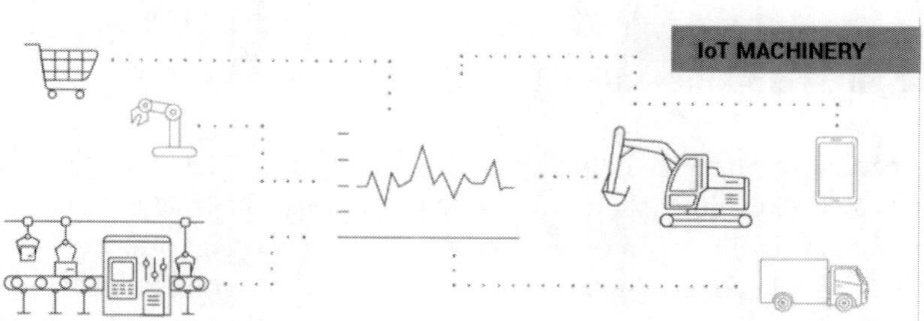

Figure 167. Deployment of Edge analytics in IoT systems to control real-time actions.

4.2.1.6. User Interface (UI)

The data gathered and processed by the IoT system must be seen, analysed, and interpreted by the user. User Interface (UI) is the method through which a user receives meaningful information. The sole visible component of the IoT system is the UI, which is nothing more than a medium for human-machine interaction. The media might be anything, including a PC, laptop, smart watch, Amazon Echo, mobile application, or website. The user may sometimes wish to continually monitor or operate the IoT system depending on the application. To boost security, for instance, one would wish to see the live video whenever he wants to at a store that has cameras placed. Similar to this, one must remotely regulate the air conditioning in the room when the outside temperature changes. All of these things are accomplished via thoughtful UI design for an IoT system. Actuators may sometimes be used to execute automated control of IoT-enabled equipment.

Figure 168. IoT user interface devices.

Modern technology offers a wide variety of interactive multimedia to reduce complex activities to simple commands on touch panels. For instance, touch panels have replaced hard switches in many of our home appliances, and this trend is spreading to almost all smart household products as shown in Figure 168. The user interface (UI) design has a stronger impact in today's global economy, often determining the choice of a certain product or appliance. The desire to possess new gadgets or smart devices that are user-friendly and compatible with standard wireless technologies exists.

5. Integration Techniques and Challenges

A potential method for improving the capabilities of electronic devices, allowing cutting-edge functionality, and extending their application areas is the integration of nanosensors. Nanosensors may be easily incorporated into electronic devices to offer real-time monitoring, sensing, and data collecting thanks to their small size and great sensitivity. To guarantee effective and dependable functioning, a number of problems are presented by this integration process. The integration methods and difficulties of nanosensors with electronic devices are introduced in this article, which also discusses issues with sensor approaches, interface and communication protocols, power supply considerations, size and form factor limitations, and security and privacy issues.

5.1. Sensor Integration Approaches

Depending on the needs of the individual application, a variety of methods may be used to combine nanosensors with electrical devices. In optical integration, sensing capabilities are provided by using the optical characteristics of nanosensors, such as their capacity to absorb or emit light. This method makes it possible to include optical nanosensors into devices, enabling distant sensing and non-contact measurements. On the other hand, electrochemical integration makes use of the electrochemical characteristics of nanosensors, including changes in conductivity or electrochemical reactions, to identify and quantify analytes. This method is especially beneficial for situations that call for quick and accurate measurements. In order to allow detection and monitoring in electrical devices, magnetic integration makes use of the magnetic features of nanosensors, such as their reaction to magnetic fields. The combination of nanosensors with biological recognition components, such as antibodies or enzymes, to detect certain biological targets is known as biosensor integration, which also creates new opportunities for medical and diagnostic applications.

5.2. Interface and Communication Protocols

To allow smooth interaction between the sensors and the host devices, interface and communication protocols must be established in order to integrate nanosensors with electronic devices. These protocols specify how the devices communicate, analyse, and interpret the data gathered by the nanosensors. Wireless or cable communication protocols might be used, depending on the application. Remote monitoring and data transfer are made possible via wireless protocols like Bluetooth or Wi-Fi, offering flexibility and simplicity. Although wired protocols like USB or I2C enable dependable and fast data transmission, they may have mobility and connection issues. The proper interface and communication protocols should be chosen based on the host device's compatibility, data transmission needs, and power consumption.

5.3. Power Supply and Energy Efficiency Considerations

Integrating nanosensors with electrical devices requires a reliable power source and effective energy management. The constraints of the host device's power supply should be considered when integrating nanosensors since they often need a reliable power source to function. In order to provide nanosensors self-sustainability, this may include investigating power harvesting strategies, such as energy scavenging from external sources like light or vibrations. To reduce the power consumption of nanosensors, ensure longer operating times, and ease the strain on the power source, energy-efficient designs and power management techniques should be used. To achieve seamless integration, it is essential to balance the energy limits of the host device with the power needs of the nanosensors.

5.4. Addressing Size and Form Factor Constraints

Constraints on size and form factor must be carefully taken into account when integrating nanosensors into electrical devices. Although nanosensors are made to be tiny, their integration should take into account the amount of room that is available inside the device without sacrificing its overall effectiveness. To fit within the architecture of the device while keeping their detecting capabilities, nanosensors should be made smaller. Additionally, its incorporation shouldn't impair the gadget's functionality or appearance. While addressing size and form factor constraints, the development of small, flexible nanosensor designs and improvements in manufacturing methods enable seamless integration.

5.5. Security and Privacy Challenges

Concerns about security and privacy are raised by the integration of nanosensors with electronic devices. Strong security measures are needed for the collection and transmission of sensitive data by nanosensors in order to guard against unauthorised access, manipulation, or data breaches. To protect the confidentiality and integrity of the acquired data, encryption techniques, secure authentication mechanisms, and data anonymization techniques should be used. The integration procedure should also take into account any vulnerabilities that may result from the interaction of nanosensors with the operating system or network connection of the host device. To build user confidence and safeguard sensitive data, it is critical to ensure the security and privacy of integrated nanosensors ("View of IoT Health Data in Electronic Health Records (EHR): Security and Privacy Issues in Era of 6G | Journal of ICT Standardization" n.d.).

By overcoming these obstacles, nanosensors may open up new possibilities for high-tech sensing, monitoring, and data collecting in a variety of industries, from consumer electronics and healthcare to smart devices and environmental monitoring.

6. Future Perspectives and Emerging Trends

Due to their small size and great sensitivity, nanosensors have the potential to revolutionise a number of industries, including consumer electronics, healthcare, and environmental monitoring. With an emphasis on developments in integration methods, miniaturisation, flexible electronics, sensor fusion, and multi-modal integration, this section examines potential futures and emerging trends in the integration of nanosensors with electronic devices. Future electronic devices might become smarter, more user-friendly, and able to provide real-time, context-aware data for better user experiences and decision-making by using these trends.

6.1. Advances in Nanosensor Integration Techniques

Significant improvements in the area of nanosensor integration continue to be made, creating new opportunities for improved sensing capabilities and seamless integration with electrical devices. To enhance the functionality, stability, and compatibility of nanosensors, researchers are investigating cutting-edge production methods and materials. In order to improve the sensitivity and selectivity of nanosensors, hybrid integration approaches that incorporate several nanomaterials or nanocomposites have showed promise ("Hybrid Nanocomposites for Nanotechnology" 2009). The size, shape, and content of nanosensors may now be precisely controlled because to developments in nanofabrication procedures such nanowire growth, nanopatterning, and self-assembly methods (Betancourt and Brannon-Peppas 2006). The expansion of the application areas and the ability to integrate nanosensors with electronic devices are made possible by these advancements in nanosensor integration methods.

6.2. Miniaturization and Flexible Electronics

Miniaturisation and the advancement of flexible electronics are key to the future of nanosensors and electronic devices. Nanosensors may be incorporated into smaller form factors, such as wearable technology, smart fabrics, or implantable devices, thanks to miniaturisation. Reduced-size nanosensors may be easily incorporated into commonplace items, allowing widespread monitoring and individualised healthcare (He et al. 2022). Additionally, the development of stretchy and flexible electronics paves the way for the incorporation of nanosensors onto novel substrates like bendable wearables, flexible displays, or conformable sensors. The possibilities for integrating nanosensors are increased by the flexibility, toughness, and conformability of flexible electronics (Gurban et al. 2023). Innovative applications in consumer electronics, healthcare, and human-machine interactions are made possible by these developments in miniaturisation and flexible electronics.

6.3. Sensor Fusion and Multi-modal Integration

In the future, multi-modal integration and sensor fusion will be used to combine several sensory modalities. The process of combining data from several sensors to get a thorough picture of the environment or object under observation is known as sensor fusion. A more precise and reliable image of the monitored parameter may be achieved by integrating the outputs of nanosensors with those from other kinds of sensors, such as optical sensors, gas sensors, or imaging sensors (Li et al. 2022). Additionally, multi-modal integration entails combining nanosensors that can concurrently detect numerous parameters, providing a comprehensive sensing solution. For instance, nanosensors that can measure temperature, humidity, and gas concentrations all at once might provide important information for applications such as smart homes or environmental monitoring (Medjahed et al. 2011). The richness and dependability of the gathered data are improved via sensor fusion and multi-modal integration, enabling more sophisticated monitoring and decision-making.

6.4. Implications for Future Electronic Gadgets

Future technological developments will be greatly impacted by the incorporation of nanosensors into electrical devices. Electronic devices may be made smarter, more user-friendly, and context-aware by the seamless integration of extremely sensitive and selective nanosensors. For instance, nanosensor-equipped cellphones may provide real-time environmental sensing, personalised suggestions based on user preferences, or health monitoring. Electronic devices with integrated nanosensors may help with early illness identification, water quality monitoring, food safety evaluation, and air pollution detection. The implementation of Internet of Things (IoT) applications, where linked devices interact and exchange data to allow a more intelligent and efficient environment, will also be aided by the integration of nanosensors into upcoming electronic products.

This progress is being driven by improvements in integration methods, miniaturisation, flexible electronics, sensor fusion, and multi-modal integration. Future electronic devices might become smarter, more user-friendly, and able to provide real-time, context-aware data for better user experiences and decision-making processes by using these trends.

Conclusion

This chapter offers a thorough discussion of the importance, potential uses, and difficulties of adding nanosensors to electrical devices. The chapter discusses a number of topics, such as the standards for choosing suitable devices, the value of integrating nanosensors with the Internet of Things (IoT) ecosystem, integration methods, and prospective developments. It starts out with an introduction emphasising the value of incorporating nanosensors with electrical devices as well as the prospective uses in consumer electronics, environmental monitoring, and healthcare. In order to provide improved sensing capabilities and real-time data collecting, it emphasises the necessity for seamless integration.

After giving an introduction of electronic devices, the chapter goes into detail about their importance and gives examples of devices that can be integrated with nanosensors. It investigates the criteria for choosing compatible devices, taking into account elements like sensor interface compatibility, processing power, data storage, wireless connectivity, power management, user interface and display, size and form factor, cost, data security and privacy, durability and reliability, scalability, interoperability, battery life, energy efficiency, sensor data processing, and customizability. The topic of regulatory standard compliance is also covered.

One key subject discussed in the chapter is the integration of nanosensors with the IoT ecosystem. In terms of sensing and data collecting, networking and communication, intelligent decision-making, miniaturisation, form factor, scalability, and deployment flexibility, it examines the significance of this integration. To offer a comprehensive knowledge of the integration process, the architecture of the IoT ecosystem and its phases are explored.

The chapter goes into further detail on how to integrate nanosensors with electrical devices and the difficulties that might arise. In addition to addressing size and form factor restrictions, security and privacy issues, it investigates sensor integration strategies, interface and communication protocols, power supply, and energy efficiency factors. Each of these difficulties is covered in depth, emphasising the difficulties and possible solutions.

The chapter finishes with a discussion of potential developments and current trends in the industry. It covers developments in miniaturisation, flexible electronics, sensor fusion, and multi-modal integration. It also covers breakthroughs in nanosensor integration approaches. Future electronic devices' consequences are examined, with a focus on the possibility for improved functionality, intelligence, and user experiences.

Overall, this chapter offers a thorough and informative analysis of how nanosensors are combined with electrical devices. It addresses a variety of issues, including the importance, selection criteria, IoT integration, difficulties, and upcoming trends. For scientists, engineers, and other professionals interested in integrating nanosensors for better sensing capabilities and the creation of cutting-edge electrical devices, it is an invaluable resource.

References

Abunadi, Ibrahim, and Ramasamy Lakshmana Kumar. 2021. "BSF-EHR: Blockchain Security Framework for Electronic Health Records of Patients." *Sensors 2021, Vol. 21, Page 2865* 21 (8): 2865. https://doi.org/10.3390/S21082865.

Alam, Tanweer, Aneta Poniszewska-Maranda, and Witold Maranda. 2021. "Cloud-Based IoT Applications and Their Roles in Smart Cities." *Smart Cities 2021, Vol. 4, Pages 1196-1219* 4 (3): 1196–1219. https://doi.org/10.3390/SMARTCITIES4030064.

Almuslem, Amani S., Sohail Faizan Shaikh, and Muhammad M. Hussain. 2019. "Flexible and Stretchable Electronics for Harsh-Environmental Applications." *Advanced Materials Technologies* 4 (9): 1900145. https://doi.org/10.1002/ADMT.201900145.

Ball, Andrew D., Fengshou Gu, Robert Cattley, Xianghong Wang, and Xiaoli Tang. 2018. "Energy Harvesting Technologies for Achieving Self-Powered Wireless Sensor Networks in Machine Condition Monitoring: A Review." *Sensors 2018, Vol. 18, Page 4113* 18 (12): 4113. https://doi.org/10.3390/S18124113.

Betancourt, Tania, and Lisa Brannon-Peppas. 2006. "Micro- and Nanofabrication Methods in Nanotechnological Medical and Pharmaceutical Devices." *International Journal of Nanomedicine* 1 (4): 483. https://doi.org/10.2147/NANO.2006.1.4.483.

Bogena, Heye Reemt, Ansgar Weuthen, and Johan Alexander Huisman. 2022. "Recent Developments in Wireless Soil Moisture Sensing to Support Scientific Research and Agricultural Management." *Sensors 2022, Vol. 22, Page 9792* 22 (24): 9792. https://doi.org/10.3390/S22249792.

Gia, Tuan Nguyen, Mai Ali, Imed Ben Dhaou, Amir M. Rahmani, Tomi Westerlund, Pasi Liljeberg, and Hannu Tenhunen. 2017. "IoT-Based Continuous Glucose Monitoring System: A Feasibility Study." *Procedia Computer Science* 109 (January): 327–34. https://doi.org/10.1016/J.PROCS.2017.05.359.

Gomes, João B. A., Joel J. P. C. Rodrigues, Ricardo A. L. Rabêlo, Neeraj Kumar, and Sergey Kozlov. 2019. "IoT-Enabled Gas Sensors: Technologies, Applications, and Opportunities." *Journal of Sensor and Actuator Networks 2019, Vol. 8, Page 57* 8 (4): 57. https://doi.org/10.3390/JSAN8040057.

Gurban, Ana Maria, Lucian Gabriel Zamfir, Petru Epure, Ioana Raluca Șuică-Bunghez, Raluca Mădălina Senin, Maria Luiza Jecu, Maria Lorena Jinga, and Mihaela Doni. 2023. "Flexible Miniaturized Electrochemical Sensors Based on Multiwalled Carbon Nanotube-Chitosan Nanomaterial for Determination of Nitrite in Soil Solutions." *Chemosensors* 11 (4): 224. https://doi.org/10.3390/CHEMOSENSORS11040224/S1.

He, Rongyan, Hao Liu, Yan Niu, Huiqing Zhang, Guy M. Genin, and Feng Xu. 2022. "Flexible Miniaturized Sensor Technologies for Long-Term Physiological Monitoring." *Npj Flexible Electronics 2022 6:1* 6 (1): 1–11. https://doi.org/10.1038/s41528-022-00146-y.

Himes, Blanca E., Lena Leszinsky, Ryan Walsh, Hannah Hepner, and Ann Chen Wu. 2019. "Mobile Health and Inhaler-Based Monitoring Devices for Asthma Management." *The Journal of Allergy and Clinical Immunology. In Practice* 7 (8): 2535. https://doi.org/10.1016/J.JAIP.2019.08.034.

Holmes, David. 2016. "Sweat-Sensing Patch for Glucose Monitoring and Drug Delivery." *Nature Reviews Endocrinology 2016 12:5* 12 (5): 247–247. https://doi.org/10.1038/nrendo.2016.56.

Huang, Haizhou, Shi Su, Nan Wu, Hao Wan, Shu Wan, Hengchang Bi, and Litao Sun. 2019. "Graphene-Based Sensors for Human Health Monitoring." *Frontiers in Chemistry* 7 (JUN): 464141. https://doi.org/10.3389/FCHEM.2019.00399/BIBTEX.

"Hybrid Nanocomposites for Nanotechnology." 2009. *Hybrid Nanocomposites for Nanotechnology.* https://doi.org/10.1007/978-0-387-30428-1.

Irkham, Irkham, Abdullahi Umar Ibrahim, Pwadubashiyi Coston Pwavodi, Fadi Al-Turjman, and Yeni Wahyuni Hartati. 2023. "Smart Graphene-Based Electrochemical Nanobiosensor for Clinical Diagnosis: Review." *Sensors 2023, Vol. 23, Page 2240* 23 (4): 2240. https://doi.org/10.3390/S23042240.

Kelechi, Anabi Hilary, Mohammed H. Alsharif, Chidumebi Agbaetuo, Osichinaka Ubadike, Alex Aligbe, Peerapong Uthansakul, Raju Kannadasan, and Ayman A. Aly. 2021. "Design of a Low-Cost Air Quality Monitoring System Using Arduino and ThingSpeak." *Computers, Materials & Continua* 70 (1): 151–69. https://doi.org/10.32604/CMC.2022.019431.

Lahlalia, Ayoub, Olivier Le Neel, Ravi Shankar, Siegfried Selberherr, and Lado Filipovic. 2019. "Improved Sensing Capability of Integrated Semiconducting Metal Oxide Gas Sensor Devices." *Sensors 2019, Vol. 19, Page 374* 19 (2): 374. https://doi.org/10.3390/S19020374.

Lakshmikantha, Varsha, Anjitha Hiriyannagowda, Akshay Manjunath, Aruna Patted, Jagadeesh Basavaiah, and Audre Arlene Anthony. 2021. "IoT Based Smart Water Quality Monitoring System." *Global Transitions Proceedings* 2 (2): 181–86. https://doi.org/10.1016/J.GLTP.2021.08.062.

Li, Cheng, Xinyu Jiang, Heng Jiang, Qinge Sha, Xiangdong Li, Guanglin Jia, Jiong Cheng, and Junyu Zheng. 2022. "Environmental Controls to Soil Heavy Metal Pollution Vary at Multiple Scales in a Highly Urbanizing Region in Southern China." *Sensors* 22 (12). https://doi.org/10.3390/S22124496.

Medjahed, Hamid, Dan Istrate, Jerome Boudy, Jean Louis Baldinger, and Bernadette Dorizzi. 2011. "A Pervasive Multi-Sensor Data Fusion for Smart Home Healthcare Monitoring." *IEEE International Conference on Fuzzy Systems*, 1466–73. https://doi.org/10.1109/FUZZY.2011.6007636.

Mohamed, Elham Farouk, and Gamal Awad. 2022. "Development of Nano-Sensor and Biosensor as an Air Pollution Detection Technique for the Foreseeable Future." *Comprehensive Analytical Chemistry* 99 (January): 163. https://doi.org/10.1016/BS.COAC.2021.11.003.

Müller-Maatsch, Judith, and Saskia M. van Ruth. 2021. "Handheld Devices for Food Authentication and Their Applications: A Review." *Foods* 10 (12): 2901. https://doi.org/10.3390/FOODS10122901/S1.

Ngo, Yen Linh Thi, Phi Luan Nguyen, Won Mook Choi, Jin Suk Chung, and Seung Hyun Hur. 2020. "Highly Sensitive Smartphone-Integrated Colorimetric Glucose Sensor Based on MnFe2O4 – Graphitic Carbon Nitride Hybrid Nanostructure." *Materials Research Bulletin* 129 (September): 110910. https://doi.org/10.1016/J.MATERRESBULL.2020.110910.

Roy Barman, Snigdha, Yu Jhen Lin, Kuan Ming Lee, Arnab Pal, Naveen Tiwari, Sangmin Lee, and Zong Hong Lin. 2023. "Triboelectric Nanosensor Integrated with Robotic Platform for Self-Powered Detection of Chemical Analytes." *ACS Nano* 17 (3): 2689–2701. https://doi.org/10.1021/ACSNANO.2C10770/SUPPL_FILE/NN2C10770_SI_003.PDF.

Saoutieff, Elise, Tiziana Polichetti, Laurent Jouanet, Adrien Faucon, Audrey Vidal, Alexandre Pereira, Sébastien Boisseau, Thomas Ernst,, Maria Lucia Miglietta, Brigida Alfano, Ettore Massera, Saverio De Vito, Do Hanh Ngan Bui, Philippe Benech, Tan-Phu Vuong, Carmen Moldovan, Yann Danlee, Thomas Walewyns, Sylvain Pétré, Denis Flandre, Armands Ancans, Modris Greitans, Adrian Mihai Ionescu. 2021. "A Wearable Low-Power Sensing Platform for Environmental and Health Monitoring: The Convergence Project." *Sensors 2021, Vol. 21, Page 1802* 21 (5): 1802. https://doi.org/10.3390/S21051802.

Sempionatto, Juliane R., José A. Lasalde-Ramírez, Kuldeep Mahato, Joseph Wang, and Wei Gao. 2022. "Wearable Chemical Sensors for Biomarker Discovery in the Omics Era." *Nature Reviews. Chemistry* 6 (12): 899. https://doi.org/10.1038/S41570-022-00439-W.

Srivastava, Sanjeev Kumar, Kah Phooi Seng, Li Minn Ang, Anibal 'Nahuel' A. Pachas, and Tom Lewis. 2022. "Drone-Based Environmental Monitoring and Image Processing Approaches for Resource Estimates of Private Native Forest." *Sensors 2022, Vol. 22, Page 7872* 22 (20): 7872. https://doi.org/10.3390/S22207872.

Tai, Jinghua, Shuang Fan, Siqi Ding, and Lishen Ren. 2022. "Gold Nanoparticles Based Optical Biosensors for Cancer Biomarker Proteins: A Review of the Current Practices." *Frontiers in Bioengineering and Biotechnology* 10 (April). https://doi.org/10.3389/FBIOE.2022.877193.

Ullah, Ata, Muhammad Azeem, Humaira Ashraf, Abdulellah A. Alaboudi, Mamoona Humayun, and N. Z. Jhanjhi. 2021. "Secure Healthcare Data Aggregation and Transmission in IoT - A Survey." *IEEE Access* 9: 16849–65. https://doi.org/10.1109/ACCESS.2021.3052850.

"View of IoT Health Data in Electronic Health Records (EHR): Security and Privacy Issues in Era of 6G | Journal of ICT Standardization." n.d. Accessed July 13, 2023. https://journals.riverpublishers.com/index.php/JICTS/article/view/11351/10685.

Wang, Donghui, Ji Tan, Hongqin Zhu, Yongfeng Mei, and Xuanyong Liu. 2021. "Biomedical Implants with Charge-Transfer Monitoring and Regulating Abilities." *Advanced Science* 8 (16): 2004393. https://doi.org/10.1002/ADVS.202004393.

Xuan, Xing, Hyo S. Yoon, and Jae Y. Park. 2018. "A Wearable Electrochemical Glucose Sensor Based on Simple and Low-Cost Fabrication Supported Micro-Patterned Reduced Graphene Oxide Nanocomposite Electrode on Flexible Substrate." *Biosensors and Bioelectronics* 109 (June): 75–82. https://doi.org/10.1016/J.BIOS.2018.02.054.

Zhang, Li, Dong Peng, Ru Ping Liang, and Jian Ding Qiu. 2018. "Graphene-Based Optical Nanosensors for Detection of Heavy Metal Ions." *TrAC Trends in Analytical Chemistry* 102 (May): 280–89. https://doi.org/10.1016/J.TRAC.2018.02.010.

About the Authors

L. Ashok Kumar
Professor
Department of EEE, PSG College of Technology
Coimbatore, Tamil Nadu, India
Email: askipsg@gmail.com

Dr. L. Ashok Kumar was a Postdoctoral Research Fellow from San Diego State University, California. He was selected among seven scientists in India for the BHAVAN Fellowship from the Indo-US Science and Technology Forum and also, and he received SYST Fellowship from DST, Govt. of India. He has 3 years of industrial experience and 22 years of academic and research experience. He has published 173 technical papers in International and National journals and presented 167 papers at National and International Conferences. He has completed 26 Government of India-funded projects worth about 15 Crores and currently 9 projects are in progress worth about 12 Crores. He has developed 27 products and out of that 23 products have been technology transferred to industries and for Government funding agencies. He has created Eight Centres of Excellence at PSG Tech in collaboration with Government agencies and Industries namely, the Centre for Audio Visual Speech Recognition, Centre for Alternate Cooling Technologies, Centre for Industrial Cyber-Physical Systems Research, Centre for Excellence in LV Switchgear, Centre for Renewable Energy Systems, Centre for Excellence in Solar PV Systems, Centre for Excellence in Solar Thermal Systems. His PhD work on wearable electronics earned him a National Award from ISTE, and he has received 26 awards in the National and in International levels. He has guided 92 graduate and postgraduate projects. He has produced 6 PhD Scholars and 12 candidates are doing PhD under his supervision. He has visited many countries for institute industry collaboration and as a keynote speaker. He has been an invited speaker in 345 programs. Also, he has organized 102 events, including conferences, workshops, and seminars. He completed his graduate program in Electrical and Electronics Engineering from the University of Madras and his post-graduate from PSG College of Technology, India, and his Master in Business Administration from IGNOU, New Delhi. After completing of his graduate degree, he joined as a project engineer for Serval Paper Boards Ltd., Coimbatore (now ITC Unit, Kovai). Presently he is working as a Professor in the Department of EEE, PSG College of Technology. He is also a Certified Charted Engineer and BSI Certified ISO 500001 2008 Lead Auditor. He has authored 19 books in his areas of interest published by Springer, CRC Press, Elsevier, Nova Publishers, Cambridge University Press, Wiley, Lambert Publishing and IGI Global. He has 11 patents, one Design patent and two Copyrights to his credit and also contributed 18 chapters in various books. He is also the Chairman of the Indian Association of Energy Management Professionals an Executive

Member in the Institution of Engineers, Coimbatore, an Executive Council Member in the Institute of Smart Structure and Systems, Bangalore, and an Associate Member in CODISSIA. He is also holding prestigious positions in various national and international forums and he is a Fellow Member in IET (UK), Fellow Member in IETE, Fellow Member in IE and Senior Member in IEEE.

K. M. Janani
Research Scholar
Department of EEE, PSG College of Technology
Coimbatore, Tamil Nadu, India
Email ID: jananiaanandh@gmail.com

Ms. K. M. Janani completed her Postgraduate M.Tech degree in Solid State Technology in 2018 from Indian Institute of Space Science and Technology, Trivandrum. She has 4 months of experience as an Assistant Professor in Shree Eshwar College of Engineering, Coimbatore, under the Department of Electronics and Communication Engineering. She also has 3 years of experience as a Junior Research Fellow under the Department of Science and Technology sponsored project in PSG College of Technology, Coimbatore. Presently she is working as a Senior Research Fellow, and pursuing her Full-time PhD in the area of Electrochemical Biosensors. She has qualified Graduate Aptitude Test for Engineering (GATE) with a percentile of 96 and an All India Rank of 6087 and the Indian Engineering Olympiad (IEO) with a City Rank of 1 and an All India Rank of 254 in 2016. She has worked on the synthesis and fabrication of 2D materials like MoS2, WS2, Cu2O and Graphene and nano FET modelling of Junctionless Field Effect Transistors. Two of her current research include the Design and Fabrication of an Electrochemical Biosensor to Estimate the Blood Cholesterol Level and the Development of a Virtual Reality-Based Rehabilitation Device. She also has hands-on experience with 3D Printing, Raman Spectroscopy, Atomic Force Microscopy (AFM), Optical Microscopy, UV-Vis Spectrophotometer, Probe Station, Thermal Evaporator, Spin Coater, and software like MATLAB, TCAD and COMSOL.

Index

L

M

N

O

P

Q

R

S

T

U

W

Z